THE COMET OF 1577: ITS PLACE IN THE HISTORY OF ASTRONOMY

BY

CLARISSE DORIS HELLMAN, M. A.

SUBMITTED IN PARTIAL FULFILLMENT OF THE REQUIREMENTS
FOR THE DEGREE OF DOCTOR OF PHILOSOPHY
IN THE
FACULTY OF POLITICAL SCIENCE
COLUMBIA UNIVERSITY

NUMBER 510

NEW YORK
1944

PRINTED IN THE UNITED STATES OF AMERICA

PREFACE

THIS book has been in preparation for nearly fourteen years, which accounts for any variation in the style of writing.

At the time the work was begun, the problem of citing titles seemed best met by quoting exactly the source of the information. Following this plan has led to apparent inconsistencies. However, in the case of books not available, it has relieved me of considerable responsibility.

Much of the material gathered could not be used to its best advantage in this book. I am looking forward to writing more on the same subject and on others closely related to it in the future.

My thanks are due to the many scholars and librarians who have helped me. It is possible to mention only a few of them by name, but each individually has my gratitude. The entire Columbia University Library staff and the faculty of political science at Columbia University all have been most helpful.

I am especially grateful to the late Professor Frederick Barry who guided me closely for almost twelve years, and to Professor Lynn Thorndike, whose suggestions and criticism have been so important both for the details and the general plan of the book. Dr. George Sarton, editor of *Isis,* gave the work its original impetus and by his continued creative encouragement greatly influenced my thought.

Many of the readers of my bibliography on the comet of 1577, published in 1934 at the suggestion of Dr. Sarton, have furnished valuable material. This is especially true of Professor Quido Vetter of Prague, who also aided me in locating copies of many of the tracts summarized. Mrs. Carl Goldmark Jr. informed me of the tracts available in Vienna. Professor Jan Schilt and Dr. Carl Boyer focused my attention on some of the astronomical and mathematical problems which needed clarifying. Dr. Alexander Pogo has been most obliging in obtaining information for me from the Harvard College Library.

Last but not least I want to extend my thanks to my own family;—to my father, Dr. Alfred M. Hellman, who first aroused my interest in the history of science, and who made it possible for me to buy the much needed photostats and books; — to my mother, who has always been ready to take over the care of my household or my children when the need arose; — and to my husband, Morton Pepper, who read the entire book in manuscript and smoothed out many awkward passages, and who, if he were not on active duty as a lieutenant in the United States Naval Reserve, would now be reading proof.

New York City
March 20, 1944

TABLE OF CONTENTS

INTRODUCTION

THE purpose of this dissertation is to illustrate the effect of the comet of 1577 on astronomical thought. It was written in the hope that it would appeal to historians of science, who, by a careful analysis of different phases of thought in various fields of endeavor, will eventually produce an integrated picture of the development of intellectual achievement. I want to show that the observations of the comet of 1577 were instrumental in bringing about a change in the opinions regarding comets, and that they mark an increased interest in and a greater number of observations of comets and a decided advance in cometary theory.

It is known that from earliest times comets were objects of wonder. Records of comets go back as far as the beginning of the third millennium B. C.,[1] when a comet was observed in China; and interest in comets has continued through Chaldean, Babylonian, Egyptian, Greek, Roman, and medieval and modern times up to the present. The early observations of comets furnished but little data, but toward the close of the fifteenth century of our era they became more accurate and more numerous.

In the first half of the sixteenth century a growing number of astronomers recorded their opinions of comets and their observations of different aspects of the phenomena. When a nova appeared in 1572, astronomers were forced to doubt the immutability of the heavens and became watchful for some other test of their suspicions or conclusions. This test was furnished by the comet of 1577, which was observed in most of Europe and in Asia by a very large number of men with improved instruments and awakened curiosity. The conclusions of these men affected, not only the theory of comets, forcing the abandonment of the notion that comets were atmospheric phenomena, but affected also the formulation of new systems

1 Pingré, I, 245.

of the world, encompassing a break-down of the Aristotelian point of view, and imparting an impetus to the acceptance of the Copernican doctrine.

In order properly to place the comet of 1577 in relation to the development of cometary theory, it is necessary to sketch historically that development up to 1577. This has been done in the first two chapters, and has become more than a preliminary survey. I have tried to include all writers who were cited by authors on the comet of 1577. Much of the material included has never before been considered from the point of view of cumulative knowledge. Obviously, the survey is based upon secondary works, but in many cases these were checked against the originals. Especially in instances where there was disagreement between reputable secondary works, it was found necessary to refer to the primary sources. On at least one point, the dates of Arrian and Posidonius, a change in interpretation was noticed in the literature and it was interesting to see what caused this change and to weigh the different sources in order to decide on the proper interpretation. Although most of the points taken up have been discussed in one or more secondary works, chapters I and II should be valuable with respect to cometary theory because they put together a great many fragments which have never before been joined in a continuous narrative.

In order to determine what was added to cometary theory by the observations of and the literature about the comet of 1577 and to appreciate the tremendous stir it caused, I have gathered a bibliography, as complete as possible, of tracts and treatises dealing with that phenomenon and have selected therefrom representative samples of authors and works to analyze and to discuss. This discussion provides a concrete picture of the state of astronomy, with special reference to comets, in the last quarter of the sixteenth century. By doing this I have been enabled to judge what changes in thought were taking place as a consequence of the appearance of the comet of 1577, what persons or group of persons were adding materially to the knowledge of comets, and how this new body of knowledge was being received.

It may safely be said that after the observations of the comet of 1577, in spite of the data accumulated from the excellent observations of the comets of 1580 and 1585, little of importance was added to the *theory* of comets until Halley's epoch-making prediction.

CHAPTER I

COMETARY THEORY TO THE END OF THE FOURTEENTH CENTURY

By the year 1577 a great many beliefs and superstitions concerning comets had grown up and taken hold, and steps had already been taken to dissipate many of them. To arrive at the state of European thought regarding comets in the year 1577, it is not necessary to write a complete history of the observations of comets, a task admirably done [1] more than a hundred and fifty years ago. For the purpose of this dissertation it is sufficient to touch upon the high spots, and it is unnecessary to discuss Chinese observations because they were not incorporated into the body of western knowledge.

Almost all that is known concerning the early observations of comets made along the Nile and in the Tigris-Euphrates basin [2] has been preserved by Greek and Latin authors,[3]

1 Pingré. Gundel's article is an example of a more recent though not nearly so complete treatment of the subject. It is divided into sections dealing with the name, the literature, popular belief, "katasterismen", astrology, form, orbit, defense, theory, observations, and lists of early comets. There seems not to be a recent treatment of the history of comets and the development of cometary theory.

2 Pingré, I, 36-41, gave a good summary of most of the available information concerning those observations.

3 The twentieth century researches in Babylonian and Egyptian history, even when they dealt with astronomy, have added nothing to our knowledge concerning observations of comets by these early peoples. Gundel, 1154, remarked that Babylonian "texte" furnish only primitive astrological prophecies but lack the astronomical data of time and place. Kugler, I, 1, mentioned comets merely as having been observed. Thompson recorded the beliefs that "When a comet reaches the path of the Sun, *Gan-ba* will be diminished; an uproar will happen twice..." (Thompson, II, 1 (No. 88)); and also that "When a star shines and its brilliance is as bright as the light of day, in its shining it takes a tail like a scorpion, it is a fortunate omen, not for the master of the house, but for the whole land." (Thompson, II, lxviii (No. 200)). The information about comets in Neugebauer's *Astronomische Chronologie* (Berlin, 1929) was all taken from Carl, Pingré, or Biot. Various astronomical texts, such as Langdon's and Fotheringham's *The*

Seneca's *Natural Questions* in particular being widely read. The question of the trustworthiness of these later commentators, in transmitting such information, does not affect the traditions which were the heritage of the sixteenth century writers on comets.

Seneca (4 B.C. to 65 A.D.) was sure that the Egyptians had not worked out any cometary astronomy. His reasons were that though Eudoxus had imported into Greece from Egypt knowledge of the motions of the planets, and Conon [4]

Venus Tablets of Ammizaduga (1928), fail to mention comets. Nor did Jastrow mention them in his *Aspects of Religious Belief and Practice in Babylonia and Assyria* (1911). However, in his *Die Religion Babyloniens und Assyriens* (Jastrow, II, 689-690), he described the same text as Thompson (No. 200), but did not say that the observed phenomenon was a comet, but called it a meteor. He went on (Jastrow, II, 696) to a school text and recorded the following belief about comets: "Ein Komet [deutet] auf feste Preise". In his foot-note Jastrow told of various other significations ascribed to comets, referring to Thompson's No. 88, which he interpreted as meaning a lowering of prices. These prophecies indicate that the direction in which a comet disappeared, the constellations through which it moved, its position in relation to the sun, and the position of its head, as well as its color, were observed. However, no "theory" of comets or data from any specific observations were set forth. Writing in 1934, Antoniadi, 98, gave Stobaeus, Diodorus, and Seneca as sources for his information on Egyptian cometography. He also cited an edition, not specified, of Hermes' *Écrits*, but the source for these is undoubtedly secondary. (See Antoniadi, 45.) Also ascribed to Hermes is the recording of a belief in the evil signification of comets, which bring destruction in their wake and are called Seth. (Hermes (1936), 216). This God is also identified with the constellation now termed the Great Bear. Archaeological treatises, describing the work and results of explorations, including works by Sir William Flinders Petrie, A.M. Blackman, Norman de Garis Davies, H. C. Rawlinson, many of which I have examined, do not seem to deal with comets, although perhaps a reader of cuneiform and hieroglyphics might find some material in these and in the texts published as a result of archaeological expeditions including those by Petrie and the University of Pennsylvania Museum. Perusal of astronomical bibliographies and periodicals has unearthed no pertinent material. After examining many works on archaeology, ancient astronomy and astrology, and ancient history, I have come to the conclusion that the Greek and Latin writers are the best sources for the history of comets among the Babylonians, or "Chaldeans" as classical writers termed them, and Egyptians.

4 Conon of Samos (third century B. C.) died young, before Archimedes. His books on astronomy contained the Chaldean observations of eclipses. See Sarton, I, 173, and Clarke, 355.

had made a record of solar eclipses observed by the Egyptians, neither of them had mentioned comets. Diodorus of Sicily, who about 30 B.C. completed his encyclopaedic history, said that the Egyptians held themselves responsible for the astrological knowledge for which the Chaldeans were famous. He ascribed prediction of comets to the Egyptians,[5] and to the Chaldeans, saying that the latter could foretell comets and other so-called atmospheric phenomena by carefully observing the five known planets.[6]

Seneca also mentioned the opinions concerning comets held by the Chaldeans. He quoted Epigenes[7] and Apollonius of Myndus,[8] whose own ideas on comets will be described below, as saying that they had studied among the Chaldeans.[9] Apollonius of Myndus said, according to Seneca, that the Chaldeans placed comets among the wandering stars (planets) and that their orbits had been determined. On the other hand, Epi-

5 Diodorus (1933-9), I, 279, wrote that "as a result of their long observations they [the Egyptians] have prior knowledge of earthquakes and floods, of the risings of the comets, and of all things which the ordinary man looks upon as beyond all finding out."

6 Diodorus (1933-9), I, 449-451. Diodorus' is the fullest extant account of the Chaldean observations of comets.

7 Fabricius, book III, chapter V, section VIII, and chapter XX, section XI, said that Epigenes was from Byzantium and that he was praised by Censorinus and Pliny. Fabricius distinguished between the Epigenes from Byzantium and the Epigenes from Rhodes who was mentioned by Varro and Columella as having written on rustic matters. According to Zedler, VIII, 1398, Epigenes flourished at the time of Alexander the Great. Rehm, in Pauly-Wissowa, VI, 65-6, assigned Epigenes to the pre-Alexandrian period. He identified the Epigenes from Byzantium with the one discussed by Varro.

8 Dreyer (1906), 190 note 3, and Pingré, I, 53 note b, said that Apollonius was a contemporary of Alexander the Great. Seneca, book VII, chapter III, said that Apollonius was skilled in casting horoscopes. Clarke, 352, stated that there may be some confusion in the text of the *Natural Questions* between Apollonius of Myndus and Apollonius of Tyana. This is most unlikely because the floruit of Apollonius of Tyana, who did not die before 97 or 98, was considerably after that of Seneca who died in 65.

9 Seneca's reference was to the fourth century B. C. and the term "Babylonians" might properly be substituted for "Chaldeans".

genes was quoted by Seneca as asserting that the Chaldeans had ascertained nothing regarding comets and thought them "fires produced by a kind of eddy of violently rotating air".[10] The variance in these descriptions of Chaldean comet-astronomy shows how little information concerning those early observations survived sufficiently long to have any influence on the development of cometary theory. The story of Chaldean observations of comets formed only the shadowy and sometimes changing background of a picture, the foreground of which was dominated by Aristotelian opinions. However, the shadow was persistent, and Joannes Stobaeus in the second half of the fifth century of our era, in that part of his anthology which is known as *Eclogae physicae et ethicae,* also mentioned the Chaldean opinions of comets.[11] He said that the Chaldeans believed that comets are other planets, stars which are hidden for a period, because of their distance, and which appear when they descend toward the earth; and also that they are called comets by those who do not know that they are true stars, which only seem to be annihilated when they return to their own region.

Information concerning the observations of comets by the early Greeks [12] is almost as scarce as that dealing with their predecessors. It, too, must be culled from the works of later commentators, and its most important source is Aristotle (384-322 B.C.), who gave his views on comets in his *Meteorologica,*[13] after first stating and refuting, as involving impossibilities, the views of his predecessors,[14] Anaxagoras,[15] Democ-

10 Seneca, book VII, chapter III.

11 Stobaeus, I, 226-7. A translation is given by Guillemin, 39. Weidler, 41, cited this passage from Stobaeus as well as (pp. 41 and 53) those from Seneca.

12 See Pingré, I, 42-60. Pingré, writing in the eighteenth century, still considered it necessary to refute Aristotle.

13 Aristotle (1923), $342^{b}25$-$345^{a}10$.

14 Rehm (1907), 374, remarked that Aristotle mentioned his sources only when he opposed them.

15 Anaxagoras of Clazomenae was born about 499 B. C. and died about 428 B. C. at Lampsacus.

ritus,[16] the Pythagoreans, Hippocrates of Chios,[17] and Aeschylus.

Aristotle said that Anaxagoras and Democritus thought that comets were conjunctions of planets, and he quoted the Pythagoreans as saying that a comet is one of the planets appearing at great intervals of time and rising but little above the horizon, which is also the case with Mercury. He added that Hippocrates and his pupil Aeschylus expressed a similar view, except that they said that the tail does not belong to the comet, but is assumed by it in certain parts of its course, when the moisture attracted to the comet reflects " our sight " to the sun.[18] He, himself, believed that a comet is not one of the planets, because all planets appear in the zodiac and comets have been observed outside. Furthermore, he noted that more comets than one have been observed at one time. If a comet were due to reflection, as believed by Hippocrates and Aeschylus, Aristotle would have expected the comet to be sometimes visible without a tail. He said that no planet had been observed except the known five, and that all of them had often been visible above the horizon at the same time; at which time, as well as

16 Democritus of Abdera, the Greek atomist, flourished about 420 B. C. The sentiments ascribed to him by Bodin, 309, that comets are the souls of men, were possibly figments of Bodin's own imagination.

17 Hippocrates flourished in Athens about 450-430 B. C.

18 Aristotle (1923), 343^a4-343^a20, gave their explanation of a comet thus: " It appears at greater intervals than the other stars because it is slowest to get clear of the sun and has been left behind by the sun to the extent of the whole of its circle before it reappears at the same point. It gets clear of the sun both towards the north and towards the south. In the space between the tropics it does not draw water to itself because that region is dried up by the sun on its course. When it moves towards the south it has no lack of the necessary moisture, but because the segment of its circle which is above the horizon is small, and that below it many times as large, it is impossible for the sun to be reflected to our sight, either when it approaches the southern tropic, or at the summer solstice. Hence in these regions it does not develop a tail at all. But when it is visible in the north it assumes a tail because the arc above the horizon is large and that below it small. For under these circumstances there is nothing to prevent our vision from being reflected to the sun."

when some planets were obscured in the neighborhood of the sun, comets had appeared. Aristotle also pointed out that a comet could appear elsewhere than in the north at summer solstice. Many were known to have appeared in the south, and he spoke of the comet which appeared in the west at the time of the earthquake in Achaea and the tidal wave. This was undoubtedly the comet of 371 B.C.,[19] at which time the towns of Helice and Bura were swallowed by the sea. The coincidence of the two events was often mentioned, as in tracts on the comet of 1577, as evidence of the effects of comets.

One of the best known observers of the comet of 371 B.C. was Ephorus of Cyme or Cumae.[20] Seneca mentioned him in an effort to refute the argument that comets are formed by a combination of stars, saying that it was easy to strip Ephorus of his authority, that he was merely a chronicler and not a man of scrupulous honor. Seneca implied that Ephorus fabricated falsehoods in order to enliven his tales, giving, as an example, the description of the splitting up into two stars of the comet of 371 B.C. Seneca thus preserved an important bit of information,[21] which centuries later could be fitted into a physical theory of comets.

19 Aristotle described the comet of 371 B. C. as appearing "to the west in winter in frosty weather when the sky was clear, in the archonship of Asteius". It set before the sun on the first day, but on the next it was a little behind the sun, setting immediately, its light extending like a leap over a third part of the sky, so as to be called a "path". The comet vanished after it had receded as far as Orion's belt. The word "leap" is Aristotle's and has a physical interpretation, meaning the space covered by a jump.

20 Ephorus was a historian and geographer and a pupil of Isocrates. According to Barber, the reasonable dates for Ephorus' life are about 405 to 330 B. C. Bostock and Riley were probably mistaken when they gave 408 B. C. as his approximate floruit. See Sarton, I, 146-7; Bostock and Riley, I, 371 note 7; Diodorus (1933-9), II, 339; and Barber, especially 3 and 132. See Marx, especially 3-23 and 250-1.

21 Seneca, book VII, chapter XVI. In the nineteenth century Biela's comet was seen to divide, so Ephorus' observation is no longer considered an impossibility, even though it is not interpreted as testifying to the formation of comets by planetary conjunction. It is easy to see how such a phenomenon might lead a man to believe that comets are formed by a union of stars.

Aristotle set forth as an argument against his predecessors, both those who believed that a comet is one of the planets and those who believed that comets are a coalescence of the planets, the " fact," which he accepted on the authority of the Egyptians and thought he had himself observed, that some of the fixed stars acquire a tail. Furthermore, he continued, comets seen in his time had faded away gradually [22] without leaving any star behind, although Democritus had insisted that certain stars were seen when comets dissolved. Democritus' theory, Aristotle asserted, would require such always to be the case. Also, on the authority of the Egyptians and from observations of the occultation by Jupiter of one of the stars of the Gemini, he stated that conjunctions and occultations take place without the formation of comets.

Aristotle then presented his own views on comets, phenomena which he thought inaccessible to observation, and an explanation of which he considered satisfactory if free from impossibilities. It is this explanation which was presented time and again by writers on comets, including many who wrote on the comet of 1577, and which was disproved by observations of that comet. Because of its importance in the history of cometary theory, it is best to quote it exactly. The passage reads as follows:[23] " We know that the dry and warm exhalation is the outermost part of the terrestrial world which falls below the circular motion.[24] It, and a great

22 " vanished without setting, gradually fading away above the horizon " (Aristotle (1923), 343b16-343b17), which Pingré, I, 26 note a, interpreted as meaning without a heliacal setting.

23 Aristotle (1923), 344a9-344b18.

24 In the *De Mundo* (Aristotle (1914), 392a32-392b5) this region is described as follows: "After the Ethereal and Divine Element, which we have shown to be governed by fixed laws and to be, moreover, free from disturbance, change, and external influence, there follows immediately an element which is subject throughout to external influence and disturbance and is, in a word, corruptible and perishable. In the outer portion of this occurs the substance which is made up of small particles and is fiery, being kindled by the ethereal element owing to its superior size and the rapidity

part of the air that is continuous with it below, is carried round the earth by the motion of the circular revolution. In the course of this motion it often ignites wherever it may happen to be of the right consistency, and this we maintain to be the cause of the 'shooting' of scattered 'stars'. We may say, then, that a comet is formed when the upper motion introduces into a gathering of this kind a fiery principle not of such excessive strength as to burn up much of the material quickly, nor so weak as soon to be extinguished, but stronger and capable of burning up much material, and when exhalation of the right consistency rises from below and meets it. The kind of comet varies according to the shape which the exhalation happens to take. If it is diffused equally on every side the star is said to be fringed, if it stretches out in one direction it is called bearded.[25] We have seen that when a fiery principle of this kind moves we seem to have a shooting-star: similarly when it stands still we seem to have a star standing still. We may compare these phenomena to a heap or mass of chaff into which a torch is thrust, or a spark thrown. That is what a shooting-star is like. The fuel is so inflammable that the fire runs through it quickly in a line. Now if this fire were to persist instead of running through the fuel and perishing away, its

of its movement. In this so-called Fiery and Disordered Element flashes shoot and fires dart, and so-called 'beams' and 'pits' and comets have their fixed position and often become extinguished." A brief sketch like the present is not concerned with the authenticity of the *De Mundo*. It is sufficient that it was part of the Aristotelian tradition which was the possession of the men who observed the comet of 1577. Capelle (1905b) considered the work to be from the first half of the second century after Christ and to be founded on two works of Posidonius. The citation given above was mentioned by Capelle (1905b), 536. Fabricius, book III, chapter VI, section XIII, considered Posidonius a possible author of the *De Mundo*. Instead of depriving the above quoted passage of value in describing the historical development of cometary theory, Capelle's contention serves to place Posidonius historically as a transmitter and codifier and shows the form of the Aristotelian conceptions concerning comets at the beginning of our era.

25 Writers on the comet of 1577 used this same classification. For example, see Dasypodius' book (item 33 of appendix, below), Busch's book (item 21), and Rocca's book (item 91).

course through the fuel would stop at the point where the latter was densest, and then the whole might begin to move. Such is a comet—like a shooting-star that contains its beginning and end in itself.

" When the matter begins to gather in the lower region independently the comet appears by itself. But when the exhalation is constituted by one of the fixed stars or the planets, owing to their motion, one of them becomes a comet. The fringe is not close to the stars themselves. Just as haloes appear to follow the sun and the moon as they move, and encircle them, when the air is dense enough for them to form along under the sun's course, so too the fringe. It stands in the relation of a halo to the stars, except that the colour of the halo is due to reflection, whereas in the case of comets the colour is something that appears actually on them.

" Now when this matter gathers in relation to a star the comet necessarily appears to follow the same course as the star. But when the comet is formed independently it falls behind the motion of the universe, like the rest of the terrestrial world. It is this fact, that a comet often forms independently, indeed oftener than round one of the regular stars, that makes it impossible to maintain that a comet is a sort of reflection, not indeed, as Hippocrates and his school say, to the sun, but to the very star it is alleged to accompany—in fact, a kind of halo [26] in the pure fuel of fire."

Aristotle accepted, without discussion, the fact that comets, when frequent, foreshadow wind and drought, which, he said, " must be taken as an indication of their fiery constitution." [27] This conclusion followed logically upon his theory of the constitution of comets:[28] " For their origin is plainly due to the plentiful supply of that secretion [wind and drought]. Hence the air is necessarily drier and the moist evaporation is so dis-

26 Aristotle later described haloes as reflections by condensations of air and vapor.

27 Aristotle (1923), $344^{b}20$-$344^{b}21$.

28 *Ibid.*, $344^{b}21$-$345^{a}5$.

solved and dissipated by the quantity of the hot exhalation as not readily to condense into water.— . . .—So when there are many comets and they are dense, it is as we say, and the years are clearly dry and windy. When they are fewer and fainter this effect does not appear in the same degree, though as a rule the wind is found to be excessive either in duration or strength. For instance when the stone at Aegospotami fell out of the air —it had been carried up by a wind and fell down in the daytime—then too a comet happened to have appeared in the west. And at the time of the great comet [371 B.C.] the winter was dry and north winds prevailed, and the wave was due to an opposition of winds . . . Again in the archonship of Nicomachus [341-340 B.C.] a comet appeared for a few days about the equinoctial circle (this one had not risen in the west), and simultaneously with it there happened the storm at Corinth."

The fact that comets are rare phenomena and appear more frequently outside than inside the tropic circles was explained by Aristotle as in part due to the solar and stellar motion, which both caused the hot principle to be secreted and dissolved it when it was gathering, and in greater part due to the fact that the stuff was collecting in the Milky Way. He thought that the latter was composed of the same matter as comets.

Additional valuable information concerning the early Greek theories of comets was given by the Greek author Plutarch, whose period of activity probably did not begin before the deaths of his older contemporaries, Seneca and Pliny. Plutarch's recapitulation of theories about comets brought their history down to a later date than did that of Aristotle, whose theory was included. Plutarch wrote as follows: " Some of the Pythagoreans say, that a comet is one of those stars which do not always appear, but after they have run through their determined course, they then rise and are visible to us. Others, that it is the reflection of our sight upon the sun, which gives the resemblance of comets much after the same manner as images are reflected in mirrors. Anaxagoras and Democritus, that two or more stars being in conjunction by their united light make a comet. Aristotle, that it is a fiery coalition of dry exhalations. Strato,[29] that it is the light of the star darting through a thick cloud that hath invested it; this is seen in light shining through lanterns. Heraclides, native of Pontus,[30] that it is a lofty cloud inflamed by a sublime fire. The like causes he assigns to the bearded comet, to those circles that are seen about the sun or stars, or those meteors which resemble pillars or beams, and all others which are of this kind. This way unanimously go all the Peripatetics, believing that these meteors, being formed by the clouds, do differ according to their various configurations. Epigenes,[31] that a comet arises from an elevation of spirit or wind, mixed with an earthy substance and set on fire. Boëthus,[32] that it is a fantasy presented to us by inflamed air. Diogenes,[33] that comets are stars. Anaxagoras, that those styled shooting stars fall down from the ether like sparks, and therefore are soon extinguished. Metrodorus,[34] that it is a

29 Straton of Lampsacus flourished about 288 B. C., becoming head of the Lyceum. His works show the influence of both Democritus and Aristotle, attempting to reconcile them. In his main field, physics, he developed Aristotelian physics. (Sarton, I, 152.)

30 Heraclides was born in Heracleia on the Black Sea about 388 B. C. and probably died between 315 and 310. He was a pupil of Plato and Aristotle and was the originator of the geoheliocentric system, which was again introduced by Tycho Brahe. See Sarton, I, 141, 125, and Duhem, I, 410-418.

31 The theories of Epigenes as set forth by Seneca will be given below. Pingré, I, 56-7, quoted Stobaeus on certain people who thought comets earthly vapors, which had risen and been ignited. This theory, Pingré said, was that of Epigenes and Apollonius, even though Stobaeus did not definitely say so.

32 A Stoic philosopher (Zedler, IV, 409).

33 This is probably Diogenes of Apollonia, a younger contemporary of Anaxagoras. This Diogenes wrote a book on nature. See Sarton, I, 96 and Fabricius, book II, chapter XXIII, sections I and XVII. The statement by Pingré, I, 55 note f, that the Diogenes mentioned by Plutarch seems to have been he who was "le Chef de la secte Ionique après Anaxagore" has little value, but probably confirms the selection of Diogenes of Apollonia.

34 This is probably Metrodorus of Chios, a pupil of Democritus and teacher of Anaxarchus. This Metrodorus wrote about atmospheric phenomena. See Pauly-Wissowa, XV, 1475-6, Metrodorus 14) M. con Chios. See also Fabricius, book VI, chapter IX, section XXX, and Pingré, I, 46 note m.

forcible illapse of the sun upon clouds which makes them to sparkle as fire. Xenophanes,[35] that all such fiery meteors are nothing else but the conglomeration of the enfired clouds, and the flashing motions of them." [36] Although Plutarch showed no reverence for the cometary theories of Aristotle and the Peripatetics, he did show that those theories had other adherents such as Heraclides.

At the time of Aristotle, Apollonius of Myndus [37] believed, according to Seneca,[38] that "a comet is not one star made up of many planets, but that many comets are planetary", and that a comet "is not an illusion nor a trail of fire produced on the borders of two stars, but is a distinctive heavenly body, just as the sun or the moon is. Its shape is not limited to the round, but is somewhat extended and produced lengthwise. On the other hand its orbit is not visible. It cuts . . . the upper part of the universe, but only emerges when at length it reaches the lowest portion of its course. There is no reason to suppose that the same comet reappears;[39] . . . Comets are as varied as they are

35 Xenophanes of Colophon flourished about 540 B. C. He is the reputed founder of the Eleatic school.

36 This section, "Of Comets and Shooting Fires, And Those Which Resemble Beams" is in the *Placita philosophorum* which is included in the *Moralia*. See Plutarch (1893), 317-8 (*De Placitis Philosophorum*, book III, chapter II.) The translation quoted above is by John Dowel, which is given in Plutarch (1883), III, 149-150. Clarke, xlvi-xlvii, cited Diels as questioning the genuineness of this "wretched epitome" and assigning it to the middle of the second century. To a contemporary and friend of Plutarch, Favorinus of Arles (d. ca. 133), not mentioned in Plutarch's section on comets, has been ascribed belief either in the cometary theory of Apollonius or in that of Democritus. See Pingré, I, 63 note c, who cited Aulus Gellius' book XIV, chapter I; Gellius; and Legré, 226. However, in order to find in Gellius' chapter any statement linking Favorinus with a theory of comets, an interpretation of the word "errones" as applying to comets as well as or instead of to planets is necessary. Quite possibly this interpretation was given in sixteenth and seventeenth century editions of Gellius' work.

37 See footnote 8 above.

38 Seneca, book VII, chapter XVII.

39 It was in this connection that Seneca spoke of the comet in the reign of Nero "which has redeemed comets from their bad character". A four-

numerous. They are unequal in size, unlike in colour.[40] Some are ruddy without any light; others are bright with a pure clear light; others are flame-coloured, but the flame is not a pure thin flame, but is enveloped in a mass of smoky fire. Some are blood-stained and threatening, bringing prognostication of bloodshed to follow in their train. They wax and wane like other planets. They are brighter when they come down toward us, and show larger from a nearer point, smaller when they depart from us, and dimmer when they retire to a greater distance." Apollonius also said, according to Seneca,[41] that stars are opaque but comets are not.

Epigenes,[42] who was mentioned by Seneca with Apollonius of Myndus, may well be discussed at this point, even though there is doubt whether he lived at the time of Apollonius or shortly before Seneca. According to Seneca,[43] Epigenes supposed that the greatest influence in determining the motions of the heavenly bodies was exercised by Saturn, whose power of contracting and massing the atmosphere explained the phenomena of thunder and lightning, beams and torches. Epigenes separated comets into two classes, which Seneca described thus: "One kind sheds its light on all sides without changing its position; the other extends a loose kind of fire in one direction, after the fashion of hair, and passes through among the stars; . . . The former variety . . . are usually low down, and arise from the same causes as beams and torches, that is, from a distempered thick atmosphere that carries in it many of the earth's exhalations, both dry and moist. . . ." Seneca continued by saying that Epigenes "supposes comets to be formed pretty

teenth century manuscript objected to Seneca's statement on the grounds that Nero was a bad ruler and hence the comet was a sign of great evil. See Thorndike, III, 582.

40 Boll (1918), 26, pointed to this citation as proving that the Babylonians observed the colors of comets.

41 Seneca, book VII, chapter XXVI.

42 See footnote 7 above.

43 Seneca, book VII, chapters IV, VI, VII, VIII, IX.

much in the same way as fires excited by whirlwind. There is this one difference, that those whirlwinds are pressed down to earth from a higher region, while these others are raised from earth to the upper regions. . . . Epigenes afterwards goes on to speak of the comets that, he says, have a more definite resemblance to stars, traversing an orbit and passing through the zodiacal signs. He attributes their origin to the same causes as produce those that he called lower comets, the only difference being that the earth's exhalations in this case contain many dry elements, and therefore seek the higher region, and are driven by the north wind toward the more exalted portions of the heavens. . . . He believes that when all the moist and dry exhalations of the earth unite, the mere discord of the different bodies turns the air into whirlwind. Then the force of that wind as it revolves sets fire by its rapid motion to all that it embraces in itself, and raises it on high. The gleam of the fire that is thus extracted remains as long as there is sufficient nutriment; when the fuel fails, the fire subsides too. . . ." Seneca, in an effort to disprove the theories he was setting forth, reasoned thus: " Let Epigenes, therefore, make his choice of the two alternatives: if the force is small, it cannot reach so high; if it is great and violent, it will the more quickly break up. But further, according to the opinion of people like Epigenes, these lower comets do not mount higher because they have too much earthiness in them. Their weight keeps them in the neighbourhood of earth . . . ."

Next, Seneca set forth the arguments of Artemidorus of Parium,[44] who urged that the five planets are not the only stars

44 Seneca, book VII, chapters XIII and XIV. Artemidorus of Parium or Parion in the Troas is known only through the mention given him in book I, chapter IV and book VII, chapters XIII and XIV of Seneca's *Natural Questions*. See Fabricius, book IV, chapter XIII, section IX; Kauffmann; and Gundel, 1170. Delambre (1817), I, 18, citing Weidler, seems to have confused Artemidorus of Parium with Artemidorus of Ephesus. Since no dates are known for Artemidorus of Parium, he may as well be discussed here as elsewhere. His teachings lean on one side toward those of Democritus and Anaxagoras and on the other toward that of Apollonius of Myndus. In connection with this last point see Rehm (1922), 12-3.

with erratic courses, merely the only ones which have been observed. Others are unknown either because of their faintness or because their orbit is so placed that they are visible only at its extremities. Thus, new stars are seen, mixing their light with that of the fixed stars, but brighter than is usual in stars. Seneca questioned this reasoning by asking why one could not then say " either that all the stars move or that none of them does ". Besides, considering the crowd of stars which Artemidorus assumed, stars would meet each other often, whereas comets are rare. Seneca discussed and criticized in highly vituperative phraseology not only the views of Artemidorus concerning comets, but also his entire scheme of the world.

At the time of Aristotle's death there lived a young man, Zeno of Citium on the island of Cyprus,[45] who thought that stars united their rays to create the image of an elongated star. However, his and his school's opinions did not make lasting headway against the growing strength of Aristotelian tradition.

The authoritative power held by Aristotle's opinions grew gradually. It was increased by the work of Posidonius and Arrian of the first century before Christ and the second after, respectively. Posidonius [46] was a pupil of Panaetius of Rhodes, who rejected the predictions of astrology and the Stoic concept of complete sympathy throughout the cosmos.[47] Panaetius and

45 Zeno, the founder of the Stoic school, lived about 336 to 264 B. C. See Seneca, book VII, chapter XIX.

46 Posidonius was born in Apamea, Syria, about 135 B. C. and died in 51 B. C., probably at Rome. A Stoic philosopher with Neo-Platonic tendencies, an encyclopaedist, geographer and astronomer, he founded a school in Rhodes in 103 B. C., was an instructor of Cicero and a friend of Pompey, and exercised great influence on Roman thought. Capelle (1905b), 531, considered Posidonius' influence on philosophy as a whole comparable only with that of Aristotle. The primary sources for fragments of his work are Cleomedes and Strabo. See Sarton, I, 204; Wilamowitz (1902), 185-6; Reinhardt (1921); Clarke, 363; Delambre (1817), I, 260; Duhem, I, 244, 282; Bostock and Riley, I, 149 note 2; Fabricius, book III, chapter XV; Cicero, book I, chapter III.

47 Panaetius was born about 180 B. C. and flourished in Rome and abroad. He died in 110-109. He is thought to have been responsible for the diffusion

others, who believed that a comet is the mere counterfeit of a star, had, according to Seneca,[48] considered "whether all seasons of the year are equally fitted to produce comets, and whether all quarters of the sky are equally suitable for their creation. They have inquired, too, whether they can be formed in all regions through which they can pass, and have discussed other points of a like kind. . . ."

Posidonius' physical theories were in close harmony with those of Aristotle[49] and were in turn upheld by Arrian.[50] Seneca wrote of his cometary theories thus: "But other fiery appearances remain for a considerable time, and do not break up until all the fuel on which they fed has been used up. Here belong the strange sights recorded by Posidonius—pillars and shields all ablaze, and other flames of marvelous strangeness. . . . They bring down sudden fire from the heights of heaven, sometimes producing a flash which is gone in a moment, sometimes compressing the air, which is forced into a glow; . . ."[51] Although the views of Posidonius were closely followed by Arrian, it is possible that the theories differed in respect to the formation of comets in the north. One interpretation of Posidonius' views leads to the conclusion that he thought that thick air was necessary for the formation of a comet and that in the north the air was not thick.[52] On the whole, Posidonius was not so much the originator of theories in the field of atmospheric

of Stoicism among the Romans. See Sarton, I, 193; Clarke, 362; Fowler, especially 36-7, quoting Cicero; Kaussen, 20-2.

48 Seneca, book VII, chapter XXX.

49 Duhem, I, 244; Reinhardt (1921), 135 (with special reference to meteorology); Rehm (1922), 35 ff.

50 See Capelle (1905a), 627-635, where the similarity between Arrian and Posidonius was noted, although, when writing the article, Capelle thought that Posidonius was a follower of Arrian. See also Capelle (1908), 612, 615, 616 ff., 632 ff.; Ringshausen; Seneca, book VI, chapters XXI and XXIV; and Capelle (1913), 337-340.

51 Seneca, book VII, chapter XX.

52 Capelle (1905a), 630-1.

physics as he was the adopter and adapter of those of others which he learned about in his journeys.[53] Posidonius made at least one observation himself, namely that of a comet visible during a solar eclipse although previously concealed by the proximity of the sun.[54]

Original or adopted, Posidonius' theories of comets had wide influence.[55] They can be reconstructed by a careful perusal of Seneca's seventh book.[56] Posidonius thought that a comet's tail was due to reflection, whereas Apollonius of Myndus considered it a part of the comet, a point noted by Seneca in arguing against the theories of the latter.[57] Like Aristotle, Posidonius thought that comets were signs of the weather, and sought the explanation of this relationship in the fiery and dry nature of the comets.[58] He taught that comets are of the same substance as the more fleeting light phenomena of the upper atmosphere, differing from them by their longer duration and their participation in the revolution of the heavens, to which is added, in some instances, a moderate motion of their own.[59] The different kinds are called after their shapes.[60] They are made out of dense air which is separated from the earth's atmosphere and ignited by the friction of the heavens revolving about it and which then follows the circular motion of the heavens. There are more comets than are visible, but they are lost in the sun's rays when they are near that body. They become visible during solar eclipses and sometimes their tails are

53 *Ibid.*, 635; Rehm (1922), 26-30, 38.

54 Seneca, book VII, chapter XX.

55 See Malchin, 21-3; and particularly Rehm (1922). Edelstein, 322-3, minimized the influence of Posidonius. However, as far as cometary theories are concerned, Posidonius furnished one step in their development, and thus his theories became a part of the body of knowledge concerning comets.

56 Rehm (1922), 20 ff.

57 *Ibid.*, 17; Seneca, book VII, chapter XXVI.

58 Rehm (1922), 18, 25.

59 *Ibid.*, 31-3.

60 *Ibid.*, 31 and note on 31-2.

visible after sunset. The nucleus is globular like a star, but the tail is a transparent streak of light. They burn so long as they find nourishment in the aetherial region, and upon this depend the duration, the proper motion, and the fluctuation of their brightness.

Posidonius adduced the following [61] reasons for his conclusions. Comets are transitory and consequently are not stars. They last longer than other luminous phenomena because their motion is higher in the warm region of the aether. The dependence on fuel and the formation from such relatively heavy substance are evinced by the fact that comets, although they appear in all parts of the heavens prefer the poles, in as much as they appear there or at any rate strive to get there. This they do because there fuel is accumulated, and there it is easier than at the equator for the heavy mass of the comet to take part in the heavenly revolution. The classification of comets with atmospheric phenomena is justified by the influence of comets on the weather. Their appearance coincides with drought, their disappearance with heavy rains, so that one must infer that comet fire eats much dry matter.

Clearly as Seneca pictured the theories of Posidonius, competent authorities have found reason to suspect the existence of a middleman between those two.[62] Whether he was Asclepiodotus remains uncertain.[63] Certainly, he can have had no effect on the reception of Seneca's great work, which not only summarized the old theories of comets but also expressed a definite opinion on the subject. Seneca made one concession to his contemporaries, when he classified comets under " meteorology "

61 *Ibid.*, 33-5.

62 Rehm (1922), 4-6; Reinhardt (1921), 137, 139.

63 Asclepiodotus seems to have been a Greek student of meteorology and military science, but his exact identity remains a mystery. Seneca, book II, chapter XXVI and book VI, chapter XVII, said that he was a pupil of Posidonius. See Asclepiodotus, especially the introduction, 230-8; Müller, K. K.; Rehm (1922), 4-5, 15; and Reinhardt (1921), 137.

as was done by Posidonius or his school. Otherwise he would have classified them under "caelestia" whereas almost all his precursors had placed them under "sublimia." [64] Seneca stated the different ideas about comets which had existed before his time and showed why they were not tenable. By disproving them, he was leading up to the arguments which seemed to him to be the logical conclusions concerning comets. This does not mean that he offered his ideas as the final word on the subject, but merely as the best which could be arrived at in the light of the existing knowledge.

When Seneca gave his classification of comets, he spoke of certain luminous phenomena thus: "Those that have a longer career and a stronger fire which follows the motion of the heavens, or those that pursue an orbit of their own, are regarded by the Stoic philosophers as Comets: . . . Different kinds of these are *pogoniae* (bearded), *lampades* (torches), and *cyparissiae* (like cypress trees), and all the rest of them: they have a thin tail of fire. It is doubtful whether beams (*trabes*) and the rare barrel-meteors (*pithitae*) should be placed in this category or not. . . ." [65] Further along, he described three types of cometary theories, the "reflection" theory, the "planet" theory, and the "eddy-of-air" theory, by direct reference to the authors of those theories.[66] After rejecting the theories of his predecessors, Seneca began the exposition of his own, but even there he harked back to those of Apollonius, of Aristotle and Panaetius, and of his own school, that of the Stoics, with none of whom he agreed.

In the twenty-second chapter Seneca said, ". . . I cannot think a comet is a sudden fire, but I rank it among Nature's permanent creations. . . ." [67] And in the following chapter he continued, "In none of the ordinary fires in the sky is the route

64 Rehm (1907), 389, 378.

65 Seneca, book I, chapter XV.

66 Book VII. See Rehm (1922), 7.

67 Seneca, book VII, chapter XXII.

curved; it is distinctive of a star (planet) that it describes a curve in its orbit. Whether other comets had this circular orbit I cannot say. The two in our own age at any rate had. Again, everything kindled by a temporary cause quickly gives out. . . . No fires have any considerable duration unless their strength is inherent. I mean the divine fires which the universe maintains eternally, because they are its parts and works. These, I say, are always active; they have an orbit the even tenor of which they preserve, and they are uniform. . . . I said a moment ago that no fire could be lasting which arose from some defect in the atmosphere. I have now to add further, that it can by no means be fixed and steady. . . . But a comet has its own settled position. For that reason it is not expelled in haste, but steadily traverses its course; it is not snuffed out, but takes its departure. If it were a wandering star (*i. e.*, planet), says some one, it would be in the zodiac. Who, say I, ever thinks of placing a single bound to the stars? or of cooping up the divine into narrow space? These very stars, which you suppose to be the only ones that move, have, as every one knows, orbits differing one from another. Why, then, should there not be some stars that have a separate distinctive orbit far removed from them? What reason is there why there should not be passages into the heavens at some part of them? But if you are convinced that every star (planet) cannot but touch the zodiac, then I say the comet might have such a wide orbit that at some point it may coincide with the zodiac. This is not necessary, but it is possible." [68] The ideas thus expressed by Seneca were cited in the early fifteenth century by Jacobus Angelus of Ulm.[69]

Seneca continued his argument through the remaining chapters of his book,[70] building up a theory, thus: ". . . Do you suppose that in this great and fair creation, among the countless stars that adorn the night with varied beauty, never suffering

68 *Ibid.*, book VII, chapter XXIII.

69 Thorndike, IV, 83. See below, in chapter II.

70 Seneca, book VII, chapters XXIV-XXXII.

the atmosphere to become empty and sluggish, there are only five stars that are allowed to move freely, while all the rest stand still, a fixed, immovable crowd? Should any one here ask me: Why, then, has their course not been observed like that of the five planets? my answer to him shall be: There are many things whose existence we allow, but whose character we are still in ignorance of. . . . Why should we be surprised, then, that comets, so rare a sight in the universe, are not embraced under definite laws, or that their beginning and end are not known, seeing that their return is at long intervals? . . . The day will yet come when the progress of research through long ages will reveal to sight the mysteries of nature that are now concealed. . . . The day will yet come when posterity will be amazed that we remained ignorant of things that will to them seem so plain. . . . The heavenly bodies may not stand or turn away. All advance; once the signal is given they start on their race. Their career will end only with their existence. . . . Men will some day be able to demonstrate in what regions comets have their paths, why their course is so far removed from the other stars, what is their size and constitution. Let us be satisfied with what we have discovered, and leave a little truth for our descendants to find out." These last sentences may not seem optimistic, but they do show a grasp of the situation, a scientific reluctance to commit himself on insufficient data, and confidence in future scientists. Seneca further said, "The whole concord of the universe is a harmony of discords. You say a comet is not a star, because its form does not correspond to the type, but is unlike other stars. You can see, no doubt, how very like that star that returns to its place after thirty years is to this which revisits its haunt within the year! Nature does not turn out her work according to a single pattern; she prides herself upon her power of variation. . . . She does not often display comets; she has assigned them a different place, different periods from the other stars, and motions unlike theirs. . . . Their appearance has, in truth, an exceptional distinction; they are not cribbed and cabined within narrow bonds, but let loose to roam freely, to range over the region of many stars."

Seneca put more emphasis on weather prognostication [71] from the appearance of comets than did Aristotle. He used Aristotle's statements, concerning weather forecasting by comets, to bolster up his own contention that a comet is a star. He said that a comet warned of rain and wind, not in the immediate future, but throughout the year, and he concluded as follows: "Hence it is plain that the comet has not derived prognostications from its immediate surroundings to reveal for the immediate future, but that it has them stored up and buried deep within by the laws of the universe." Aristotle had argued that the slowness of a comet's motion was proof of its being heavy and containing much earthy matter, and that comets' orbits are usually toward the poles. Seneca tried to disprove both contentions. He said that the course of comets is sluggish but that they have further to go, and in regard to their being borne down because of their weight, he pointed out that they are not borne down but around, that the most recent comet "was elevating its orbit when it faded from sight", and that the comet in the reign of Claudius "first appeared in the north, and continued without intermission to rise straight up to a higher elevation until it disappeared." The ideas of Seneca were closely modeled on those of Posidonius, and also bear a resemblance to those of Apollonius of Myndus. Seneca, however, exhibited a critical ability as well as a breadth of vision and imagination, especially with regard to the closed orbit of a comet and a notion of periodicity, which have made his work unique.

A Stoic, not mentioned by Seneca, Marcus Manilius, lived in the Augustan era and wrote a poem on astronomy,[72] which, although it added nothing to the study or science of comets, was

71 Thorndike, I, 103, said on the subject of prognostication that "... Seneca accepts natural divination in well-nigh all its branches: sacrificial, augury, astrology, and divination from thunder" and that "He believes that all unusual celestial phenomena are to be looked upon as prodigies and portents."

72 *Astronomicon Libri Quinque.* The end of book I deals with comets.

an organ of transmission. It was read throughout the Middle Ages and was a factor in determining the thought of the men of the sixteenth century.[73] Manilius adhered to the Aristotelian theory of the earthly origin of comets. According to him, the risen vapors were easily ignited and the force of fire was everywhere. The consequences of the appearance of a comet, as stated by him, were much more severe than any that we have seen outlined before his day, and much closer to the predictions which accompanied many of the treatises on the comet of 1577. He thought that comets brought drought, death, and pestilence in their train.

Although Pliny the Elder's encyclopaedic, but indiscriminate, *Natural History*[74] appeared about 77 A. D., it is not known whether its author had read Seneca's *Natural Questions*.[75] In any case, with respect to comets, Pliny followed the work of Aristotle, rather than the work of his own contemporary, who was not mentioned as a source. Pliny introduced the subject of comets by describing their general appearance " as if shaggy

73 There are many extant manuscripts. In 1579, Scaliger re-edited the poem. See Manilius. This rendition went through several editions.

74 Observers of the comet of 1577 must have been well acquainted with Pliny's *Natural History*. Complete editions of that work now preserved in the B. M. include the editions printed as follows: 1507 in Vercelli; 1511 and 1514 in Paris (both edited by N. Maillard); 1513 in Venice; two in 1516 in Paris; 1518 in Hagenau (a reprint of the 1497 Venice edition); two in 1524 in Cologne by the press of Cervicornus; 1525 in Venice; 1525, 1530, 1539, 1549 and 1554 in Basle by the Froben press; 1532 in Paris; a 4-volume edition in 1536, -35, -38, in Venice (a variation in this edition has the imprint of 1540 on the third part); 1543 in Paris; 1548, 1553, 1561-2, 1563, 1587 in Lyons; 1559 in Venice; 1582 and 1599 in Frankfort-on-the Main; 1582 and 1582-93 in Heidelberg. Of particular importance to the astronomers and astrologers of the sixteenth century were two commentaries on book II. One of these was by Jacob Ziegler and was published in Basle in 1531 (Thorndike, V, 387-8 note 44; B. M. catalogue) and in Cologne in 1550 (B. M. catalogue). The other was by Jacob Milich and was first printed at Hagenau in 1535, then again at Schwäbisch Hall in 1538 and at Frankfort in 1543 (Thorndike, V, 387) and again in 1563 (B. M. catalogue).

75 See Clarke, xlviii-xlix.

with bloody locks, and surrounded with bristles like hair." [76] On his description depended his classification, which, though similar to that employed by Seneca, was more detailed. It was this: "Those stars, which have a mane hanging down from their lower part, like a long beard, are named Pogoniae. Those that are named Acontiae vibrate like a dart with a very quick motion. . . . When they are short and pointed they are named Xiphiae; these are the pale kind; they shine like a sword and are without any rays; while we name those Discei, which, being of an amber colour, in conformity with their name, emit a few rays from their margin only. A kind named Pitheus exhibits the figure of a cask, appearing convex and emitting a smoky light. The kind named Cerastias has the appearance of a horn; . . . Lampadias is like a burning torch; Hippias is like a horse's mane; it has a very rapid motion, like a circle revolving on itself." [77] In addition to these classes, Pliny told of a white comet with silver hair so brilliant that it could not be looked at and having the aspect of a Deity in human form. Other comets described by Pliny included one having the appearance of a fleece, surrounded by a crown, and one where the appearance of a mane was changed to that of a spear. He also said that the shortest period during which a comet had been visible was seven days and the longest a hundred and eighty.

Many of the terms employed by Pliny in classifying comets continued in use for many centuries. The Greek names given above have their Latin counterparts. For example, the term "Pogoniae" would be the Latin "barbati," "Acontiae" would correspond to "jaculi," "Xiphiae" to "ensis," "Discei" to "orbis," "Pitheus" to "dolium," "Cerastias" or "ceras" to "cornu," "Lampadias" or "lampas" to "fax" and "Hippias" to "equus".

76 Pliny, Book II, chapters 22 and 23. Pliny devoted less space to the stars and heavens than to terrestrial phenomena. Thorndike, I, 94, thinks that this difference in emphasis was due to Pliny's being less a believer in astrology than in magic. Delambre (1817), I, 288-9, summarized Pliny's chapters on comets.

77 Pliny, book II, chapter 22.

Pliny thought that some comets remained stationary but that others moved like planets, or at least changed their positions with reference to the fixed stars. He believed that most of them were seen toward the north, particularly in the Milky Way.[78] He quoted Aristotle as saying that comets foreshadowed wind and heat and also that several comets might be seen at one time, but added that he knew of no one else who had so observed.[79] Pliny also said that comets were visible in the winter months and about the south pole, but that they then had no rays coming from them. He added that hairs are sometimes attached to planets and stars. Pliny stated that " Comets are never seen in the western part of the heavens." If he had this from Aristotle, it was probably his interpretation of Aristotle's way of saying that they faded away gradually.[80] Pliny evidently gave more credence to the influence of comets on the future [81] than did Aristotle and Seneca, for he said that they were regarded as terrifying and were not easy to expiate, and he recalled recent comets and events to uphold those ideas. To read the portents, Pliny thought it " important to notice towards what part it [the comet] darts its beams, or from what star it receives its influence, what it resembles, and in what place it shines." The deductions to be made from these observations were that "If it resembles a flute, it portends something unfavourable respecting music; if it appears in the parts of the signs referred to the secret members, something respecting lewdness of manners; something respecting wit and learning, if they form a triangular or quadrangular figure with the posi-

78 Pliny, book II, chapter 23. Here Pliny was in accord with Seneca, book VII, chapter XXI, quoting the Stoics. On the other hand Aristotle, see above, spoke of comets appearing elsewhere than in the north.

79 Pliny, book II, chapter 23.

80 See note 22 above and Pingré, I, 26 note a.

81 Thorndike, I, 97, said that "Aside from the question of the control of human destiny by the constellations at birth, Pliny's general theories of the universe and of the influence of the stars upon terrestrial nature are roughly similar to those of astrology."

tion of some of the fixed stars; and that some one will be poisoned, if they appear in the head of either the northern or the southern serpent."

With his usual procedure of including all available bits of information, Pliny next told of a temple in Rome dedicated by the Emperor Augustus to a comet which appeared during the games he was celebrating not long after the death of his father, Caesar. Pliny cited Augustus as saying that this comet was visible beneath the Great Bear, for seven days, rising about the eleventh hour, or about an hour before sunset, and shining brightly. It was supposed, by the common people, he said, to indicate the entrance of the soul of Caesar among the Gods. Pliny believed that Augustus interpreted the auspicious omen as being produced for himself, and added that it had truly proved salutary for all.

In conclusion, Pliny added that "Some persons suppose that these stars are permanent, and that they move through their proper orbits, but that they are only visible when they recede from the sun.[82] Others suppose that they are produced by an accidental vapour together with the force of fire, and that, from this circumstance, they are liable to be dissipated."[83] Thus Pliny added nothing new to the theory of comets, nor indeed, was he very explicit about their origin, constitution, or motion. His importance in this history of the theory of comets rests on his great popularity in the sixteenth century, to which he transmitted considerable learning concerning comets, as well as a terminology which persisted before, during and after that century.[84]

Arrian, who wrote a book on meteorology and a monograph on comets, lived in the second century of our era.[85] His works

82 This is similar to the opinion expressed by Seneca, when discussing the Stoics, book VII, chapter XIX.

83 Pliny, book II, chapter 23.

84 See note 74, above, for a list of sixteenth century editions of the *Natural History*.

85 A fragment of his writings on comets has been preserved by Stobaeus, I, 227-8 (book I, chapter 28). Arrian was also mentioned in the works of

show the influence of Aristotle and of Posidonius and include observations not mentioned by Aristotle. He combated the theory that comets are formed in the air, and tried to prove that they announced neither good nor bad,[86] refusing them both astrological and meteorological meaning.[87] He considered them condensations of air pressed out of the atmosphere and ignited in the lowest layer of the aether, next to the air, which revolve with the aether and have existence only so long as the inflammable matter lasts,[88] a theory taken from Posidonius. Because only fragments of Arrian's works survived, his influence on sixteenth century observers was indirect.

No history of any astronomical subject, prior to the rise of physical astronomy, can be complete without mention of Ptolemy, who flourished about the middle of the second century of our era. Ptolemy's influence on various branches of astronomy and his coordination of them are too well known to require comment here. As far as the development of the theory of comets is concerned, this influence took two directions; first

Photius, as writing on the nature of comets and attempting to prove that they announced neither good nor evil. See Delambre (1817), I, 315. Many authoritative works, including Sarton, I, 184, and Capelle (1905a) and (1905b), have assigned Arrian to the first half of the second century before our era. The year after the appearance of the above cited articles by Capelle, there appeared a short note, Wilamowitz (1906), commenting on Capelle (1905a) and placing Arrian definitely in the second century of our era. In 1913, Capelle (1913) considered Arrian to have been a follower of Posidonius and compared the work of the two men on that basis, considering similarities to be due to Arrian's use of Posidonius' work. Capelle (1913), 345 note, cited Wilamowitz (1906), thus showing why he had changed his opinion. Rehm (1922) and Reinhardt (1926), 381 note 1, and Reinhardt (1921), 136, unquestioningly accepted the changed chronology. Aside from the reasons given by Wilamowitz-Moellendorff, it seems logical to make the change in chronology to account for the fact that Seneca did not mention Arrian although he gave the subject matter of Arrian's work, taken from that of Posidonius, whose influence can be seen throughout the *Natural Questions*. The work of Arrian and Posidonius came down to the Middle Ages together, unaffected by the relative dates of the lives of the two men.

86 Delambre (1817), I, 315.

87 Rehm (1922), 24.

88 Capelle (1905a), 626-7.

through the indirect pressure put upon that theory to conform to the mathematical structure portrayed by Ptolemy, although, indeed, comets are not mentioned in the *Almagest;* and in the second place through the appeal of Ptolemy's sanction of astrology and divination. This second direction of Ptolemy's influence on cometary theory was the more direct. He put astrology on a firm basis, and the rules he laid down were substantially the same as those followed by many of the writers on the comet of 1577.

The *Tetrabiblos,* also known as the *Quadripartitum* or *De iudiciis* was written by Ptolemy,[89] although Posidonius may have been the source of much of it.[90] It widely influenced writers in the declining Roman Empire and throughout the Middle Ages,[91] and thus bears directly on the heritage of sixteenth century astronomers. It was in the *Quadripartitum* that Ptolemy associated with each planet one or more of the elemental qualities, hot, cold, dry and moist,[92] as was habitually done in sixteenth century cometary tracts. Ptolemy asserted that the influence of the stars was not inevitable and that those events not arising from the motion of the sky can be altered by applying opposite remedies.[93] It was partly this notion which was the basis of the sixteenth century prayers to avert the " consequences " of comets. The end of the second book of the *Quadripartitum* dealt with meteorological phenomena, including comets.[94] There Ptolemy laid down rules for weather prediction and there too he declared that unusual celestial phenom-

89 Boll (1894), 180.

90 Thorndike, I, 111.

91 Boll (1894), 127; Thorndike, I, 115-6.

92 Thorndike, I, 113-4.

93 *Ibid.*, 112.

94 Ptolemy (1541), 457-8. The translation of the first two books of the *Quadripartitum* in the above edition is that of Camerarius and was published first in 1535. Translations into English are numerous and were printed as early as 1535 (?) (see B. M. catalogue). A translation into English from the Greek was printed in London in 1822 (Ptolemy (1822)).

ena portended definite events, the appearance of comets announcing wind and dirt. This is the only mention of comets by Ptolemy; but the hundredth paragraph of the *Centiloquium*, long attributed to that author, dealt with comets, and, as far as most sixteenth century astronomers were concerned, had the authority of Ptolemy.[95] According to this paragraph, if comets appear in the cardinal points at a distance of eleven signs from the sun, the king or prince of some kingdom will die. If the comet appears in a succeeding place things will go well with the treasury of that kingdom, but it will change its governor. If in a place which has passed the meridian, there will be sickness and sudden deaths. If the comets move from west to east, a foreign enemy will invade the country; if they do not move, the enemy will be from the provinces.[96] The sentiments here expressed were repeated again and again in the following centuries.

The years that passed between Ptolemy's death and the fifteenth century were not productive of any new cometary theory. Brilliant comets were observed and recorded, but the observations were perfunctory and contained barely sufficient data to identify those of the comets which have been proved to be periodic.

Among the early Christian writers often cited by sixteenth century authors was Origen (ca. 185-254), the Greek theologian, exegete and encyclopaedist. In his work, *Against Celsus*, he expressed his opinions on comets. Discussing the star which was seen at the birth of Christ,[97] Origen remarked that new

95 The *Centiloquium* or *Karpos* was ascribed to Ptolemy in medieval Latin manuscripts but is probably spurious. See Thorndike, I, 111, and Boll (1894), 180-1. It is surprising that in the sixteenth century there was so little questioning of the authenticity of this work. Morshemius, in a work printed in 1558, distinguished Cardan as having questioned the authenticity of the *Centiloquium* (Thorndike, V, 403), and Pontus de Tyard, in his *Mantice ou discours de la verité de divination par astrologie* (likewise printed in 1558) seems to have ascribed the *Centiloquium* to Haly (Thorndike, VI, 107).

96 Ptolemy (1541), 504.

97 See Thorndike, I, 436-461, especially 456-7, and Origen, I, 461-2 (*Against Celsus*, book I, chapters LVIII-LIX).

stars partaking of the nature of the celestial bodies which occasionally appear, such as comets and meteors, indicated such events as the changes of dynasties or the outbreak of wars. He inclined toward the opinion of Chaeremon the Stoic, whose *Treatise on Comets* he cited, that comets also sometimes appeared when something good was to happen. Origen also said that there was no prophecy connecting a comet with a particular kingdom or a particular time.

Hephaestion of Thebes in Egypt flourished a little more than a century after Origen. He was a Greek astrologer who knew and made frequent use of Ptolemy's *Tetrabiblos.*[98] His work was read in the sixteenth century, Camerarius editing fragments of it in 1532.[99] Hephaestion believed in seven kinds of comets, five of which were named after the planets, color being the basis of comparison.[100] The twenty-first through the twenty-fifth chapters of the first book of his astrological compilation deal with the subject of comets and meteors.[101] They show the direct influence of the *Tetrabiblos,* especially of the last part of book II, from which Hephaestion took his information.

Contemporary with Hephaestion, lived Ammianus Marcellinus, the Latin historian, whose *Rerum gestarum libri XXXI* contain a brief exposition of the different opinions held by philosophers concerning comets.[102]

The conception of comets as fatal omens, despite occasional references to comets as forerunners of good, persisted together with faith in astrology. A generation after Hephaestion, Synesius of Cyrene,[103] who was versed in astronomy and geome-

98 Boll; Engelbrecht, 28-9.

99 Boll; Engelbrecht, 13-4; B. M. catalogue.

100 Boll (1918), 27.

101 Engelbrecht, 24, 82-102.

102 Pingré, I, 62; Marcellinus, 401-2 (book XXV, chapter X).

103 Synesius, Bishop of Ptolemais, was born in Cyrene, probably between 370 and 375. See Sarton, I, 388-9; Thorndike, I, 540-4; Druon, 9-56; v. Campenhausen.

try, and had faith in astrology,[104] wrote a work, *Praise of Baldness,* in which he objected to calling comets stars, and also said that they were evil portents, foretelling the worst public disasters, and that they could be appeased by diviners and soothsayers.[105]

Among the men destined to keep alive the different theories of cometary science was the Byzantine author, John Laurentius Lydus (490-ca. 565). He spoke of comets in his *De Mensibus* [106] and in his *De Ostentis,*[107] following the Aristotelian tradition, and saying that comets were below the moon and were not stars, but were formed of earthly exhalations ignited in the aether.[108] He repeated the classification of comets which had grown up since Aristotle's time, naming and describing nine divisions which he said were taught by Aristotle, or ten as taught by Apuleius Romanus. His divisions were " hippias, xiphias, pogonias, docias, pithus, lampadias, cometes, disceus, typhon, cerastes." Citing Ptolemy as his authority, Lydus gave the name of an additional type of comet, " salpinx ".[109] He discussed the differences between comets, giving not only the names but also the characteristics of the different types and telling with which planets they were associated.[110] For example,

104 Thorndike, I, 542-3. See also Kolbe, B.: *Der Bischof Synesius von Cyrene als Physiker und Astronom,* Berlin, 1850, which was referred to by FitzGerald, I, 104.

105 Thorndike, I, 543. Synesius, II, 257 (*A Eulogy of Baldness,* chapter 10), reads in part as follows: " It is not even pious, in my opinion, to call these [comets] stars, but if you wish to call them so, this much at least is clear, that hair is an evil, inasmuch as even in a star it produces a perishable form. And whenever these comets appear, they are an evil portent, which the diviners and the soothsayers appease. They assuredly foretell public disasters, enslavements of nations, desolations of cities, deaths of kings, nothing small or moderate, but everything that exceeds the disastrous."

106 Book III, chapter 41; book IV, chapter 73.

107 Chapters 4, 11-16; *Diarium Tonitruale.*

108 Lydus, 46-7 (*De Mensibus,* book III, chapter 41).

109 *Ibid.,* 101-2 (*De Mensibus,* book IV, chapter 73). Apuleius Romanus is, doubtless, the same as Apuleius of Madaura. See Pauly-Wissowa, II, 249.

110 Lydus, 285-290.

Hippias is of the nature of Venus and received its name because of its speed, and Xiphias is connected with Mercury. Moreover, Lydus gave the predictions or significations attendant upon the different comets.[111]

Reiteration of the conception of comets as harbingers of evil came from Isidore of Seville (ca. 560 or 570-ca. 636), who, despite his denunciation of any attempt to predict future events from the stars and his hostility to astrologers, was ready to assert that comets signified revelations, wars and pestilences.[112]

Less than forty years after Isidore's death, the Venerable Bede (673-735)) was born in or near Jarrow, Durham. A Benedictine theologian, an historian and a scientist, his knowledge of science, chiefly because of his knowledge of Pliny, was superior to that of Isidore.[113] However, in his *De Natura Rerum,* Bede showed himself to be of the same opinion as Isidore by regarding comets as portents.[114] He said that comets were stars with fiery hair, appearing unexpectedly and portending changes of rule, or pestilence or wars or winds or heat. He believed that some of them move like planets, that others are stationary, and that nearly all are in some part of the north, usually in the Milky Way. The shortest time during which one remained visible was seven days, he said, the longest eighty. He thought that sometimes the streamers are strewn among the planets and stars, but that a comet never appears in the western part of the sky.[115] In the *Historia Ecclesiastica,* Bede's most important book, notices of comets are mostly confined to a statement of the time of the comet's appearance and a summary of the events which might be said to follow as a result. In 678

111 In his *Diarium Tonitruale* he made the statement for May 27th, that if it thunders there will be ominous signs and a comet will suddenly appear (Lydus, 331).

112 Thorndike, I, 632-3, citing Isidore's *De natura rerum,* XXVI, 15, and *Etymologies,* III, 71, 16; Wedel, 28.

113 Sarton, I, 510-1.

114 Thorndike, I, 635; Wedel, 29.

115 Bede (1843-4), VI (*Opuscula Scientifica, Et Appendix*), 111.

a comet appeared in August and lasted three months. It rose in the morning showing a lofty column of radiant flame.[116] Such notices furnished significant data to later cataloguers and to astronomers who wished to identify past appearances of periodic comets. The two comets appearing about the sun in January 729 were dealt with at considerable length.[117] One of them appeared before sunrise, the other after sunset, signifying destruction for east and west. Their tails were turned toward the north. They remained nearly two weeks and were followed by great disasters. Bede is supposed to have said that comets never move toward the south, supporting the contention that the Star of Bethlehem could not have been a comet because it led the Wise Men south from Jerusalem to Bethlehem.[118]

A somewhat younger contemporary of Bede, who lived in a different part of the world, was destined to become an authority on comets, not because of his scientific learning but because of his importance as a theologian of the Greek Church. This was John of Damascus. Part of his work was known in the Latin Middle Ages and he was cited in several sixteenth century tracts on the comet of 1577. In his *Exposition of the Orthodox Faith* he said that comets "are signs of the death of kings, and they are not any of the stars that were made in the beginning, but are formed at the . . . time by divine command and again dissolved." [119] He seemed to consider the star of the Magi a comet.[120] He was cited by Jacobus Angelus at the beginning

116 *Ibid.*, III, 57 (*Historia Ecclesiastica*, book IV, chapter XII). In the following book, chapter 24, the comet of 678 is again mentioned, with its attendant results, and announcement is made of the appeararnce of comets in 729. However, chapter 24 is a summary and is not included in Bede (1843-4), but can be found in Bede (1896), I, 352-360.

117 Bede (1843-4), III, 290-3.

118 According to a fifteenth century treatise on the comet of 1468, cited by Thorndike, IV, 419.

119 John of Damascus, 24. Robinson, 5, said that John of Damascus was consistent, presumably with his theology, when saying that, as signs, comets were created and dissipated by God.

120 John of Damascus, 24.

of the fifteenth century as believing comets to be special divine creations that lasted only a short time,[121] Jacobus evidently having in mind the passage quoted above. Liberati, in his book on the comet of 1577, quoted John of Damascus as saying that comets are made by God to signify the death of kings, princes and important people, and that they announce winds, earthquakes and great tempests on the sea.[122]

The fear of comets continued through the Middle Ages.[123] However, the Arabic astronomers do not seem to have paid much attention to those bodies. For example, Alfraganus (Al-Fargani), who flourished in the ninth century, made no mention of comets in his elementary work on astronomy, commonly called *Elementa astronomica.*[124] Albategni, who lived in the last half of the ninth century and the first half of the tenth, likewise failed to mention comets in his astronomical work.[125] The first outstanding Arabic name in cometary history is that of the Muslim astrologer, Albohazen (Albohali or Haly or Abenragel), who flourished in the first half of the eleventh century. He seems to have written a tract devoted to comets. Certainly sixteenth century astrologers thought he had done so,[126] and he was cited in sixteenth century cometary tracts. His treatment of comets seems to have been purely astrological.[127]

121 Thorndike, IV, 83.

122 Item 67b of appendix, below, A_4v.

123 Thorndike, I, 673, specifically mentioned the tenth century as a time when men "resorted to enchantments, auguries, and other forms of divination".

124 See Alfraganus (1669) or Alfraganus (1910).

125 See Albategni.

126 See the following work by M. Frytschius: *Catalogus prodigiorum, miraculorum atque ostentorum, tam* [in] *coelo quam in terra, in poenam scelerum etc. Additus est: Tractatus Albohazenhalij de cometarum significationibus per XII. signa zodiaci.* Nuremberg, 1563. This was listed by Rosenthal, catalogue 168, item 1098. A copy can be found in the B.M. See also Zinner (1934), 89, and Thorndike, VI, 490, for reference to this work.

127 See Haly. The 41st chapter of the eighth part was entitled "De uisione cometarum", and gave a series of comet significations.

Besides Albohazen, a commentator, also known as Haly, lived in the first half of the eleventh century. He was an Egyptian, 'Ali ibn Ridwān, who wrote, among other commentaries, on Ptolemy's *Centiloquium* and *Quadripartitum*.[128] In the second half of the eleventh century, Averroes wrote a commentary on Aristotle's *Meteorologica,* which was printed as early as 1474.[129] In addition, there must have been other Arabic translations of and commentaries on this work of Aristotle's, and these furnished the basis of some of the early Latin translations or commentaries.[130]

In the second half of the eleventh century a very bright comet attracted widespread attention but no scientific observation. The comet of 1066, later found to be one appearance of Halley's comet, aroused wide popular interest because of its brilliance, and, being observed in northern Europe, was quickly linked with the Norman conquest.[131] The comet was represented in the Bayeux tapestry, and was mentioned in Ingulph's chronicle and in Orderic Vitalis' *Ecclesiastical History*.[132] A treatise by Jerome of Sancto Marcho, printed in the early years of the sixteenth century but probably compiled before 1505, gave the Norman conquest as an illustration of the effects of comets.[133] A monk, Oliver of Malmesbury, is supposed to have predicted the destruction of his country upon seeing the comet on April 24, 1066.[134]

Despite the dearth of new cometary observations, men of learning continued to make pronouncements concerning comets.

128 Sarton, II, 343; Zedler, XII, 325. The B. M. has a copy of the 1484 printed edition.

129 In Padua. See B. N. catalogue. There was also an edition in Lyons in 1542. See B. M. catalogue.

130 Hellmann (1917), 5, in addition to the commentary by Averroes, mentioned one by Alfarabius (ca. 900).

131 Pingré, I, 373-8.

132 Robinson, 6-7.

133 Thorndike, IV, 704 ff.

134 Sarton, I, 720-1; D. N. B., XLII, 140; Pingré, I, 378.

Peter Abelard (1079-1142) did not believe that comets were new stars.[135] Contemporary with Peter was William of Conches who accepted comets as omens caused by the will of God and did not attempt a natural explanation of the events which followed them.[136]

In the first half of the thirteenth century, Michael Scot, renowned as a translator, philosopher and astrologer, was active in Spain and Sicily.[137] Scot, whose literary fame in the Middle Ages is based on his astrological writings, was well acquainted with Aristotle's *Meteorologica,*[138] but his astronomical work showed the influence of Al-Fargani.[139] His astrological and meteorological *Liber particularis* showed the influence of Isidore, Roman tradition, Aristotle's *Meteorologica,* ecclesiastical writers, and bits of Arabic learning.[140] In his *Liber Introductorius* he included divination from comets under the subject of aeromancy,[141] and although he seems not to have been particularly interested in comets, he is too important an astrologer to ignore.

A belief in comets as signs of slaughters and important events on earth was shown in his *De Legibus* by William of Auvergne, who was bishop of Paris from 1228 until his death in 1249.[142] He also spoke of the belief, which he called "universal," that comets forecast political changes and the death of kings. He

135 Thorndike, II, 7.

136 *Ibid.*, 50-65, especially 57-8. Professor Thorndike says that William wrote (p. 60) "... non est ergo stella sed ignis iuxta voluntatem creatoris ad aliquid designandum accensus."

137 See Haskins (1922); Haskins (1927), 272-298 (revised from Haskins (1922)); Sarton, II, 579-582; Thorndike, II, 307-337. The first known date in Scot's career is August 18, 1217, when he completed his translation of Al-Bitrogi's *On the Sphere* (Haskins (1927), 273).

138 Haskins (1927), 284, 285.

139 *Ibid.*, 288.

140 *Ibid.*, 291-2.

141 Thorndike, II, 320.

142 *Ibid.*, II, 371.

did not consider the Star of Bethlehem a comet. Further repetition of the belief that comets signified pestilence, famine or war, was given later in the same century by Vincent of Beauvais.[143]

Robert Grosseteste[144] (ca. 1175-1253), Bishop of Lincoln, like most Christian authors, exempted man from the control of the stars, partly on account of his free will and rational soul.[145] He did not believe that the stars were of the same nature as the spheres.[146] He thought that the stars were originally generated from the four elements,[147] and had become unchangeable and incorruptible.[148] He did not believe a comet to be a new star because nothing changes in the region above the moon, and for the same reason he considered the comet sublunar; what was new, was the appearance of the comet in the sublunar world.[149] His theory of the generation of stars made it reasonable for him to believe that comets were elementary and yet closely associated with the heavenly bodies. Refuting those who held that a comet was a planet or star and its tail a reflection of the sun's rays from the planet or star, Grosseteste said that the tail was not always extended "back toward the sun." [150] However-

143 *Ibid.*, II, 469.

144 Grosseteste was praised by his countrymen Matthew Paris, who observed the comet of 1239 (Pingré, I, 403), and Roger Bacon, Grosseteste's pupil (Thorndike, II, 436-7). The latest edition of Grosseteste's writings is Ludwig Baur's. See Grosseteste. The *De Cometis* is on pages 36-41 of the second part of that work. Baur's criticism of the *De Cometis* is on pages 69*-72* of the first part. In 1933, using a manuscript preserved in Florence and not known to Baur, S. H. Thomson re-edited the *De Cometis* (see Thomson). The principal difference is one of order, giving the work a more logical sequence. There are, also, several sections not present in the Berlin manuscript used by Baur.

145 Thorndike, II, 446.

146 Grosseteste, 32 ("De generatione stellarum").

147 *Ibid.*, 33.

148 *Ibid.*, 35-6.

149 *Ibid.*, 36-7; Baur (1917), 71; Thomson, 23.

150 "In oppositum solis", Grosseteste, 40; Thomson, 22.

ever, he does not seem to have arrived at any general statement concerning comets' tails. Citing Ptolemy as his authority, Grosseteste gave the following names to the different types of comets: Veru, Cenaculum, Pertica, Miles, Dominus Aschone, Maculia or Aurea, Argentum, Rosa, and Virga.[151] He expressed the usual opinion that a comet is sublimated fire, separated from terrestrial nature and assimilated to celestial nature.[152] The heavenly bodies cause this separation and assimilation, each comet having a particular star which draws it. The star, even if fixed, must be related to a particular planet, and consequently the comet is under the rule of a planet. Due to the action of the celestial bodies, particles of a spiritual sort assimilated to the celestial natures are incorporated in every earthly object. When a comet is generated, these fiery particles are carried on high. This is the beginning of a more general release of the spiritual nature and of the consequent corruption of the terrestrial objects and compounds concerned, namely those that are ruled by the planet which controls the comet and those in the region whence the comet was sublimated. However, it is difficult to determine exactly where the comet has most influence, and opinion in this matter may be governed by the greatness of the alarm of those who see the comet. Grosseteste was trying to find natural causes for the commonly believed consequences of the appearances of comets![153]

Another treatise, a *Summa philosophiae,* in which a theory of comets was given, has been wrongly attributed to Grosseteste.[154]

151 This is true in some of the manuscripts. See Grosseteste, 37 note. Baur, editor, 71*, in his criticism, expressed the opinion that the introductory paragraphs taken from Ptolemy and Haly may be by Grosseteste himself. Thomson, 19, called the section naming the comets "an undisguised paraphrase of a short section of Ptolemy's Almagest, describing nine *stellae cum caudis,* . . .", a passage which I have been unable to locate. Indeed, as stated above, Ptolemy did not treat of comets in the *Almagest.*

152 Thorndike, II, 446-7; Baur, editor, 71*; Baur (1917), 71-2; Grosseteste, 38.

153 Baur (1917), 72.

154 It is included in Grosseteste, 275-643, as an apocryphal treatise. It contains a passage mentioning the comet of 1264, nine years after Grosseteste's

The author of the *Summa* differed from Grossteste in that he did not attempt to explain naturally the opinion that comets signify disaster to whole regions. He held that their appearance was caused, not by chance or nature, but by the will of God alone, and by the ministry of intelligences.[155]

The great activity of the theologian, philosopher and scientist, Albert the Great (ca. 1193-1280), included an interest in astrology and astronomy. He was not a great astronomer and, after fluctuating between the ideas of Al-Bitrogi and Ptolemy, he finally accepted the Ptolemaic system. He wrote a *De Meteoris,*[156] which was a commentary on Aristotle's, in which he discussed comets and why they signified wars and the deaths of rulers rather than poor men.[157] This was cited by Jacobus Angelus in the beginning of the fifteenth century as giving nine chief effects of comets,[158] and as expressing the opinion that a comet was produced by the projection of planetary light.[159] He observed the comet of 1240, which lasted six months.[160] Also

death, and, unless this passage was inserted later, this work cannot be by Grosseteste, except in the sense that it might be interpreted as representing his teaching or as being an incomplete work finished by someone else. See Thorndike, II, 448.

155 Thorndike, II, 452, citing Grosseteste, 586.

156 This was printed in Venice in 1488 (Davis and Orioli, catalogues 78, 93; Hellmann (1917), 19). See Albertus Magnus, 477 ff. The third tract of the first book of the *Libri Meteororum* deals with comets (Albertus Magnus, 499-508). In it Albert recounted past theories of comets.

157 Thorndike, II, 583. See Albertus Magnus, 507-8 (Liber I *Meteororum,* Tractatus III *De Cometis,* Caput XI *Et est* digressio *quare cometae significant mortem potentum et bella*).

158 Thorndike, IV, 83; Angelus, B_1 v-B_4 r. Although Albert discussed the meaning of comets in his *De Meteoris,* he did not list their nine chief effects there. Possibly Angelus found them in another work by Albert.

159 Collard, 85. See Albertus Magnus, 500 (Liber I *Meteororum,* Tractatus III *De Cometis,* Caput IV *De opinione eorum qui dixerunt cometen esse vaporem adhaerentem planetae, sicut sol in mane cernitur cum colore vaporis*). Albert, however, believed "quod cometes nihil aliud est quam vapor terrestris grossus, . . ." (Albertus Magnus, 502).

160 Sarton, II, 937; Pingré, I, 403-4; Albertus Magnus, 504 (however, not mentioning the comet's duration).

attributed to Albertus Magnus is a *Speculum Astronomiae.*[161] In it the events signified by comets are classed under the head of revolutions, that is, what God was going to accomplish in a given year, using the stars as his instruments.[162]

Thomas Aquinas, who also wrote a commentary on Aristotle's *Meteorologica,* closely followed Aristotle's theory of comets with which he agreed.[163] Aquinas' work was based on the Greek, and explained Aristotle's text. In his *Summa Theologica,* quoting St. Jerome, Aquinas mentioned comets among the fifteen signs preceding the Lord's coming to Judgment,[164] an opinion also held by Albertus Magnus.[165]

A comprehensive treatise by the Dominican brother, Giles, was written on the occasion of the appearance of the comet of 1264.[166] It is divided into ten sections and deals with the essence, motion and signification of comets. Giles was familiar with the literature on the subject. He cited in detail Aristotle's *Meteorologica,* mentioning Albert's commentary on it, and Seneca's *Quaestiones naturales.* He also referred to the last sentence of the *Centiloquium* and cited Haly's commentary. He mentioned the works of John of Damascus, Isidore, and Albumasar. Giles was well acquainted with Grosseteste's *De Cometis.*[167] He referred to the latter's mention of those who held that a comet's tail was due to the reflection of the sun's light from a star, and with whom, as was shown above, Grosseteste disagreed. Giles gave a rather complete résumé of Grosseteste's theory of

161 Thorndike, II, 693-4.

162 *Ibid.*, II, 700-1.

163 Thomas Aquinas (1886), 348-356.

164 Thomas Aquinas (1921), 88.

165 Robinson, 7, says that the opinion was handed on by Albertus Magnus, who, however, was scarcely in a position to "hand on" the opinion of his pupil!

166 Thorndike, II, 453. Professor Thorndike has kindly made available his draft of the Latin of Giles, taken from a rotograph of the manuscript. Sarton, II, 946, 961, believes it possible to identify this Giles with Giles of Lessines.

167 Giles spoke of Grosseteste as "a man in our times".

comets, which, as far as their generation was concerned, Giles did not believe differed much from Aristotle's theory. Realizing that he was superseding Aristotle's investigation, he listed nine kinds of comets.[168] He also said that Isidore had distinguished thirty kinds, Aristotle two, Ptolemy three, Seneca none (believing all comets to be the same), and the Arabs five (confusing them with the planets). When quoting Aristotle's statement that comets have been seen outside the zodiac, he pointed out that such was the case with the comet of 1264. This, according to Giles, was seen in France in the east, before sunrise, from before August 1st to the beginning of October.[169] He saw it north of the zodiac in Cancer, then south of that circle, in Gemini between the dog and Orion. In addition to the diurnal motion, it had a retrograde motion. During two solar months, he saw it move 40° in latitude although scarcely 3° in longitude. Like Grosseteste, Giles made no general statement concerning the direction of comets' tails, but stated that he observed the tail of the comet of 1264 first on one side, then on the other. While not offering a theory of his own, Giles adequately summarized the existing ones.

Roger Bacon referred to the comet of 1264 more than once in his writings. A brief tract, *De cometis,* that remains in manuscript, appears to have been suggested by it.[170] In the *Opus Maius,* Bacon gave the date of the comet as July 1264, called it horrible and said it had been generated by virtue of Mars and moved towards Mars, which was in Taurus at the time although the comet started in Cancer.[171] Citing Albumasar, Bacon

168 These were like Grosseteste's except that Giles had a comet called "Nigra" and none called "Virga". Further along in his treatise, Giles gave Grosseteste's list.

169 Pingré, I, 407-8, said that the comet attracted wide attention in France from the middle of July to October, and was seen in China; Sarton, II, 984, that it was visible in China.

170 Little, 379.

171 Bacon (1897), 385 (*Operis Majoris Pars Quarta. Astrologia*). See also Little, 379-380.

pointed out the effects of the positions of the planets, among which effects were pestilences and comets. Because of the nature of Mars, Bacon thought that the comet of 1264 portended discord and wars. He followed Grosseteste, specifically, in believing that a comet consisted of fiery vapor sublimated and assimilated to celestial nature. In the *Tractatus Brevis*,[172] an introduction to his edition of the *Secretum Secretorum* which he attributed to Aristotle, Bacon mentioned the comet of 1264 in a discussion of meteors and comets. He gave Ptolemy's *Centiloquium* and Aristotle's *Meteorologica* as his sources and took the occasion to state again that comets not only are formed of inflamed vapor but are sublimated by virtue of a certain planet or fixed star whose motion they follow. Citing Algazel's *De Naturalibus,* he added that comets are in the sphere of fire above the air.

The voluminous astrological work of Guido Bonatti, who died in 1297, is divided into ten or twelve treatises.[173] In the tenth, the relationship between comets and wind and drought is mentioned.[174] At the opening of the fifteenth century, Bonatti was cited by Jacobus Angelus as noting nine kinds of comets, and also as mentioning a comet in the Arabic year 663.[175] Bonatti was frequently cited in the sixteenth century.

Also often cited in that century was Leopold of Austria, who probably flourished about the middle of the second half of the thirteenth century. An astronomer and meteorologist, he wrote a treatise, which he called a compilation, on astronomy. It was astrological with an astronomical basis, and, like Bonatti's

172 Bacon (1920), 1-24, especially 9-12.

173 The *Liber Astronomicus* was first printed in Augsburg in 1491 (Thorndike, I, xx). See Bonatti (1491). There are ten treatises in the 1491 edition. See also Thorndike (1916), 254; Thorndike, II, 638, 826.

174 Bonatti (1491), EE_5.

175 Thorndike, IV, 93, 85-6. The Arabic year 663 was interpreted in the manuscript edition of Jacobus Angelus' tract (preserved at Erfurt) as meaning 1262, and in the incunabulum edition (in the Cornell University library) as 1260, but it probably meant 1265.

work, was divided into ten treatises. The authors of sixteenth century cometary tracts were particularly interested in Leopold's treatment of the types of comets as ruled by the different planets.[176] He began by stating the Aristotelian theory of the origin of comets and continued by saying that there were nine varieties. These he named and divided into groups according to the planets which ruled them, those belonging to Mars, for example, bringing terror. He stressed the importance of the position of the comet at the time of its appearance in determining its meaning, and went into great detail concerning the significance attached to the appearance of a comet in each of the twelve houses.

In 1302, but with later additions, Henry of Malines or Henry Bate (1246-1310 or later) finished his *Speculum divinorum et quorundam naturalium.*[177] This is an encyclopaedia of science and philosophy, and is divided into twenty-three parts.[178] Citing many authors, it attempts a compromise between Platonic and Aristotelian theories.[179] It mentions comets in part II, chapter XV,[180] and deals with them in part XVIII, chapter XV, where

176 This treatment can be found at the end of the fifth tract in the first edition, Ratdolt's in 1489 (Leupold, f-f_4).

177 Wallerand, (7), (23). Bate influenced later astrologers down to the sixteenth century, but this particular work received little notice from later philosophers (Sarton, II, 994-5). Sarton, II, 994, gave the alternate dates 1244 or 1246 for Bate's birth; Wallerand, (7), gave only 1246. The most complete account of Bate's life and works is that published in 1931 by Wallerand, (7)–(23), who also gives a list of previous treatments of the subject. The *Speculum* remained in manuscript until Wallerand published its dedication, table of contents, and parts I and II (see Bate).

178 Sarton, II, 995; Wallerand, (22)–(23).

179 Wallerand, (23) note 56; Sarton, II, 995. This becomes apparent from reading Bate's table of contents (Bate, 3-32). Part XXIII, chapter XXIV, has the title: "Antiquarum opinionum quarundam circa praemissa quaedam correctio simul et sententia Platonis et Aristotelis in idipsum reductio finalis." Bate, 32.

180 Bate, 171-2. "Rursus, id quod de lacteo dicit Philosophus, primo *Meteorologicorum*, in aere scilicet ipsum consistere, atque causatum esse ex fumosa sive spumosa exhalatione seu concretione subtili in directo quidem

Bate described the comet of 1264 as unusually large and visible from June 25th to the early part of October.[181]

Cecco d'Ascoli (d. 1327), in his *Commentary on the Sphere,* discussed the comet called Milex or Miles, which he said was not the cause of the period of darkness during Christ's passion. Although the comet Milex supposedly presaged religious change and injury to rulers, Cecco did not believe it would cut off the sun's light nor that it would be found at the altitude necessary to interpose.[182]

About the middle of the century there flourished an Englishman, Robert of York, also known as Perscrutator. In his work on weather prediction, which, although not printed, was known in the sixteenth century, Perscrutator seemed to avoid using the word " comet " although he was speaking of ' " stars that appear in the air " '. He did not regard comets as stars, believing them to be made of earthly vapor mixed with water to make them glow. He said that all water contains light, a statement he supported by saying that if water placed in a vase at night were stirred, light would appear. Since these stars did not appear to be burning, he thought it wrong to ascribe their luminosity to heat in the region of upper fire.[183] Like his contemporaries, Perscrutator believed that comets portended war. In 1313, a star with a great tail appeared in latitude 54°, that of York, and moved from north-east to south-west. According to Perscrutator, it indicated the defeat of the English by the Scotch, because Scotland, in his belief, was north-east of England.[184]

cuiusdam astrorum multitudinis consistente, ac totius caeli lationem consequente, quemadmodum cometes circa unum fit astrum quod assequitur, seu halo circa solem aut lunam; quamvis inquam verisimiliter et valde probabiliter dictum appareat, exquisitius tamen perscrutando contraria est ratio fortior et quod . . . ratione nobis certius est experimentum sensus."

181 Wallerand, (10)–(11) and note 20; Bate, 26.

182 Thorndike, II, 961.

183 *Ibid.*, III, 115-6, 118.

184 *Ibid.*, III, 117.

From 1315 until after the Black Death in 1348, Geoffrey of Meaux wrote numerous treatises of a minor character, mostly astrological but to some extent also medical and astronomical.[185] He discussed the comets of 1315 and 1337, and after the great plague he reviewed its astrological causes.[186] He relied on the authority of the great men who preceded him except when their writings were repugnant to the Catholic Faith, mentioning particularly John of Damascus, Aristotle, Ptolemy and Albumasar, according to all of whom a comet forecast future events. According to Geoffrey, this tenet was not against holy faith because it was a matter of disposition or inclination, not of necessity. His concern was with the events signified by the comet of 1315, which appeared some days before the feast of St. Thomas. He assumed that this phenomenon was a comet, not a true star. As to the cause of its generation, Geoffrey interpreted John of Damascus as believing that comets were not made of earthly vapors by virtue of the stars, but were newly created by the divine will as a sign of future marvels. Geoffrey, himself, was inclined to agree with others who, he said, believed in the production of comets by virtue of the conjunctions or configurations of certain planets in appropriate quarters of the sky. He pointed out that Mars was in Leo when the comet of 1315 first appeared, Saturn being in opposition in Aquarius. A double opposition was produced by Mars returning, in its retrograde motion, to Saturn. The comet, like Mars, was in the north and had a retrograde movement which Geoffrey recorded in detail, observing the comet's motion through the constellations. This nightly observation is unusual in the annals of medieval comet history.

Geoffrey, in keeping with his times, tried to fix the sphere of influence of the comet, placing it in the seventh clime but granting that the effects would also be felt in the sixth clime and in several other regions. He cited other writers as holding that the effects of a comet appearing before sunrise would be felt

185 *Ibid.*, III, 281.

186 *Ibid.*, III, 281, 285-6, 715.

quickly, but that those of a comet appearing after sunset would be delayed. On this basis, accounting for the size and duration of the comet, he estimated that the virtue of the comet of 1315, which was visible both by day and by night, would last at least two years. This comet, according to Geoffrey, signified corrupt blood and unnatural choler, the consequences being numerous robberies and dissensions and a scarcity of good faith, truth and justice. The juncture of the comet with Jupiter pointed to further ills. The fact that Mars, the lord of the coming year, was in an aquatic sign, indicated that many would drown in the sea.

Geoffrey's treatise on the comet of 1337 is similar to that on the comet of 1315.[187] In it he cited John of Damascus, at the same time accepting the belief that the comet was produced naturally in the sky by the influence of the planets. He recorded repeated observations of the comet and mentioned an eclipse [188] in connection with it. He traced the motions of Mars and Saturn from the time of the eclipse, and came to the conclusion that they were the cause of the generation of the comet, because, according to Abraham, superior planets in conjunction attract vapors from the earth more strongly when they are retrograde. Geoffrey believed it necessary to know the sign of the zodiac under which a comet was generated in order to form a true judgment from it. The comet of 1337 was observed for twelve days before Geoffrey knew of it. Then he found its position by observing the fixed stars nearest to it and drawing circles through them from the pole. He thus observed that the comet moved toward the pole, not sideways, and concluded that it had come from Gemini, not Cancer or Taurus. A method similar to Geoffrey's was employed by Maestlin when he observed that the nova of 1572 did not move.[189] The significations ascribed by Geoffrey to the comet of 1337 resembled those that

187 *Ibid.*, III, 286-7.

188 The solar eclipse of the preceding March 3rd.

189 See chapters II and III, below.

he ascribed to the earlier one. The effects were to last about two years.

Giovanni Villani, a victim of the Black Death, disagreed with Geoffrey's conclusion that the comet of 1337 proceeded from Gemini, not Cancer or Taurus. He noted two comets in that year: one in Taurus, named Ascone, lasting four months; the other in Cancer, called Rosa, and lasting two months.[190] He noted a horrible pestilence in Florence in 1340 which he connected with a comet in Virgo and the beginning of Leo in March.[191] He also remarked on the appearance of the comet Negra in Taurus in 1347. It was of the nature of Saturn and signified deaths of rulers and great mortality in the regions under Taurus and Saturn.[192]

Conrad of Megenburg (1309-1398) also wrote about the comet of 1337,[193] which he himself observed in Paris. He said that it lasted more than four weeks, was near Ursa Major, and turned its tail toward the German lands. It was in the eleventh chapter of the second section of his *Buch der Natur* that he spoke of the comet of 1337, and there he also said that the air is divided into three regions, in the highest of which comets are seen. He is particularly noted for his nomenclature of directions in the heavens: " Mittag " for south, " Sonnenaufgang " and " Sonnenuntergang " for east and west, and " Himmelswagen " for north. The word " Mittag " was used until the end of the eighteenth century, but from the sixteenth to the eighteenth, " Mitternacht " was used for north.[194] In the diagrams and texts of tracts on the comet of 1577 this usage can be observed.

190 Thorndike, III, 287 and 287 note 16.

191 *Ibid.*, III, 232.

192 *Ibid.*, III, 316.

193 Hellmann (1891), 5-13, especially 8 note 2. See also A. D. B., XVI, 648-650; Sarton, II, 593, 786; Sarton (1936) and the answers to Sarton (1936), in later issues of *Isis*, concerning the early editions of Conrad's *Deutsche Sphaera.*

194 Zinner (1934), 5.

Augustine of Trent in a work written in 1340, in connection with "the pestilence of infirmities" in that year, regarded the influence of Mars and the appearance of two comets as evil influences for that year. He warned young people to be particularly careful because of the comet in Leo.[195]

In the second half of the century, the comet of 1368 attracted much attention, calling forth an astrological interpretation of its significance from John of Legnano at Bologna, which was cited as late as 1431 in an annual prediction for that year.[196] In many ways John's treatise on the comet bears a close resemblance to the less important sixteenth century cometary tracts. His scientific observations are not so good as those of his predecessor, Geoffrey of Meaux, or of his successor, Jacobus Angelus;[197] but they do illustrate the general upward trend in the fourteenth century. First he considered what a comet is, then the different kinds of comets; and as a third step he presented a natural physical explanation of them followed by a treatment of their astrological significance in relation to the signs and planets. He then gave an astrological treatment of the comet of 1368; and finally listed some of the notable comets of the past.[198] His view was the usual Aristotelian one that a comet is neither a star nor a part of the sky. He thought that there were two ways of explaining the effects of comets, naturally as in Aristotle's *Meteorologica,* and astrologically in accordance with the signs and planets to which they were related. Using the first method he showed that the natural results of comets included winds, floods, wars, deaths of princes and religious changes, comets making men choleric so that they were consequently inclined to wars. Princes, living a more dissipated life, were especially choleric; and because they spent so much time in wars, they were exposed to death. From history, examples can be drawn to show that great changes, such as the

195 Thorndike, III, 227.

196 *Ibid.,* III, 492, 597.

197 *Ibid.,* III, 595.

198 *Idem.*

Norman conquest, were preceded by comets. According to John, a comet could not be wholly a good sign although it might bring good in one place. He was not sure whether the comet of 1368 was in Taurus or Gemini, and so gave predictions for either alternative. In the beginning, he wished to place it under Mars, which was in Taurus, but taking the comet's color into consideration, he placed it under Saturn, at that time in Sagittarius. He thus illustrated the foundations upon which fourteenth century critics of astrology based their objections, the difficulty of telling in what sign or under what planet a comet was. John's treatment of the relation between astrology and religion was none too clear. He believed that three comets under Nero marked the spread of Christianity but that this was due to supernatural divine virtue, not to the force of the planets. John's authorities, which he often stated precisely, were the *Quadripartitum* and *Centiloquium,* the works of Haly and Albumasar, and those of Michael Scot and Leopold of Austria. But he added that true Catholics should place their faith in Augustine's writings and the laws of Justinian.[199]

Two men who wrote on comets in the second half of the fourteenth century are noted for their attacks on astrology. They are Nicolas Oresme and Henry of Hesse or of Langenstein. Nicolas Oresme, at his death in 1382, left behind him, among other writings, French translations of and commentaries upon various works of Aristotle,[200] a work on divinations, first written in French in December 1361 and later translated into Latin,[201] and an early work, *Tractatus contra astronomos judiciarios.*[202] In the last named and in many others of his works Oresme criticized astrology, but with little lasting

199 *Ibid.*, III, 595-6.

200 *Ibid.*, III, 398.

201 *Ibid.*, III, 401; Curtze (1870), 17-9; Meunier, 48-58.

202 Jourdain, 145-6; Curtze (1870), 16-7. The text of this tract, with the title *Tractatus magistri Nicolai Orem contra astrologos,* has been printed in Pruckner, 227-245. Curtze (1870), 17, seems to have considered this work as covering pretty much the same ground as the work on divinations. However, it does not deal with the subject of comets.

effect.[203] He is important as one of at least three medieval scholars who considered the possibility of the motion of the earth, supposing the eighth sphere of fixed stars to be immobile.[204] His commentaries on Aristotle gave him the opportunity of expressing himself on this matter. The work on divinations is of especial interest here because in it Oresme admitted the possibility of general conjecture concerning wars and pestilences from comets. He did not believe that comets necessarily brought evil, supporting his point by Seneca's report that the comet under Octavian was beneficial.[205] Furthermore, in the fourth question of his *Quotlibeta,* Oresme conceded that comets and planet conjunctions produced changes on earth, but he doubted the possibility of accurate forecasts of these.[206]

Henry of Hesse (1325-1397) went further than Oresme towards scepticism,[207] denying prediction from comets. He supplemented Oresme's onslaughts on divination or astrology by specifically criticizing particular parts of astrology such as the belief in comets and planet conjunctions as signs or causes of future events. Upon the appearance of a comet in 1368, Henry wrote his *Questio de cometa,*[208] in which he denied that the appearance of a comet was a prognosis of future events.[209] Al-

203 Thorndike (1929), 22, wrote: " The occult sciences lost nothing of their hold upon the human mind during the fourteenth and fifteenth centuries, and continued to be inextricably combined with the natural and mathematical science of the time."

204 Thorndike (1929), 17, 141; Borchert. The other two were Franciscus de Mayronis and Albert of Saxony. In the fifteenth century, Nicholas of Cusa had the same idea.

205 Thorndike, III, 417-8.

206 *Ibid.*, III, 418.

207 *Ibid.*, III, 492.

208 Roth, 97. The text of the *Questio de cometa* can be found in Pruckner, 89-138. Pruckner also gives an excellent discussion of Henry's writings against the astrologers in general and of his *Questio de cometa* and *Tractatus contra astrologos coniunctionistas* in particular.

209 Thorndike, III, 493; Pruckner, 23. In this respect, Henry's work can be compared to Bayle's much later essay.

though Henry adopted, as his scientific premises, Aristotle's incorrect theory of the origin of comets,[210] since this was the generally accepted theory, his attack on prognostication from comets may have had influence. Henry thought that comets formed only in mid-air and so could have no effect on the earth.[211] Although winds may accompany comets and even cause them, comets do not produce winds. It is true that pestilence often follows comets, but the reason is that both are due to the same cause, the exhalation from the earth of the pestilential vapors within it.[212] Henry denied the association of a comet with a particular constellation, holding that astrological influence was not necessary to account for the natural attendants of comets.[213] He strengthened his argument by pointing out the difficulty in determining the exact spot where a comet first appeared,[214] which, as we have seen, was considered necessary in making astrological predictions from comets. Henry also discussed the size, shape and motion of comets. He thought that their circular motion was derived from the diurnal movement of the heavens, although he believed comets belonged solely to the inferior world.[215] Furthermore, he did not approve of the custom of ascribing all unusual events to comets, eclipses and conjunctions.[216] According to Henry, the comet of 1368 began on the evening before Palm Sunday and lasted three weeks. In the three days following its first appearance, it moved three or four degrees nearer the pole.[217] But Henry's observations of the comet are not given in detail and might be said to be on a par with those of John of Legnano.

210 Thorndike, III, 493.

211 *Idem*; Pruckner, 23.

212 Thorndike, III, 493.

213 *Ibid.*, III, 493-4; Pruckner, 25.

214 Thorndike, III, 494; Pruckner, 27.

215 Thorndike, III, 494, 755-7.

216 *Ibid.*, III, 495.

217 Pruckner, 23, 41.

At the close of the fourteenth century, probably between 1389 and 1395,[218] Cardinal Pierre D'Ailly wrote a commentary on Aristotle's *Meteorologica.* It was printed in Strasburg in 1504 [219] and in Vienna in 1509.[220] Although merely "an abbreviated paraphrase" [221] of Aristotle's work, it must have served to help keep alive Aristotle's theory of comets. D'Ailly's work on Manilius' astronomy [222] may have had a similar effect.

With the opening of the fifteenth century, the history of the development of theories about comets enters a new stage. Therefore, it might be well if this survey were interrupted long enough to summarize briefly the heritage of the fifteenth century observers. Up to and including the time of Seneca, a great many theories of comets were presented, in fact, nearly all the views which were to become the heritage of sixteenth century astronomers. In earliest times, the Chaldeans thought that it was possible to foretell comets, although no method of doing so was devised. Later, Apollonius of Myndus believed comets planetary, and Seneca himself envisioned a theory of comets based on that of planets, expressing a belief in comet orbits. However, he had no clear theory, merely a trust in "natural law." On the other hand, many learned men did not believe that comets were planets. Aristotle had his own, now well-known, theory of comet generation. Others had believed that comets were formed by planet conjunctions or that fixed stars might acquire tails and thus become comets. It was also suggested that comets' tails were due to solar reflection, a theory held by Tycho in the sixteenth century. Furthermore, efforts were made to classify comets. Both Aristotle and Epigenes postulated two types, in accordance with the shapes of comets. Seneca's classification of comets was more elaborate but on the same principle. Like-

218 Salembier, 369.

219 Goldschmidt, catalogue 25.

220 Poggendorff, I, 19.

221 Thorndike, IV, 102.

222 Salembier, 369.

wise, belief in comets as omens began before the beginning of the Christian era, and was increased and elaborated later on. Aristotle held that comets foreshadowed wind and drought, as was consistent with his theory of their generation. The disaster at Helice and Bura was connected with the comet of 371 B.C., and, at a later date, Manilius believed in even more severe consequences of comets. Seneca placed more emphasis on comets as weather signs than had Aristotle.

Shortly after Seneca's death, Pliny made an elaborate classification of comets, arranged by color as well as by shape. Furthermore, Pliny adhered to the Aristotelian theory of the generation of comets, which was fast becoming the only accepted one. In addition, Pliny placed increased credence in the power of comets over the future, a credence which was later strengthened by Ptolemy's astrology, in spite of Arrian's attempt to prove that comets announced neither good nor bad events.

During the twelve centuries which followed the death of Ptolemy, little was added to cometary theory. Belief in comets as evil omens was strengthened. Comets were associated with particular planets or constellations, and it became of utmost importance to determine the place in which a comet first made its appearance. However, the fourteenth century witnessed some attacks on astrology. The previously established classifications of comets were repeated and somewhat elaborated. The chief value of those years to the development of the theory of comets was in the continued observations of those phenomena and the consequent increase of data on the subject, and the repeated efforts to interpret comets.

CHAPTER II

COMETARY THEORY FROM THE BEGINNING OF THE FIFTEENTH CENTURY TO 1577

THE fifteenth century opened very propitiously as far as cometary history is concerned. Two comets appeared in the year 1402 and the first [1] of them was ably described in a treatise by James or Jacobus Angelus of Ulm. In this treatise the author was described as Iacobus Angeli, from Ulm, as a master in arts and a licentiate in medicine and as physician to Duke Leopold of Austria.[2] The information concerning the printed copies of the treatise is confusing.[3] It seems to have been writ-

1 The first comet was thought by Pingré, I, 449, to be the comet of 1661 in an earlier appearance. Pingré cited Struyck as identifying the comet with that of 1702. The second comet of 1402 was seen in Constantinople in the summer (Pingré, I, 449-451).

2 Angelus, C_6 r. (Angelus, A_2 r, speaks of Duke Leopold of Austria, Styria, and so forth.) See also Collard, 82; Thorndike, IV, 80-1. James said that he was in Paris in 1382 (Angelus, C_4 r), to which Thorndike, IV, 81, adds "very likely as a student". Collard, 82, went so far as to say that James, the son of a pharmacist, studied in Paris in 1382, thanks to a subvention from the magistrates of his native town. Certainly, James (Angelus, A_2 r) thanked the magistrates, council and citizens of Ulm for their generosity to him. According to Collard, 82, James returned to Ulm but did not remain there long, and no trace of his activity as physician to Leopold has been found. However, he was registered at the university of Vienna for the winter session in 1391 (Collard, 82-3). There were other men of the same name who should be distinguished from the author of the treatise on the comet of 1402. See Thorndike, I, 105-6, IV, 82, and Poggendorff, I, 47, and various catalogues. A Johannes Angelus edited the 1491 edition of Bonatti's *Liber astronomicus.* Joh. Engel or Angelus from Aischach, who was "Magister" in Ingolstadt, wrote a calendar for 1484 (Zinner (1934), 87). Hans Engel (Hellmann (1899), 29), who was the author of the *Dewtsche Practick* for 1488, was identical with Johannes Angelus, who came from Aichen in Bavaria, was professor of astronomy in Vienna and produced several astrological works.

3 Thorndike, IV, 80, says that the treatise is extant both in a manuscript copy preserved at Erfurt (where the author is described as Iacobus Engelhart) and in an incunabulum edition printed at Memmingen in Bavaria about 1490.

ten in 1402 and deals with the comet which appeared and caused great wonder in the beginning of February of that year.[4] It was divided into ten chapters, and the first part seems to have been based largely on Albertus Magnus' commentary on Aristotle's *Meteorologica*.[5] James told of the theory that a comet is a conjunction of several stars called planets, and is of divine origin, saying that according to Seneca this theory was held by Anaxagoras, Democritus and Apollonius, and that it was described by Aristotle.[6] James also cited Italian philosophers, called Pythagoreans, as saying that a comet was not a star but a planet, and Hippocrates [7] as saying that comets sprang from stars or planets, the tails being due to the reflection

See Angelus in the general bibliography below. This is the copy which Thorndike used (Thorndike, IV, 80-1, n. 4). The Cornell copy of the tract is not dated (except on the modern binding which gives the date as 1480), which is also true of the B. M. copy to which the Cornell copy seems to correspond (see Robert Proctor, III, 608, where the book was also assigned to Memmingen). Collard knew about the B. M. copy through Proctor's *Index* (Collard, 83), but he used the copy in the library of the Belgian royal observatory (Collard, 82). This was undoubtedly the copy described by Houzeau, 5534. Houzeau gave it the date 1480 in parenthesis, indicating that that copy also was not dated. Collard, 83, believed that the treatise was printed by Johannes Reger of Ulm about 1480 and that the 1490 *Tractatus de cometis* was the work of Jacobus' son. Collard's summary follows the Cornell copy so closely that it must be concluded either that the tract in Belgium and the tract in America are copies of the same edition of the same treatise or that there were two incunabula editions of it. Collard, 82, giving the printed work the date 1480, believed this the first printed astrological treatise on conjunctions, eclipses and comets. Collard's summary of the tract is not as reliable as that given by Thorndike, IV, 82-7.

4 Angelus, B_4 r ff.

5 Thorndike, IV, 82-3. According to Collard, 84, James said that the Arabic epoch and the western Middle Ages, up until the time of Regiomontanus, remained faithful to the Greek theory which confused shooting stars, balls of fire, and comets with thunder and lightning. This is an obvious interpolation by Collard, himself.

6 Angelus, A_2 r-A_2 v. Collard, 84, wrongly said that Angelus ascribed this theory to Aristotle.

7 "Ypocras et socii sui Nichius et Paulus qui concordant cum pictagoriciis ..." (Angelus, A_2 v).

of light.[8] He also cited John of Damascus, Seneca, Aristotle, and Albertus Magnus. He discussed general phenomena which he deemed atmospheric and due to earthly exhalations, and then [9] set forth the Aristotelian theory of the generation of comets. He mentioned the opinions of Agasel [Algazel], Alpetragius, Haly,[10] as expressed in his commentary on the *Centiloquium,* and Albumasar, coming to the conclusion that comets are elementary, not celestial. Discussing the various shapes of comets' tails, he added that the tail of the comet of 1402 was shaped like a pyramid. In the fourth and fifth chapters [11] James dealt with the various types of comets, following Bonatti's nine classifications, and giving their nine chief effects after the manner of Albertus Magnus.

Beginning with chapter six, James discussed the comet of 1402.[12] The comet was first visible in Swabia about the beginning of February, 1402, and was still visible in Ulm on March 15th, when it set at the point where the sun sets at the time of summer solstice. It was slightly larger than Venus as a morning star and was similar to it in color but not so bright. The tail was for a time white, not very long,[13] directed upwards, and, at first, pointed south, then north. It grew in thickness and brightness so that by March 15th it appeared as long as a lance and pyramidal in shape.[14] When the comet was first seen, Mars was in the last division of Aries, or approximately there. Although he did not see the comet then, James believed that it was there

8 Angelus, A_2 v. Collard translated this passage as meaning that the Italian philosophers believed a comet to include only one star.

9 Angelus, A_5 v-B_1 v (chapter III).

10 This must be the Egyptian, 'Ali ibn Ridwān. See Chapter I, note 128 above.

11 Angelus, B_1 v-B_2 r and B_2 r-B_4 r.

12 Angelus, B_4 r-B_4 v. A French translation of this passage was given by Collard, 86-7.

13 Collard, 87, said "puis fort longue".

14 Angelus, B_4 r.

at the same time and that Mars and the comet were in conjunction in the third and last division of Aries.[15] So on that occasion, at some time it was true that a circle passing through the pole of the zodiac and through the middle of the sky [zenith] passed through the centers of both; nay rather, on the 22nd of March it [the comet] was seen at the second hour near the sun at the distance of a lance to the north.[16] It was evident, therefore, that the comet moved contrary to the sun, across a great distance from the north toward the south, where it finally disappeared on the feast of Passover on March 26th or 27th. The comet seems to have been lost in the rays of the sun. Traces of it appeared in the east before sunrise, when James saw three long, thick hairs, and in the west after sunset, when he saw one hair. Since it was thought that both Mars and the sign Aries are of a hot and dry nature, and that a comet originates from a hot and dry exhalation, the comet's influence was supposed to be characterized by the above mentioned qualities.

Next, James gave the astrological significations of the comet of 1402.[17] Setting forth the general rules for prediction from comets, based on their density or rarity, on the lands over which they appeared, and the causes of their generation, he described the three "dispositions in the air" which, in Swabia, had pre-

15 Angelus' phrases, "in vltima facie" and "in tertia facie seu vltima arietis", were translated by Collard as meaning "in the last phase" and "in the third and last phase of Aries". There is no authority for this meaning, especially since constellations don't have phases. Thorndike, IV, 83, retains the original Latin word "facies". It seems safe to use the translation "division", suggested by Professor Thorndike, since the expression "third and last" would imply a division into three, giving each division the value of 10°. Professor Thorndike furnishes the following two citations: 1. "Facies autem signorum sic distinguuntur secundum quod unum quodque signum dividitur in tres partes equales quarum quelibet vocatur facies et quelibet earum constat ex decem gradibus." (Guido Bonatti, II, 18 [wrongly numbered 17 in Venice, 1506, edition]) 2. "De faciebus. Scias quod in quolibet signo sunt tres facies . . .". (Haly, I, 3).

16 Collard, 87, misconstrued this passage as meaning that the comet had the length of a lance.

17 Angelus, B_4 v-C_8 r (chapters VII and VIII).

ceded the comet of 1402.[18] The first of these occurred on the eve or day of St. Paul's conversion in 1399, when there was a universal and uniform light in the air for a considerable time after sunset, when it was too dark to see indoors without a light. It was not from the moon, the sun, a star, or any reflection of light, but must have been due to a widespread exhalation causing a bright flame or fire throughout the horizon. The second phenomenon was in the autumn of 1400, when a flame as long as a lance with a head like a calf and growing narrower toward the tail came from the west toward the east through the air and vanished in the pinnacle of a house while James was watching. Similar fires were seen throughout Swabia in the twilight at the same hour. The third disposition was the appearance of thunderstorms at the beginning of April, 1401, which lasted to the end of August. Much of the growing food was destroyed. James concluded that these phenomena were forerunners of the comet and bore witness to the continual rising of earthly exhalations, and that the comet's influence would be felt in the places where these dispositions had appeared [Swabia]. Astrologers judge a comet's effects by the direction of its tail. On that basis James concluded that Spain, France, England, Scotland and southern Germany were among the countries threatened by the comet. Referring to Bonatti, James concluded that if this comet was of the type "cenaculum" it presaged want and religious wars; if of the type "pertica," drought; if it was related to Mars it signified wars. James thought that its appearance in Aries meant that it would affect the Italians. He recalled that the comet of 1382 was followed by several conjunctions and subsequent ills in 1385, and he was fearful of what might happen after the conjunction of Saturn and Jupiter three years later on January 12th, 1405. He cited Albumasar as saying that such a conjunction signified changes and serious and dangerous events.

18 The passage dealing with these "dispositions" was reprinted by Thorndike, IV, 664-5.

In the ninth chapter James gave examples of the significations of comets from the time of Nero to the beginning of the fifteenth century, and showed his belief that the effects of a comet were not carried out in the first year after the comet's appearance but over a period of six years.[19] In the final chapter he tried to show that faithful Catholics ought not to mind the significations of this or other comets, but commit everything to God, and tried to allay the fears previously aroused.

The comet of 1402 was hailed in Wales as a favorable sign for the rising of Owen Glendower.[20] According to Simon de Phares, Gilles de Louviers, a canon of Paris, predicted the comet of 1402 as well as one in 1399, and according to the same authority, Peter of Monte Alcino based a prognostication on the comet which was in the twenty-eighth degree of Aries under Mars on February 25th, 1402.[21] Blasius of Parma was another to whom was ascribed the prediction of the comet, Verru, of February 25th, 1402.[22] In his *Judgment of the Revolution of the Year 1405,* Blasius predicted comets among other events which he considered gloomy.[23] The comet of 1402 was also mentioned in the *Chronique de Jean Stavelot,* where the comet was said to have been seen from February 25th to March 19th.[24] In Austria, the same comet was the subject of a tract by Friedrich von Drosendorf.[25]

Angelus' detailed treatise on the comet of 1402 was followed in the middle of the fifteenth century by increased observation and mention of comets, showing a new stage in cometary history. Among the observers was Herman Schedel (1410-1485),

19 A major portion of this chapter was reprinted by Thorndike, IV, 662-4.

20 Thorndike, IV, 87.

21 Phares, 236, 241; Thorndike, IV, 80.

22 Thorndike, IV, 78, on the authority of Simon de Phares.

23 Thorndike, IV, 76.

24 Collard, 87.

25 Zinner (1938), 14.

the physician.[26] Another was Nicholas Comes or Niccolò de Comitibus of Padua. In 1450 he addressed a treatise on weather prediction and other astrological judgments to Malatesta de Malatestis, in which comets were mentioned among the more apparent signs by which unskilled persons might foresee weather changes.[27]

Halley's comet appeared in 1456 and was visible almost throughout the month of June.[28] A *Judgment* based thereon was written at the university of Vienna.[29] Platina, probably the noted humanist (1421-1481), wrote on that comet,[30] and it was also observed in Nuremberg.[31] Two comets were said to have appeared in Rome in that year and Pope Calixtus III ordered prayers; but it is not certain whether he was influenced in this by the comets.[32] At any rate, according to Nicolaus e Fara, living in Budapest, the two comets were thought by John of Capistrano to be signs of Christian victory.[33] In 1456, possibly because of the comet of that year, William de Bechis of Florence addressed a treatise on comets to Piero de' Medici. He maintained a conservative religious position, saving freedom of the will, and held that comets were signs, not causes.[34]

One of the many fifteenth century authors who associated earthquakes with comets and discussed the two kinds of phenomena in the same tracts was brother Matthew of Aquila, professor of sacred theology of the order of Celestines. He wrote on the earthquake of 1456,[35] and, among its important

26 Zinner (1934), 19.

27 Thorndike, IV, 250-2.

28 Pingré, I, 459-465.

29 Thorndike, IV, 413.

30 Proctor and Crommelin, 46.

31 Zinner (1934), 66.

32 Thorndike, IV, 414.

33 *Idem.*

34 *Ibid.*, IV, 298, 417. William de Bechis began his studies at the Augustinian convent in Padua in 1433 and later (1470) became bishop of Fiesole.

35 Thorndike, IV, 416-7.

antecedents, mentioned a comet appearing May 18th. In his treatise he devoted much space to comets. A follower of the Aristotelian ideas concerning their generation, he considered them not only signs but also causes of coming evil, because their hot, putrid vapors contaminated the air. He distinguished three comets in 1456, two before and one after the earthquake which took place on December 5th between the tenth and eleventh hours of the night.

The well known Italian humanist, Giovanni Pontano [36] (1426-1503) was another of the men who wrote on Halley's comet in its 1456 return.[37] In his *Meteororum Liber* he devoted two sections to comets in general.[38] At the close of his commentary on the *Centiloquium,* which he attributed to Ptolemy, Pontano wrote a considerable discussion on comets. In this he adhered to the Aristotelian theories, and discussed the astrological implications of comets.[39] That his works were still read at the time of the comet of 1577 is evident from the fact that part of the second section on comets in his *Meteororum Liber* was printed in Bazelius' *Prognosticon nouum.*[40]

As a result of the appearance of the comet of 1456 many questions were put to Peurbach, who replied that comets had appeared before, and recalled the paths of the comets of 1402 and 1433.[41] A manuscript tract on the comet of 1456 has been attributed to him. He tried to measure the distance of the comet, an attempt which has been called the first of its kind.[42] Regiomontanus, too, saw the comet of 1456, but he did not

36 See Tallarigo, especially I, 482-512 (chapter IV, "L'Astrologia"), and Boffito.

37 Proctor and Crommelin, 46.

38 Pontanos (1902), I, 195-7, 215-7.

39 Pontanos (1519), folios 90-3.

40 Item 10 of appendix, below.

41 Zinner (1938), 23. See also Thorndike, IV, 413, note 2.

42 Zinner (1938), 26, calls Peurbach's work comparable to Toscanelli's.

make continuous observations of it nor of the comet of the following year.[43]

The most important astronomical observations of the 1456 appearance of Halley's comet were undoubtedly those by Paolo dal Toscanelli[44] (1397-1482). Toscanelli's observations were preserved in manuscript only, until the nineteenth century.[45] Indeed, they did not become as well known as they deserved and had little influence on the progress of cometary science. His manuscript on comets, as would be expected, showed evidence of belief in astrology. There were included both a prediction by Pietro Bono Avogaro on June 17, 1456 of the effects of the comet and a similar one of his own.[46] Toscanelli also observed the comets of 1433, 1449-50, 1457 and 1472. His observations of the comet of 1433 were so accurate that they furnished all the data necessary for calculating the comet's orbit.[47] His observations of the comet of 1449-50 were far superior to those by others, and his observations of Halley's comet in 1456 are a valuable addition to the knowledge of the motion of that body.[48] In 1457, two comets appeared, and again Toscanelli was at work.[49] Finally, the comet of 1472 furnished the object of new activity on his part.[50] It is because of the accuracy and detail of his observations that he falls in line with Geoffrey of Meaux, Jacobus Angelus and Regiomontanus as giving evidence of gradually increasing wealth and accuracy in the observation of comets. These are part of the necessary background for any improvement in theory, and, together with the introduction of

43 *Ibid.*, 40, 212.

44 See Uzielli. Uzielli, 308-385, was republished, with additions. See Celoria.

45 Celoria, 3.

46 Thorndike, IV, 432, questions Celoria's conclusion that Toscanelli turned from faith in astrology to distrust of it.

47 Celoria, 5-10.

48 *Ibid.*, 10-36.

49 *Ibid.*, 37-46.

50 *Ibid.*, 46-57.

the use of the theory of parallax in regard to comets, render the fifteenth century outstanding in the progress of cometary astronomy, although, of course, many observers after 1456 continued to make incomplete and inaccurate observations of comets.

When in 1458 King Alfonso died and Lorenzo Bonincontri di San Miniato lost his wife and two of his sons in a pestilence, the latter ascribed the loss to the baneful influence of two comets,[51] undoubtedly the two observed by Toscanelli in 1457. Simon de Phares said that Perre de Graville, a Norman master at Paris, made a prediction from a comet in 1465, but possibly the date should be 1456.[52] After the comet which appeared in 1468, Martin, archdeacon of Sora and canon of the church at Zagreb, wrote a work containing specific predictions such as the death of George of Podiebrad. Martin favored the contention of some, that they saw the comet three days before September 22nd, but in his title he used this date for the comet's appearance.[53] According to Martin [54] the comet first appeared near the front paws of the Bear and then traversed Leo and Virgo. He believed that the comet originated from exhalations and discussed what constellations caused it, by what planet it was governed, and the evils it would bring. He made use of the *Quadripartitum* and maintained that the belief that comets meant death to kings was supported by experience as well as by authority. Reviewing the evils following past comets, Martin proclaimed that the comet of 1468 was a threat to Pope Paul II. In 1472, Martin (probably the same one though then called a parish priest of Buda) addressed a judgment on the comet of that year to Matthias Corvinus.[55] An anonymous

51 Thorndike, IV, 405, 407.

52 *Ibid.*, IV, 418. Pingré, I, 466, mentioned for 1465 only a comet visible in Japan and China.

53 Thorndike, IV, 419-421; Zinner (1938), 109, who probably took his information from Thorndike.

54 Thorndike, IV, 421-2.

55 *Ibid.*, IV, 424.

treatise addressed to Pope Paul II was called forth by the comet of 1468. This treatise inquired into the cause and material of this comet and of comets in general. It followed Albertus Magnus and twice cited Leopold of Austria, but limited the prognostications to generalities.[56] John Lichtenberger, a German astrologer, said, in another work, that he had written a treatise at Speyer on the comet which appeared in Gemini September 22, 1468.[57]

The comet of 1472, which is generally dated in January, seems to have inspired more treatises than did the comets of 1456 and 1468 [58] and probably had many more observers. John of Glogau, in his astrological *Summa,* called that comet a great comet.[59] Because of it Pietro Bono Avogaro wrote a treatise on comets [60] and Henry Sutton wrote a prognostication concerning the phenomenon of 1472, which he called a tailed or bearded star.[61] Angelo Cato de Supino of Benevento, who called himself both a philosopher and a physician, wrote a treatise on the comet.[62] He called it " Pogonias " and said that it had been unequalled in the past 1500 years.[63] He said that it first appeared on January 7th but that the material for it had begun to conglutinate on January 1st; that at first it was scarcely visible but that by the eighth day it equalled a second magnitude star in size; and that by the fourteenth day it was

56 *Ibid.*, IV, 418-9, 700-1.

57 *Ibid.*, IV, 422.

58 *Idem.* Certainly, more are preserved, but this may be partly due to the increased use of printing, partly to the later date and consequently greater chance of survival.

59 Thorndike, IV, 422, 453.

60 *Ibid.*, 422.

61 *Ibid.*, 428.

62 The treatise was printed by Sixtus Riessinger in Naples in 1472 and was addressed to Don Juan of Aragon, son of King Ferdinand. It was also written for the glory of the university of Naples. Thorndike, IV, 425. The treatise was also mentioned by Riccioli, II, 1.

63 Thorndike, IV, 427.

almost as large as the moon, with a tail 36° in length by 4° in breadth, sweeping out a sixth of the sky, a degree being estimated as 5000 miles [64] at the distance of the comet. Angelo added that the comet was forecasting great ills, especially for kings, and drew general conclusions as to its effect. He ended by adding that God if He wished could change any or all the evils threatened by the comet.[65] Another tract on the same comet was, according to Simon de Phares, written by Laurens Hutz. He wrote a prediction from the comet which appeared on St. Agnes' eve [January 20th] in 1472 and addressed it to Louis XI.[66]

A Pole, John de Bossis, lecturer on astrology at Bologna in 1471 to 1472, wrote a treatise on the comet of 1472 at the request of Bishop Nicholas. In this treatise,[67] John gave twelve efficient causes of the comet, which included the positions and conjunctions of the planets on certain dates. He rejected the opinions of the ancients but gave the usual explanation that comets were exhalations from the earth rising into the upper air and refined by the action of "superiors" or the influence of heavenly bodies and assimilated to the nature of the sky so as not to be quickly consumed. He said that the present comet was generated in the fifteenth degree of Libra, its movement governed both by the diurnal motion of the primum mo-

64 Angelo thought that comets circulated where the sphere of air and fire meet. See Thorndike, IV, 427. It is difficult to believe that Angelo meant to give the value of 5000 miles to a degree at that distance. It leads to the surprisingly accurate belief that comets are approximately 300,000 miles away, and is entirely out of line with his general reasoning. Angelo does not seem to have made any great departure in theory, and it seems reasonable to suppose that if he had intended to place comets at such a great distance, greater than the moon's, he would have said so, more especially since his spheres of air and fire must have been sublunar. Angelo has not given any basis for his estimate, nor value for his mile, unless Professor Thorndike omitted it from his resumé.

65 Thorndike, IV, 427-8.

66 *Ibid.*, IV, 425.

67 *Ibid.*, IV, 422-4.

bile and by the virtue of the fixed star whose nature first attracted it. The varying velocity of the comet was due to its attraction by fixed stars of the nature of Mars or Saturn in the neighborhood of which planets its movement was retarded. Its exceedingly rapid movement at one time was due to the virtue it absorbed from its dominant fixed stars and planets. John also described the tail (not its direction), the course and the color of the comet and its significations. He devoted a whole page to the Pope, whom the comet threatened, and also predicted danger to the emperor.

In Bamberg the comet of 1472 was observed by Antonius von Rotenhaus. His chief interest was in the meaning of the comet. His treatise was nearly a word for word transcription of 1456 observations, and even copied Regiomontanus' data for that comet.[68] An author, whose name may possibly be Laurentius Cerastius of Viterbo, discussed the following seven points about the comet of 1472 :[69] what a comet is in kind; what the present comet should be called; in what sign it appeared; when its effects would begin; what it signified; how great would be the evils from it; and how long they would last. He besought God to ward off all evil, although one evil announced by the comet he would tell only orally, not in writing. Among the many judgments on the comet of 1472, were one by master Valentinus Zathor, and one by an astrologer in the town " Newmarckt " near Nuremberg.[70] An anonymous printed work on the comet of 1472 was published in Rome in that year.[71] Its author, possibly Nicholas Hartman, attributed the comet to the conjunction of Mars and Saturn in the third degree of Gemini in 1471. He believed that the comet would not injure Pope Sixtus IV, although it was dangerous to those who had Libra in the ascendant of their nativities. About the

68 Zinner (1934), 66 ff.

69 Thorndike, IV, 430.

70 *Idem.*

71 *Ibid.*, IV, 429.

time of the appearance of the comet of 1472, but without mentioning it, Ralph of Rudesheim, a licentiate in sacred theology, considered questions concerning comets, such as whether a comet, originated from matter of the elements, gave true significations of the death of a prince or other notable events.[72] Probably the leading astrologer to busy himself with the comet of 1472, if indeed he was the author of the tract in question, was Conrad Heingarter, also known as Thurecensis and Turicensis.[73] Rather than make predictions of particular events,

72 *Ibid.*, IV, 429-30.

73 *Ibid.*, IV, 357; Hain 15512, 15513; Vollbehr, 21. According to Wickersheimer, 325-6, "il existe un *Tractatus de cometis* imprimé, dont le médecin zurichois Conradus Türst passe généralement pour l'auteur et dont on pourrait se demander s'il ne devrait pas être attribué à Conradus Heingarter. . . . L'auteur ne se nomme pas et fait connaître seulement sa qualité de "Thurecensis phisicus", mais quelques bibliographes . . . ont cru pouvoir avancer que le prénom de ce médecin était Conradus. Il est bien certain que "Thurecensis phisicus" a ici le sens de "Züricher Stadtphysicus".... Or cette qualité n'a jamais appartenu à Conradus Heingarter et par conséquent le *Tractatus de cometis* n'est pas de lui. Cet ouvrage reconnait-il pour auteur Conradus Türst qui fut médecin de Maximilien d'Autriche, ou le Franconien Eberhart Schleusinger de Gassmansdorf, comme l'a écrit... le Dr Friedrich Hegi? (4). [(4) Friedrich Hegi. Neues zur Lebensgeschichte Dr. Konrad Türsts. *Anzeiger für schweizerische Geschichte*, XI (1912), p. 280 et suiv.]..." Hegi, who, having attributed another astronomical tract to Schleusinger spoke, 284, of the tract on the comet of 1472 as 'Dessen "tractatus de Stellis, Cometis, earumque jndiciis [sic] et seorsim de illa, quae A. 1472 Tiguri apparuit"...'. Hegi said that it was printed twice but seemed inclined to think that the printer's date should read 1482, identifying this with the previously discussed tract, for "Augenscheinlich war Schleusinger 1472 noch gar nicht in Zürich, dagegen 1482." Hegi had previously, 283, said that Türst first became city physician in 1489. Hegi did not even suggest Türst as the author of the tract on the comet of 1472. According to Zinner (1934), 66 ff., Schleusinger observed the comet of 1472 in Zürich and attempted to determine its size, but did not claim to have made his own observations and may have taken his measurements from Regiomontanus. According to Ludendorff, there were two incunabula editions and one later edition of Schleusinger's book. Both Zinner and Ludendorff were undoubtedly dealing with the tract attributed in this chapter to Heingarter. Wolf (1877), 182 and note 4, likewise attributed the tract to Schleusinger, and Wolf (1849), 102-3, attributed the following title for the tract to Johann Jacob Wagner: *Eberhardi Schleusingeri de Garmanstorf Franconiae, Artium et Medicinae Doctoris, Physici*

persons and places, Conrad noted the general effects of the comet, and was inclined to consider its influence good. Like Aristotle, he believed comets were not heavenly bodies but were generated from the earth by the action of the stars. This action, according to Conrad, was under the influence of Saturn. Mars caused comets to fly in the air. He did not think that the tail of the comet of 1472 moved with a movement like Mars in its epicycle, or, indeed, that planets had an epicycle.[74] Much of his work shows a tendency to defend astrology, thus implying a certain opposition to it or scepticism concerning it.[75] The first part of his tract on comets deals with generalities, such as what comets are and how they are generated, what are their motion, color, types, what they signify and so forth. The second part is concerned with the comet of 1472 in particular, is largely astronomical rather than astrological, and discusses the comet's size, length, distance from the earth, its motion and that of its tail, as well as its astrological implications.[76]

The greatest contributions to cometary theory which resulted from observations of the comet of 1472 were those of Johannes Müller (1436-1476) of Königsberg, better known as Regiomontanus. Regardless of the tendency of modern historians[77] to detract from the value of his contributions to

Tigurini, Tractatus de Cometis, speciatim de Cometa A. C. 1472. Beronae (Beromünster) 1473. Zinner (1938), 157, attributed the tract on the comet of 1472 to Schleusinger. See also Thorndike, V, 362 note 136. Heingarter seems to have been a German, born in Zürich but active in France. He was a physician as well as an astrologer, and many of his works are both medical and astrological. The only one of his treatises in print seems to be that on the comet of 1472 (Thorndike, IV, 359). It was printed in 1472 by Helias or Heliae of Louffen at Beromünster and reprinted in 1474 by Hans Aurl in Venice or Rome (Thorndike, IV, 359; B. M. catalogue; Hain 15512, 15513; Vollbehr, 21). Wolf (1877), 182 and note 4, called this the earliest printed comet tract. It was later translated into Italian (Thorndike, IV, 360). Pingré, I, 239, mentioned an edition, printed in Basel in 1556, probably that edited by Gratarolo (see Thorndike, V, 600).

74 Thorndike, IV, 360.

75 *Ibid.*, IV, 366-7.

76 *Ibid.*, IV, 360, 692-4.

77 Thorndike (1929), 142-150 (chapter VIII, "Peurbach And Regiomontanus: Their Great Reputation Re-examined").

mathematics and astronomy by bringing to light the work of earlier men, it must be conceded that he made advances in the observation of comets which profoundly influenced later tracts on comets, an influence all the more powerful because of the Regiomontanus "legend".[78] There were many editions, all posthumous, of Regiomontanus' very important work on comets,[79] the scope of which is indicated by the titles of the sixteen problems into which it is divided. These are:[80] (1) "To present a preliminary investigation of the distance of a comet from the earth." (2) "To discuss the variant appearance of a comet in the circle of altitude." (3) "In another manner, to complete the same." (4) "To finish what the preceding sections dealt with, with other arguments." (5) "To know the true place of a comet in the ecliptic with the aid of an instrument." (6) "To measure the variant appearance of a comet in longitude." (7) "To examine whatever apparent latitude a comet has." (8) "To investigate the variant aspect of a comet in the circle of altitude otherwise than above." (9) "To interpret accurately the apparent position of a comet." (10) "To

78 It is not necessary to go to the extreme length which prompted Günther (A. D. B., XXII, 579) to say "Die . . . dem Jahre 1472 entstammende Abhandlung "De cometae magnitudine longitudineque ac de loco ejus vero problemata XVI" erörtert mit vollster Sicherheit alle Fragen, die vor Erfindung des Teleskops überhaupt vernünftigerweise gestellt werden konnten; jedenfalls ist von M. [Müller] bis auf Kepler . . . . nichts Besseres über die Kometen beschrieben worden."

79 See note 78 for the title. It was printed, more than fifty years after Regiomontanus' death, in Nuremberg in 1531 (Crawford library catalogue, 104), and again in Basle in 1544 and 1548 (Will, III, 280), and in Leyden in 1618 (in item [108] of appendix, below). Ziegler, 44, listed a 1544 Nuremberg edition, edited by Schöner. He may have had in mind the following work: *Scripta . . . J. Regiomontani, de Torqueto, Astrolabio armillari, Regula Magna Ptolemaica, Baculoque, Astronomico, & observationibus Cometarum, aucta necessariis, J. Schoneri . . . additionibus.—Item Observationes motuum Solis, ac Stellarum . . . Item libellus . . . G. Purbachii de Quadrato geometrico.* [Edited by J. Schoener] *Norimbergae*, 1544. (B. M. catalogue; H. C. L., Astr. 655.44 A) According to Zinner (1938), 258-9, this work contains Regiomontanus' observations from 1457-75 and Walther's from 1575-1604 [misprint for 1475-1504]. See Thorndike, V, 365 and note 148.

80 Translated from the Latin in item [108] of appendix, below.

measure the distance of a comet from the center of the world and from the center of sight." (11) "To instruct easily how many miles lie between the center of a comet and the center of the earth or the center of sight." (12) "To discover with skill the diameter of a comet observed with an instrument." (13) "To compare the diameter of the body of a comet with the semidiameter of the earth by a fixed proportion." (14) "Finally to measure the amount of material of a comet." (15) "To inquire skillfully the length of a comet's tail." (16) "To investigate, thereafter, the thickness of such a tail." By observing the parallax and consequently the distance from the earth, the rotation time, the apparent diameter, and the size of comets, Regiomontanus, uninterested in the meaning of the comet of 1472,[81] strongly encouraged scientific observation of comets. His failure to attempt any prediction from the comet was a significant departure, indicating a shift of interest and a preoccupation with positive astronomy, and illustrating an attitude exhibited by Tycho a century later.[82] The *Scripta clarissimi mathematici M. Joannis Regiomontani* contain a section on comets by Regiomontanus.[83] His observations of the comet of 1472 were preserved by Jacob Ziegler [84] and Hagecius [85] in

81 Zinner (1934), 77; unlike his contemporaries, Regiomontanus does not seem to have written about the meaning of comets, and his work on the comet to 1472 does not discuss the "meaning" of that comet. According to Zinner (1938), 176, Regiomontanus, in his letters, made the point that he had not mixed his comet works and his astrological works.

82 See chapter III, below, especially note 12.

83 See note 79 above; Zinner (1938), 259.

84 *Jacobi Ziegleri Landaui Conceptionum in Genesim mundi, et Exodum, Commentarij.... Ad haec, Ioannis De Monte Regio libellus, de Cometa.... Basileae, apvd Joannem Oporinum.* The *De Cometa* is on pages 172-4. See Zinner (1938), 267-8. According to Zinner (1938), 156, 268, the work was printed in 1548, but according to the B. N. catalogue it was printed in 1540. At least one authority, Zinner (1938), 157-8, 173, 267, is of the opinion that Ziegler's edition of the *De Cometa* was copied from a passage in the *Tractatus de Cometis*, assigned above to Conrad Heingarter, and that Regiomontanus was not its author. Zinner attributed the *De Cometa* to Schleusinger. Ziegler and Schleusinger were acquainted. However, Zinner did not wish to

the sixteenth century. Regiomontanus saw the comet January 13th, under the sign of Libra, among the stars of Virgo, and continued to observe it until the end of February, and looked for it even later.[86] He found that it moved most rapidly toward the middle of that period, when it traversed four signs in one day.[87] He noted the place of the comet at each observation, and stated toward which stars the tail pointed, remarking that in one night, from sunset to after sunrise, it pointed in four directions, making a circle.[88] Measuring differences in the comet's position in relation to Spica, he found that the comet's parallax could not exceed 6° and that the comet was at least 8,200 German miles away.[89] Regiomontanus has been considered the first

detract from the importance of the *De Cometa* in forwarding the understanding of comets as heavenly bodies. See also Thorndike, V, 370.

85 Hagecius (1574). The volume is a quarto of 176 pages and is 227 mm. high. On the verso of the title-page is a list of the tracts included in the book and among them is one entitled *Ioannes de Monte Regio de Cometa anni 1475*. This treatise occupies pages 146 to 149. The date is an error on the part of Hagecius or his printer. The error seems to have originated in the Ziegler edition (see Zinner (1938), 258). The comet discussed was undoubtedly that of 1472, although several later writers have used the date 1475. See Pingré, I, 68, 234, 472, 475, and 477 where the error was reported due to Ziegler. The Crawford library catalogue, 216, gave the date 1475 without question, as did also Gassendi, 74, and Riccioli, II, 1. There seems to have been no comet in 1475.

86 The statement by Regiomontanus, in Hagecius (1574), 146, reads: "Idibvs Ianuariis, anno Domini 1475. [sic] visus est Cometa sub Libra, cũ stellis Virginis: ... donec cum stellis Ceti occasus Eliacus nobis ipsum occultaret, in vltimis diebus Februarij.... Tamen propter figuram eius ad Solem, & maximè in plagis septentrionalibus, non nisi parum in fine suae apparitionis videri poterat versus meridiem in diebus Aprilis, si motu suo regularitatem seruasset."

87 Regiomontanus in Hagecius (1574), 146.

88 *Ibid.*, 147.

89 *Ibid.*, 147-8. He wrote: "Considerando itaque maximam diuersitatē aspectus capitis cometae à Spica stella sibi vicina, quae iuxta possibilitatem omnibus difformitatibus reductis, maior comprehendi non poterat quàm 6 graduùm, instrumentis congruis ad hoc ordinatis: ad·quam aspectus diuersitatem necessariò sequitur, corpus Cometae à superficie terrae distitisse in nonecupla distantia ad semidiametrum terrae: quae ad minus est octo millia & ducenta milliaria: ..." He used the value 913 German miles for the semi-

to apply the method of parallaxes to the observation of comet distances, opening a path for later observers,[90] although his own value for the parallax of the comet of 1472 was too great. Using a cross-staff,[91] Regiomontanus found the diameter of the head of the comet to be 11′ and that of the coma to be 34′, and from these observations and his calculation of the comet's distance from the earth, he determined its actual size, the diameters of the head and of the coma being respectively 16 [92] and 81 German miles.[93]

diameter of the earth. He also made the more general statement: "Distantia verò cometae à terra comprehenditur ex diuersitate aspectus ipsius cometae, vel alicuius eius partis, ad aliquod Astrum sibi vicinum."

90 Pingré, I, 68. Janssen, I, 120, thought that Regiomontanus was the first western astronomer to determine the distance, size and rotation time of a comet. Fiedler, 10, made no reservation as to "western" astronomer. Presumably Peurbach's efforts, if any, to measure a comet's distance were based on parallax also. See note 42, above.

91 Regiomontanus has been credited (Janssen, I, 120) with the invention of this instrument, but the credit should go to Levi ben Gerson, who made the discovery in 1325 (Darmstaedter, 58; Zinner (1938), 159). Zinner (1938), 159, called the cross-staff a development from an instrument used in the time of Hipparchus, but spoke of Regiomontanus' use of it. Duhem, IV, 40, while attributing the instrument's invention to Levi ben Gerson, said that Regiomontanus used it to such advantage that it was sometimes attributed to him. Duhem, V, 203, said that Levi ben Gerson so improved the cross-staff that it could be used to observe the apparent diameters of the sun and moon.

92 Using Regiomontanus' values for the comet's distance and the apparent diameter of its head, the computed value of the diameter of the head is approximately 26 German miles. Since the value of 81 miles for the diameter of the coma checks with the other figures, it can safely be supposed that 16 was a misprint for 26. Indeed, from Regiomontanus' own computations it is apparent that the error was an arithmetical one, and consequently probably a misprint, for he wrote: "Cùm enim diameter capitis cometae vndecim minuta circuli magni chordaret : vt instrumẽtis deprehendebatur : quorum minutorum chorda est vndecim minuta, & medium fere : illa ergo chorda, vt dictum est in prima parte, multiplicetur in distantiam cometae à terra : id est octo millia & 200 miliaria : & producentur 94 millia, & 300 : quae diuidantur per finum totum, scilicet 3 millia & 600 minuta, exhibunt inde 16 miliaria : quae est quantitas diametri capitis cometae : . . ." (Hagecius (1574), 148-9). $94300 \div 3600 = 26+$.

93 Regiomontanus in Hagecius (1574), 148-9. For a discussion of the value of a German mile see chapter III, note 28 and chapter IV, note 124, below.

There are some records of the appearance of a second comet in 1472, but these probably erroneously distinguish different observations of the comet seen by Regiomontanus.[94]

Bernard Walther, the wealthy Nuremberg citizen who befriended Regiomontanus on the latter's arrival in Nuremberg, and who may have observed the comet of 1472 with him,[95] continued to make observations after Regiomontanus' death. He observed the comet of 1491,[96] and discussed the effects of the comet of 1501, adhering to the idea that comets were formed from vapors from the earth.[97] He is said to have been the first to measure the distance of a planet from two well known stars [98] and he took practical notice of the effect of refraction, observing the sun when it appeared to be outside the ecliptic, although he did not investigate the laws of refraction.[99] However, although he made a great many observations, he was by no means the intellectual equal of his protegé.

In the closing years of the fifteenth century and the early years of the sixteenth, comets were mentioned several times. Writing in 1492 Marsilio Ficino expressed the opinion that comets were caused by celestial beings.[100] Johann Werner, a disciple of Regiomontanus, wrote a *Judgment* on the comet of 1500, but, unfortunately, it has not been printed and there is little available information concerning it.[101] Gregorius Reisch, in his *Margarita philosophica,* which was first printed in 1503 although dedicated in 1495, looked upon comets as earthly ex-

94 Pingré, I, 476.

95 Berry, 88.

96 Zinner (1934), 13; Pingré, I, 478.

97 Zinner (1934), 66 ff. Zinner (1938), 167, said that Walther observed the comet of 1500, which is probably the date Zinner (1934), 66 ff. meant.

98 Dreyer (1890), 345. This must, of course, be the angular distance on the celestial sphere.

99 Dreyer (1890), 336.

100 Thorndike, IV, 563.

101 *Ibid.*, V, 351.

halations which signified coming sterility, pestilence, and seditions.[102] Agostino Nifo of Sessa, in a work *On the Causes of Our Calamities,* which was finished in 1504 although first printed in 1505, devoted the second of four books to comets, thus giving support to the belief in their future significance.[103] In 1506 a bright comet [104] had several observers, none particularly important to this discussion.

Observation of the position and direction of a comet's tail was quite common by the time the comet of 1577 appeared, but not so common as to occur without comment. One of the men who helped to popularize observations concerning the common characteristic of comets' tails, namely that they are always opposed to the sun, and who was frequently cited by writers on the comet of 1577, including Tycho himself,[105] was Girolamo Fracastoro.[106] He studied in Padua, where possibly he and Copernicus exchanged ideas.[107] His fame as an astronomer rests

102 *Ibid.,* V, 139, 141.

103 *Ibid.,* V, 71-2, 162-3.

104 Pingré, I, 481-2. Struve, I, 785, listed a tract on the comet of 1506 by Hansen Virdung von Hassfurt, who undoubtedly should be identified with Johann Virdung of Hassfurt, who, according to Günther (A. D. B., XL, 9) wrote on that comet. See Thorndike, IV, 456-7, V, 401; Hellmann (1924), 30; and the articles by Birkenmajer (Isis, 19(2) : 364-378, June, 1933) and Thorndike (Isis, 25(2) : 363-371, September, 1936; Isis, 26(2) : 321, March, 1937; Isis, 34(4) : 291-3, Spring, 1943); and Hellmann (1924), 30.

105 Brahe, IV, 137 (*De Mvndi Aetherei . . . Phaenomenis,* chapter 7), where the observations of tails were mentioned, and Brahe, II, 379, (*Progymnasmata,* Part II), where Fracastoro's observations of comets were mentioned in passing, and section IV of the German work on the comet of 1577 (item 20a of appendix, below), where the direction of the tail is mentioned.

106 Abraham, 12, on the authority of Professor Barbarani of Verona, and Singer, 1, on the authority of Professor Roberto Massalongo, gave 1478 as the date of Fracastoro's birth. Poggendorff, I, 787; Dreyer, editor, VIII, 456, and Dreyer (1906), 296; Rossi, 34; Tiraboschi, VII, part III, 293, and many other sources, gave the date 1483. Greswell, 456, used 1484. Fracastoro died in 1553.

107 Singer, 2. Copernicus' stay in Padua took place between the years 1501 and 1506 (Favaro (1881), 39), during which time Fracastoro was teaching

on his *Homocentrica*. . . .[108] Its aim was to supplant the Ptolemaic epicycles and eccentrics with a system of concentric spheres somewhat like the system of Eudoxus.[109] Fracastoro's system required seventy-nine spheres, six of which were for the fixed stars.[110] The last sublunary sphere was similar to the one assumed by Seneca and others to account for the motions of comets,[111] for Fracastoro, like others of his time, thought comets sublunar.[112]

In the section on the sun,[113] Fracastoro described several

logic there. Favaro (1881), 43-5, citing Bailly (*Histoire de l'Astronomie moderne en Europe*, II, Paris, 1805, 19-20) and Libri (*Histoire des sciences mathématiques en Italie, depuis la renaissance des lettres jusqu'à la fin du dix-septième siécle* [sic]. III, 2nd edition, Halle, 1865, 100), advanced the suggestion that Fracastoro was a precursor of Copernicus in forming a new system of the world. But, according to Favaro (1881), 44, Siegmund Günther (*Studien zur Geschichte der mathematischen und physikalischen Geographie*, I, *Die Lehre von der Erdrundung und Erdbewegung im Mittelalter bei den Occidentalen*. Halle, 1877, 37-28 [sic]), unaware of the simultaneous residence in Padua by the two famous contemporaries, did not agree on this point. Favaro (1881), 42, 43, thought that there was no doubt that Copernicus had been a pupil of Fracastoro. Favaro thought that although Copernicus had studied logic in Cracow he had to review it and preferred Fracastoro to the other lecturers on the subject. Nevertheless, it is difficult to accept such a statement. In view of Fracastoro's youth (see note 106, above), it seems unlikely that he had any great reputation at the time when Copernicus was in Padua. But, on the other hand, Copernicus, several years Fracastoro's senior, had achieved considerable repute (see Thorndike, V, 408) and already had attended several universities before enrolling in the medical school at Padua. Without disputing the possibility of the two men having heard of each other, it seems doubtful that either of them had as yet reached conclusions he could impart, or, indeed, that Copernicus would have been influenced by the youth, Fracastoro.

108 *Homocentrica. Eiusdem de causis criticorum dierum per ea quae in nobis sunt*. Venice, 1538. This was often reprinted. The first part has the title *Homocentricorum sive de stellis liber unus*.

109 Singer, 6.

110 Berry, 121; Rossi, 97. See also Thorndike, V, 490.

111 Dreyer (1906), 300.

112 Fracastoro, 43r, 44r (*Homocentricorvm, Sive De Stellis, Liber Vnus, Ad Pavlvm III, Pontificem Max.*) See also Thorndike, V, 490.

113 *Ibid.*, 42r-44v (*Homocentrica*, section III, "De Sole", chapter 23, "De diuersitate aspectus in quadraturis, & in motu veloci & tardo."

comets, namely that of 1472, which he called "pogonias", and which was seen in February of that year, and three which he observed and for which he also gave data. These were the comet first seen September 8th or 9th, 1531;[114] the comet of 1532,[115] which he said was visible from September 22nd to December 3rd, and was three times the size of Jupiter, and the motion of which he tried to resolve into its component parts; and finally the comet of 1533 which he saw on July 1st between the Pleiades and the stars in the horn of Aries.[116] Then he made the important observation, as yet not accepted without comment, that comets' tails are always turned away from the sun,[117] drawing this conclusion from his observations of the last three comets and what he had read about that of 1472.[118] This observation, alone, would make Fracastoro worthy of mention here.

An astronomer and mathematician of great repute, who proved vastly important in shaping the history of cometary theories, and who was a contemporary of Fracastoro, was Peter Apian, professor of mathematics at Ingolstadt, astronomer to Charles V and Ferdinand I. He did not reject the Aristotelian theory of the generation of comets, but he did much toward its later invalidation.[119] He observed the comet of 1531,

114 Was this the same comet seen in August by Cardan? Should the word "September" read "August"? See Pingré, I, 489-491.

115 Pingré, I, 493-4. Fracastoro gave both December 3rd and December 8th as the last day when the comet was visible, but the "8" seems to have been an error in printing.

116 Fracastoro, 43v; Dreyer (1906), 300; Dreyer, editor, II, 457; Delambre (1819), 389-390; Rossi, 112-3.

117 Dreyer (1906), 300; Dreyer, editor, II, 457; Dreyer (1890), 166.

118 Fracastoro, 44r, where the following sentence occurs: "Obiter autem nec silebimus vnum, quod commune fuit his tribus cometis, dignum (vt arbitror) relatu: omnes enim comam seu barbam proijcere è directo semper in oppositam Soli partem, vt si Sol in aequinoctiali fuisset versus orientem, barba in aequinoctiali versus occidentem protendebatur, & quantum Sol vnam in partem deflexisset, tantum in oppositam barba illa semper & ipsa sese vertebat: quod & ille etiam cometa fecisse legitur, qui anno 1472 apparuit..."

119 Günther (1882), 57, 61-2.

the second comet of 1532, and the comets of 1533, 1538 and 1539.[120] He noticed the position and motion of a comet and saw that its tail always pointed to that part of the sky which was opposite the sun.[121] His observations may not have been very accurate,[122] but his remarks on the direction of a comet's tail are historically important. There has been considerable dispute over the priority of the observation of this trait of comets.[123] Although it is the common opinion that Apian was the first to observe the fact, as he set forth in his *Astronomicum Caesareum,* which appeared in Ingolstadt in 1540, the observation had been previously made by Fracastoro as is evinced by the latter's *Homocentrica* which appeared in Verona in 1538.[124] Nor had observations concerning the direction of a comet's tail been lacking during antiquity and the Middle Ages.[125] However, although Fracastoro's work was not unknown to his immediate successors,[126] the credit for the obser-

120 *Ibid.,* 59; Dreyer, editor, II, 457. Olbers attempted, without much success, to compute the elements of the orbit of the comet of 1533 from Apian's observations. Olbers, 126, said that Apian observed the comet of 1533 only four times and gave, for each time, only the longitude and latitude of the comet, not the observation itself, which he did give for the comets of 1531 and 1532. Apian's observations of the comets of 1531 and 1532, the latter being given in Apian's *Practica* for 1532, were analysed by Günther (1882), 56-7, 59. Apian's observations of the comets of 1533 and 1539 were also used in later orbit computations (Günther (1882), 60).

121 Pingré, I, 68; Günther (1882), 57.

122 Bruhns, 506; Pingré, I, 491.

123 Günther (1882), 60-1.

124 Dreyer, editor, III, 398. However, Apian had observed the direction of comets' tails and commented on it in his early tracts on comets. See Günther (1882), 57.

125 Pogo (1934), 444. Pogo cited especially Grosseteste's *De Cometis* and the observations made in China of the comet of 837 taken from Biot's article in the *Connaissance des temps pour l'an 1846* (1843), Additions, p. 79. See chapter I, above, especially note 150. For the most part, statements concerning the direction of a comet's tail had been specific, such as "it pointed to such and such a constellation" or "it pointed east", not general, such as "it pointed away from the sun".

126 See the discussion of Tycho Brahe's German work on the comet of 1577, in chapter III, below.

vation was given to Apian by their contemporaries, for example by Gemma Frisius.[127] It is even possible that Apian's books were the first to have diagrams of the direction of a comet's tail.[128] The first of these diagrams appeared on the title-page of his *Practica* for 1532,[129] and illustrated Apian's observations of Halley's comet in August, 1531. The second diagram is the woodcut on the title-page of the tract on the comet of 1532 and represents observations of that comet.[130] In these diagrams the axis of the tail is prolonged and passes through the sun.[131]

Apian seems to have been aware of the method of parallax for determining distance and made many measurements of terrestrial distance, using the quadrant and the cross-staff.[132] Although acquainted with Regiomontanus' work, he does not seem to have achieved any measurements of a comet's actual distance,[133] but he did observe the positions of comets in the same way that he observed the positions of other heavenly bodies.[134] He also noted that comets in the neighborhood of the sun disappear in the rays of that body and later reappear.[135] And, like his contemporaries, he made predictions from the appearances of comets.[136]

For at least three reasons Apian's work is particularly important as a background for the observers of the comet of 1577.

127 Pogo (1934), 443.

128 *Ibid.*, 443-4 and Plates I and II.

129 *Practica auff dz. 1532 Jar.* See the description of this book in Ortroy (1901), 109-111.

130 *Ein kurtzer bericht d'Obseruation vnnd vrtels / des Jüngst erschinnen Cometen / jm weinmon vñ wintermon dises XXXII. Jars.* See the description of this book in Ortroy (1901), 302-4.

131 See Pogo (1934) and Delambre (1819), 392-3.

132 Günther (1882), 39-40, 58.

133 *Ibid.*, 56, 58.

134 *Ibid.*, 59.

135 *Idem.*

136 *Ibid.*, 64-5.

In the first place, Peter Apian had a son, Philip, born in 1531, who taught at Tübingen, and who must have been acquainted with his father's work and have had the opportunity to disseminate that knowledge.[137] Among Philip's pupils was Michael Maestlin,[138] who later succeeded him.[139] Philip Apian does not seem to have made any observations on comets, but he did observe the nova of 1572 which he considered a comet and placed in the region of heavenly bodies.[140] A second instance of Apian's influence is the impression made by his *Astronomicum Caesareum* on the Landgrave of Hesse Cassel.[141] In the third place, Gemma Frisius, the father of Cornelius Gemma,[142] edited Apian's *Cosmographicus Liber*.[143] Gemma's works were widely read and held in high esteem.[144]

Another scientist in the first half of the sixteenth century to point out the fact that comets' tails are pointed away from the sun was Jerome Cardan.[145] His achievements in the fields of

137 Peter Apian himself laid the foundation of his son's studies. But the younger Apian seems to have been away from home during his most formative student years, returning home after his father's death in 1552. See Günther (1882), 82-3, 113-4.

138 See chapter III, below.

139 Wolf (1877), 266.

140 Günther (1882), 118-9.

141 See chapter II, below.

142 See chapter II, below. Cornelius Gemma, who observed the comet of 1577, spoke of the comet of 1533 in his *De Naturae divinis characterismis*. See Carl, 49.

143 For the many editions of Apian's *Cosmographicus Liber*, called *Cosmographia* in later editions, consult Ortroy (1901), 113-156; Ortroy (1920), 84-6 and 165-189; B. M. catalogue under Gemma's name; Pogo (1935), 471-5 ("Apian's Cosmographia and Gemma's De principiis").

144 Ortroy (1920), 37-45. See Pogo (1935), especially 483-5 ("Life and works of Gemma Frisius"); McColley; Seyn, II, 1796-8, where Ortroy (1920) was cited. Gemma Frisius, a well known doctor, astronomer, and geographer, was not, as has often been supposed, a pupil of Apian (Ortroy (1920), 22-3). Gemma observed the comets of 1533, 1538 and 1539 (Dreyer, editor, II, 457; Pingré, I, 496-7, 499, 500).

145 Dreyer (1890), 166, where Apian, Fracastoro, Gemma and Cardan were mentioned by name. A clear and concise summary of Cardan's theory of comets, based on the *De Subtilitate* and the *De Varietate* can be found in Rixner, II, ("Hieronymus Cardanus..."), 32-3.

philosophy, physics, and mathematics cannot be discussed here but should be borne in mind in order to realize the stature of the man who discussed comets. His theory of comets, denying the Aristotelian doctrine, is particularly interesting because it was distinctly understood by Tycho. He proposed the use of the method of parallaxes to determine whether comets were above or below the moon.[146] However, he does not seem to have given anywhere the day to day observations of a comet which were necessary to insure a fairly accurate value of the parallax, nor the necessary computations.[147] Indeed, his statement that the comet of 1532 was above the moon was based on the comparative speeds of the two bodies.[148]

In the *De Subtilitate* [149] after a demonstration of the height to which vapors can rise, Cardan argued that comets seen from Milan to be in Capricorn, are not formed of vapors, because the latter cannot rise so far, because comets would need continually renewed material and because comets last so long that the whole earth would not suffice for the fire.[150] He proposed

146 See note 151 in this chapter.

147 Pingré, I, 70, said that Cardan, although he did not pretend to be the author of the method of parallaxes, used the lack of perceptible parallax to destroy the system of "Comètes-météores". Pingré added that this at least proved that toward the middle of the sixteenth century the method had already proved successfully that the true place of comets was usually beyond the lunar orbit. The statement by Pingré is a bit too emphatic, because Tycho Brahe's observations of the comet of 1577 were the first to be sufficiently accurate to prove the supra-lunar position of comets and to establish the incompatibility of the observations with the Aristotelian standpoint. Guillemin, 47-8, was at best inaccurate when he said: "At the end of the fifteenth century, we find... Cardan remarking that comets are situated in a region far beyond the moon, founding his opinion upon the smallness or absence of parallax". Cardan did not live in the fifteenth century and it is improbable that the method of parallaxes was used successfully before the second half of the sixteenth century to prove that comets are supra-lunar.

148 See next paragraph below and note 152.

149 The *De Subtilitate* first appeared in Nuremberg in 1550. Cantor (1905), 135; Poggendorff, I, 377. It is, in reality, an up-to-date (1550) compendium of learning.

150 Cardan (1663), III, 420 (*De Subtilitate*, Book IV). Cardan did not mention a specific comet seen from Milan, although the French translation

use of the principle of parallaxes but he made no observations, at least none recorded here, nor calculations to uphold this suggestion.[151] Next he described the triple motion of comets, which he considered common to all comets. One motion was from east to west requiring 24 sidereal hours, that is, diurnal rotation, and another was from west to east, sometimes more, sometimes less, resembling that of Venus. The third motion was in latitude. He said that, from September 22nd to December 3rd, the comet of 1532 moved less than 1° per day in longitude, and he concluded that it could not have been below the moon, because then it would have moved faster than the moon.[152] He cited both Pliny and Fracastoro on the duration of comets, the latter on the comet of 1531 in the *Homocentrica.* He showed that he recognized that the direction of a comet's tail away from the sun was a trait common to all comets.[153] A third trait which he thought common to all comets was that their longitude is such that they most often accompany the sun and appear only in the evening. This idea is easily understandable when the difficulty of naked eye observations during daylight hours is considered and when it is realized that comet observations before the invention of the telescope were of necessity made when the comet was near perihelion. Cardan concluded that a comet is a globe formed in the sky and illuminated by the sun, the rays of which, shining through the comet, give the appearance of a

of the passage, given by Pingré, I, 70-1, gives that impression. This is not true of the French translation by le Blanc (see Cardan (1578), 103r, [marked 85]).

151 Concerning the determination of the distance of comets, Cardan wrote: " Sed depraehendere an Cometes in elementorum regione sit, an in coelo fiat : facilè est admodum. Nam si maiorem habeat diuersitatem, quàm Luna, in elementorum esse regione necesse est : sed si minorem, in coelo fiet procul dubio." Cardan (1663), III, 420.

152 " Ex quo patet : sub Luna eum esse non posse, nam sic motu primi orbis velociùs, quàm Luna moueretur." Cardan (1663), III, 420.

153 "Aliud verò commune Cometis est, quòd cauda semper ad vnguem, partem Soli oppositam respicit : vt cùm Sol occidit, Cometes caudam habet ad amussim Orientem versus, vt in obscura Lunae parte nobis singulis diebus videre licet." Cardan (1663), III, 420 (*De Subtilitate*).

beard or tail.[154] Thus, he said, something can be generated in the sky or else it is necessary to say, which is most true, that the sky is full of bodies invisible to us. If the air becomes dry, these bodies become visible, and consequently comets, if not the cause, at least are the sign of dryness and corruption, famine, death, and so forth.[155]

Comets as portents were considered elsewhere by Cardan, for example, in the *Encomium Neronis;*[156] and the *De Vtilitate,* where the influence of comets on those born at the time of the comet's appearance is discussed; [157] and the *Problematum Naturalium,* where the relation between comets and dryness and moisture is shown.[158] His commentary on Ptolemy's *Quadripartitum* afforded him the opportunity to set forth his own opinions in contradistinction to those previously held.[159]

Equally important with the *De Subtilitate* as a source for Cardan's theory of comets is his *De Rerum Varietate.*[160] In the first chapter he said that a comet is not made in the elementary region which is always changing. On the contrary, a comet remains a long time and neither descends because of the vapors nor ascends because of the fire. However, of these two, it is necessary that the second happen if the fire be kindled from

154 Cardan (1663), III, 420 (*De Subtilitate*). Tycho also believed that a comet's tail was due to the passage of solar rays through the comet. See the discussion of Tycho's German book on the comet of 1577, in chapter II, below.

155 Cardan (1663), III, 420 (*De Subtilitate*). The passage is not as smooth as would be expected of a description of accepted facts. Therefore, Pingré's (I, 70-2) quotations of and paraphrases from it, while giving the general meaning, are misleading.

156 Cardan (1663), I, 187

157 *Ibid.*, II, 221 (*De Utilitate,* cap. XXV "De Signis").

158 *Ibid.*, II, 621 (*Problematum Naturalium,* Sectio Prima).

159 Cardan (1554), 150-9.

160 This was first published in 1557. Cantor (1905), 135, gave the date as 1556 but Poggendorff, I, 377, gave the later date. Although both the B. M. and B. N. have copies of the 1557 edition, neither has a copy of any earlier edition. The *De Varietate* was supposed to supplement the *De Subtilitate.* See Waters, 104-5, 108-9; Thorndike, V, 570.

vapors, nor would it always be moved by that triple motion but as spears or burning stars.[161] The light, said Cardan,[162] is collected in one part of the heaven and increased and appears round. It has a tail formed by sunlight, for the light of the comet itself is not so clear as to allow the sunlight to permeate it unhindered nor yet so dense as to reflect light, the latter being a characteristic of the moon, the former of the stars. Therefore, argued Cardan, the nature of comets lies between that of the moon and that of the stars. Cardan continued by repeating his description of a comet's motion. He believed comets "happened" more easily and more often near the pole because there "the light of the sun, which is distant, hinders but slightly the generation of comets, which occurs from the scant light of the stars."[163] He again noted the direction of comets' tails, which he thought were made by the sun's rays.[164] He accounted for the long duration of comets by pointing out that they are formed from many sources of light.[165] Again, he remarked that there are more comets than are seen, many appearing at dusk and at midnight. He believed that they are seen when there are winds and when the air is finest, in the summer rather than in the winter.[166] As in the *De Subtilitate,* he said that comets are followed by dryness, deaths and so forth, but he attributed these to natural causes due to the manner in which comets are formed. For further notice of the signification of comets he referred the reader to his commentary on the *Quadripartitum*. In the *De Rerum Varietate,* he returned to the sub-

161 "Non fit cometes in elementari regione, quae semper est inconstans: cometes autem diu manet, nec etiam descendit propter vapores, nec ascendit ob ignem: horum autem duorum alterum fieri necesse esset, si ignis esset ex vaporibus accensus: nec triplici illo motu moueretur semper, sed vt iacula, aut stellae incensae." Cardan (1663), III, 1.

162 Cardan (1663), III, 2.

163 *Idem.*

164 *Idem.* "Quòd verò cauda fieri possit à radiis Solis, experimentum docet candela accensa Soli exposita, sic vt radij per illius ignem transeant."

165 Cardan (1663), III, 2.

166 *Idem.*

ject of comets, commented on the wide effects of those visible in the day, and described the one seen in March 1556.[167] He grouped comets by their appearance in the manner and by the names used by Pliny.[168]

Jean Pena, royal mathematician at Paris, held with Cardan that comets were not exhalations from the earth and that, because their tails were always turned away from the sun, they were transparent bodies through which the rays of the sun could be refracted. He further held that by the principles of optics some comets could be shown to be supra-lunar.[169]

The subject of comets as portents was treated in the first half of the sixteenth century by an Englishman, John Robyns, in his *De portentosis cometis.*[170] He held that it is not the comets themselves which produce the great effects which follow them. Saying that most astrologers ascribed the generation of comets to eclipses and revolutions of years of the world, he added that no comet is generated without the influence of the hot and dry planets, Mars, Saturn, and the sun, and that since two of these are unfortunate, comets are likely to forecast ills although they may share the nature of stars like Jupiter and Venus and portend good fortune to some extent. Because they are generated by hot and dry planets, they cannot presage rainy years. Robyns included observations on the comet of 1532 in a work on future events.[171]

Another observer of comets in the first half of the sixteenth century was Joannes Vögelin or Vogelinus. He has been called

167 *Ibid.*, III, 274-5 (book 14, chapter 69).

168 *Ibid.*, III, 276 (book 14, chapter 70).

169 Thorndike, VI, 71.

170 *Ibid.*, V, 320-1. Robyns, elected fellow of All Souls college, Oxford, in 1520, later became chaplain to Henry VIII and canon of Christchurch and Windsor. His *De portentosis cometis,* a manuscript, was addressed to Henry VIII and was the outcome of discussions with that monarch.

171 Thorndike, V, 320. The work on future events is also a manuscript addressed to Henry VIII.

the disciple of Regiomontanus,[172] probably because of his connection with the university of Vienna, where Regiomontanus and Peurbach had met, and because of his use of the method of parallaxes. He wrote on the comet of 1532 and has erroneously been said also to have observed a comet in 1527.[173] Vögelin, unlike Regiomontanus, was interested in the signification of the comet he observed, but he also tried to determine its parallax, unfortunately with poor results.[174] However, he and Regiomontanus were frequently cited together by Tycho.[175] William

172 Pingré, I, 69, 494; Riccioli, I, xxxix, II, 1. Vögelin, who was not born until the end of the fifteenth century and first became professor of mathematics in Vienna in 1528 (Dreyer, editor, VIII, 464-5), cannot have known Regiomontanus personally. However, he did print the *Spherica* of Theodosius, which Regiomontanus had proposed to do. See Thorndike, V, 356.

173 Weidler, 341; Riccioli, I, xxxix; Günther (Voegelin), 142. In regard to observations of a comet in 1527, an error has crept into the literature. Vögelin's *Significatio cometae qui anno 1532 apparuit* was first printed in Vienna in 1533 (Houzeau, 5565; Struve, I, 786; Schottenloher, IV, 376; Thorndike, V, 322), but it was reprinted in Hagecius' *Dialexis de novae ... stellae* in 1574, where the date of the comet was given as 1527 in the title of Vögelin's work at the beginning of that work (Hagecius (1574), 150). However, the date 1532 appears on the verso of the title-page of Hagecius (1574), where Vögelin's work is listed, and again on page 151, in Vögelin's own introduction, where he has written: "Huius instructione formatus obseruaui ego Cometam qui hoc anno, id est 1532 apparuit..."; and once more in chapter I, where the comet is referred to as having appeared in 1532. In the B.M. catalogue, under "Voegelin", the words "or rather 1532" are enclosed in brackets after the date 1527. Thorndike, V, 322, speaks of the tract on the comet of 1532 but rightly makes no mention of any observations of a comet in 1527. Günther (Voegelin), 142, wrongly said that Vögelin's tract was also reprinted in Tycho's *Progymnasmata.* It was merely cited there. Günther (Voegelin), 142, said that another work on comets by Vögelin was preserved in manuscript in Vienna.

174 Vögelin believed that the comet was 6140 paces (passus) or 1535 German miles from the center of the earth (Vögelin in Hagecius (1574), 159), and that the difference in position of the comet as seen by an observer on the earth's surface and from the center of the earth, measured along the ecliptic, was 32° 27′ 14″ (Vögelin in Hagecius (1574), 162). This last figure would correspond to the parallax. Other computations in the same tract had yielded slightly larger values. Brahe, VII, 212, cited Vögelin's value for it as 35°.

175 Brahe, III, 26, 27, 140, IV, 208, 447, VI, 65, VII, 68, 258.

Gilbert, at the close of the sixteenth century, was fully aware that at least some comets are supra-lunar, and knew that the nova of 1572, which he seemed to confuse with a comet, had been conclusively proved, by the observations of Tycho, Maestlin, Digges, Dee, Hagecius, Munosius and Cornelius Gemma, to be supra-lunar.[176] Nevertheless, he cited Vögelin's determination of the parallax of the comet of 1532, along with observations of comets by others, as proving that comets were sometimes sublunar.[177]

The astronomical and astrological works of many eminent men, including much by Regiomontanus, were edited[178] by Johann Schöner (1477-1547), who was a mathematician, geographer and astronomer in his own right. He observed Halley's comet in its 1531 appearance and wrote about it.[179] Schöner began to observe the comet on August 15th, and, although he made almost daily observations, noting the position of the comet and its tail, and the comet's motion, his observations, because they are not sufficiently precise, have little value. He retained Aristotle's theory of the generation of comets. In his short book he devoted more than a page to what a comet is, more than a half page to the nature of the comet of 1531, almost two pages to the general meaning of that comet, about a page to its particular significance, half a page to what regions were threatened by it, and more than half a page to a discussion of

176 Gilbert, 155, 236, 243. Gilbert's work was not published until the middle of the seventeenth century. See Thorndike, VI, 379-381.

177 Gilbert, 242, giving Vögelin's value for the parallax as 32° 27′.

178 This included the 1531 edition of Regiomontanus' sixteen problems relating to comets (Coote, 156) and possibly a 1544 Nuremberg edition of the same work (Ziegler, 44). This 1544 edition was not mentioned by Coote and may be a misprint on the part of Ziegler, who may have had in mind another work by Regiomontanus which was edited by Schöner. See note 79 in this chapter. Schöner also published observations of eclipses, comets, the planets and fixed stars by Peurbach, Regiomontanus and Walther. See Thorndike, V, 365-7.

179 Schöner. There is a copy in the C.U.L. It was probably printed in 1531 and is a quarto, 187 mm. high, of one and a half signatures. See Zinner (1934), 66 ff.; Coote, 156; Thorndike, V, 357 and 357 notes 118 and 119.

the size of the comet, which he estimated as several German miles in length.[180] Thus, he did not live up to the example set by Regiomontanus, nor add anything new in the observation of comets as his contemporaries, Fracastoro, Apian, and Cardan, had done.

Schöner read the horoscope of Nicolaus Gugler, presumably of Nuremberg, who himself wrote a brief tract on comets, probably in the second quarter of the sixteenth century.[181]

Jacobus Petramellarius of Bologna, in his prediction for 1533,[182] referred to that of 1532, stating that he had told what comets and three suns seen in Apulia would announce.[183] Other minor tracts on comets include one by Peter Creutzer on a comet in 1527 (sic), which was printed in Nuremberg, probably that year; one by Nicolaus Prueckner on the comet of 1532, which was printed in Strasburg that year;[184] one by Ludovicus Vitalis on the comet in September 1531;[185] and one on the significance of the comet of 1531 by Andreas Perlach.[186]

180 Schöner, A_{iii} v. Stating that the moon was 30,000 German miles from the earth, Schöner added that the comet was 1000 miles from the earth and that its rays were several miles in length (" etwa vil deutscher meyl ", possibly meaning an equal number of miles, a translation which, however, has not been substantiated) although not more than one and a half German miles thick. He gave no computations to show how he arrived at this value for the comet's distance, and only said that he had done so " auss gewiser kunst der zal vnd messunge ".

181 Thorndike, V, 368-371. This tract is preserved in a manuscript at Paris. Although there seem to have been two Nicolaus Guglers, probably father and son, the author of this tract seems to have been a doctor of laws, a physician, a mathematician, a judge at Nuremberg, an advocate of the imperial court, and a counselor of the king of Denmark. According to Hellmann (1924), 27, a Nicolaus Gugler wrote prognostications in 1563 and 1564.

182 This prediction is contained in a volume now in the B. M. but formerly the property of Boncompagni. See Thorndike, V, 234, 238.

183 Thorndike, V, 239.

184 See note 193 below.

185 Thorndike, V, 243.

186 *Ibid.*, V, 382.

Copernicus was one of the many observers of the comet of 1533.[187] He appears not to have been particularly interested in comets, for he mentioned them only once in his *De Revolutionibus,*[188] and his observations of the comet of 1533 are very likely his only observations of the sort. This uniqueness makes them all the more interesting, although they have little importance in the development of the theory of comets. They were mentioned by William Zenocarus in the sixteenth century.[189] Though they seem to have been irretrievably lost, it appears that they were not in agreement with those made by Apian, Cardan, Gemma Frisius and Jerome Scala, and gave rise to a controversy between Copernicus and those men.[190]

Not all observations of comets made in the first half of the sixteenth century led to advances in cometary theory.[191] Theophrastus Paracelsus [192] (1493-1541), whose influence was felt in many lands, and whose theories in medicine proved

187 Curtze (1878), 41-3.

188 Curtze (1878), 41, said that the only mention of comets in the *De Revolutionibus* occurs in book I, chapter VIII. They seem not to have been mentioned in the *Commentariolus* and are not listed in Rosen's index.

189 The passage containing Copernicus' observations of the comet was quoted in toto by Curtze (1878), 41-2. It is from Zenocarus' *De Repvblica, Vita . . . : Imperatoris, Caesaris, Augusti . . . , Libri septem,* which was printed in Ghent in 1559. According to the B. M. catalogue the "libri septem" are really only five.

190 Thorndike, V, 410; Curtze (1878), 42. Scala must have been an unimportant observer since his name does not appear in Pingré or in any of the usual bibliographies or biographical dictionaries, (A. D. B.; B. M. catalogue; Freher; Jöcher; Michaud; Poggendorff; Riccardi; Zedler). Curtze (1878), 43, said that Scala's and Cardan's observations of the comet of 1533 might yet come to light.

191 An interesting account of writings on the comet of 1531 was presented by Rauscher. He gave a historical discussion of contemporary writings on that comet by the reformers, and showed the astrological and religious aspects of their observations. The men whose writings were discussed by Rauscher include Luther, Melanchthon, Zwingli, Paracelsus and Nausea.

192 He used the name Philippus Aureolus Paracelsus Theophrastus Bambast von Hohenheim.

fruitful, wrote a tract in German on the comet of 1531.[193] In this work he sought the reason for all phenomena in God alone.[194] In a treatise on the comet of 1532,[195] Paracelsus described that comet as appearing in December and moving athwart the path of the previous comet. He described its tail as outstretched and like a switch or a broom.[196] In 1534 he again discussed the comets of 1531 and 1532.[197] In a work

193 According to Sudhoff, 13, the tract bears the title *Vsslegung des Com- || meten erschynen im hochbirg, zů || mitlem Augsten, Anno 1531. Durch || den hochgelertenn Herren || Paracelsum. rc.*, and is a quarto of eight leaves, with the date and place of publication not given, and with a crude woodcut of a comet on the title-page. Sudhoff said that undoubtedly the book was printed in Zurich. For further information see Sudhoff, 13; Wolf, III, 22-3. Wolf, III, 21; Wolf (1877), 183; Wolf (1849), 104-5, called this the first tract on comets in the German language. Wolf was corrected by Sudhoff, 13-4, as follows: 'Wenn Rudolf Wolf... dies für die erste Cometenschrift in deutscher Sprache hält, so ist er im Irrthum; es erschien z. B. schon vier Jahre früher die folgende: "Ausslegung Peter Creutzers, etwan des || weytberümptē Astrologi M. Jo. Liechtenbergers discipels, vber den || erschröcklichen Cometen, so im Westrich vñ vmbligenden grentzen er- || schinen, am xj tag Weinmonats, des M. D. xxvij. jars. zů eeren || den wolgepornen Herrn, herr Johann || vñ Philips Frantzen || beyde, Will vnd Reingrauen etc." ... am Ende "Gedruckt zů Nůrmberg durch Georg Wachter"...' In 1532 a tract by Nicolaus Prueckner entitled *Was ein Comet sey: woher er kom̃e, und seinen Ursprung habe... und sonderlich von dem Cometen erschiñen im Weinmonat des XXXII. jars* was printed in Strasburg (B. M. catalogue; Struve, I, 786; Schottenloher, IV, 376). See also Thorndike, V, 322; Hellmann (1924), 29. Apian's tract on the comet of 1532 was also in German. See note 130, above. It is evident that tracts on comets in the German language were making their appearance in the beginning of the second quarter of the sixteenth century; and it will become apparent that they increased in number in the ensuing years.

194 Wolf (1849), 105; Wolf, III, 24.

195 According to Sudhoff, 14, this tract bore the title *Ausslegung Dess Cometen vnnd Virgultae, in Hohen Teutschen Landen erschienen, durch den Hochgelehrten Herrn Paracelsum Doctorem, An. XXXII.*

196 Sudhoff, 14; Wolf, III, 25.

197 According to Sudhoff, 15, the title of this treatise is *Von den wunderbar= || lichen, vbernatürlichen zey= || chen, so iñ vier jaren ein ander nach, im̃ || hym̃el, gewülcke vnd lufft, ersehen, Von sternen, Re= || genbŏgen, Fewrregen, Plůtregen, Wilde thierer, Tra = || ckenschieffen, Fewrin mañ, mit sampt ander der= || gleychen. Auch ausslegung der zweyen || Cometen, so biss her yrrig auss = || gelegt seynd....*

which, even if apocryphal,[198] is, nevertheless, the type of tract about comets which passed for his, he spoke of the origin of comets as "ein Fatum aller Völcker," and said that they were raised on high as a sign of new lamentations and occurrences.[199] In the *Fragmenta Meteorica* a fantastic statement includes the words: "Est enim Cometa opus à spiritibus aëris confectum, sicut Magus faciem facit sui (haeredis)." [200] In the *De Tvmoribvs, Pvstvlis, Ac Vlceribvs Morbi Gallici* comets are spoken of as being transplanted because of conjunctions,[201] and elsewhere they are called portents.[202] Genuine or not, the book on meteors that appears under the name of Paracelsus should be mentioned here.[203] It does not raise the caliber of the

198 Wolf, III, 27, 37.

199 The passage reads: "Der Ursprung von Cometen ist ein Fatum aller Völcker, die under seinem gang ligend: Bedeut under ihnen als Elend und noth, oder Freyhcit, oder etwas unerhörts newes dings oder Wesens unnd seiner dienstbarkeit: Wirdt geboren auss dem Fatum der Herren, der Geistlichen, der Leyen, unnd dergleichen. Etlicher wirdt allein geboren auss dem Fato der Herren, und bedeut under ihnen den newen Murmel, derselbig sieht Herlich: Etwann allein der Geistlichen, derselbig sieht Schwentzet: Etwann allein der Bawren, derselbig ist grob. Und setzen ihre Form von dem, das sie bedeuten: Als ein Stern unnd ein Schwantz, auss ursachen, dass ein langer Schwantz werden wirdt auss dem handel den er bedeutt. Ist er breitt, ein breitter weitleuffiger handel, und wie es ein endt nimbt am Schwantz Frisch oder Frölich, also endet sich das auch. Darumb weiter von Cometen nichts anderst zu verstehen ist, dann allein dass sie von undern Fatis geboren worden in das ober, zu einer Prognostication auf die Welt eines newen geschreys und geschichts." Wolf, III, 26-7, cited the section as from *Philosophia Theophrasti Bombast ab Hohenheim*, II, 1-63, of Huser's edition (1616-1618) of the collected works of Paracelsus. The citation is copied from Wolf, but can be found in Latin in Paracelsus, II, 267 ("De Elemento Ignis Textvs V. De Cometis").

200 Paracelsus, II, 340.

201 *Ibid.*, III, 115.

202 *Ibid.*, II, 395 ("Liber Philos. De Nymph. Sylphis, Pygmaeis Et Salamandris"). See Sudhoff, 120-2, where the first (1566) edition is described. As Spunda, 113, says, "Der Komet hat bei Paracelsus dieselbe Funktion als Unheilsbote wie im Altertum und Mittelalter."

203 It first appeared in German in 1556 (Sudhoff, 115, 122), and in Latin in 1569 (Sudhoff, 175-6), although the latter does not seem to be a translation of the former.

cometary theories attributed to him.[204] Paracelsus' theories about comets, even if but slightly circulated,[205] must have been a deterrent to progress in this field. Many other scholars in the first half of the sixteenth century wrote about comets without affecting the current knowledge concerning them. Frederick Nausea, Bishop of Vienna, wrote a book *Super Cometâ hujus anni 1531* which was printed in Mainz in 1531 [206] and translated into English by Abraham Fleming in 1577 under the title *Of All blasing starrs in generall.*[207] Nausea distinguished two types of comets, natural and supernatural.[208] He stated, as the

204 In it are the following sentences: "Cometa est crescens singulare, ex nulla matrice, ex nullo semine prognatum, sed à spiritibus compositum. Hi enim futuros euentus, fortunam aut infortunium, mortem ac vitam, bellum ac annonae caritatem praesciunt...." (Paracelsus, II, 318 ("De Meteoris Lib. I, Cap. De Cometis")), and "Cometa quae apparet, ex naturali causâ non oritur: sed initium est, & vestigium, à spiritibus expressum, quod nouum quiddam, aut mutationem, aut malum publicum, aut quicquid aliud regionibus damnosum esse potest, designat. Cursus & motio ipsius dirigitur à spiritibus, vt insignes mutationes aut diuturnas afflictiones, casúsque miros alios portendant...." (Paracelsus, II, 336-7 ("De Meteoris, Lib. III, Cap. XXXIII")). The cometary theories attributed to Paracelsus have been excellently summarized by Rixner, I, ("Theophrastus Paracelsus..."), 179, as follows: "Die Kometen sind ein sonderbares Gewächs am Himmel, gleichsam ein Unkraut (Zizanium) unter den Sternen;—eine unregelmässige Erzeugung der imaginatio animae mundi majoris, ohne einen andern Samen.... Sie waren nicht in der Zahl der Sterne der ersten Schöpfung mit begriffen, sondern sind spätere zufällige Erzeugungen des Himmels aus sich selbst, nicht aus den aufsteigenden Dünsten der Erde. Einige mögen wohl auch neuere unmittelbare Schöpfungen Gottes seyn, die dann nicht etwa nur Regen und Wind, wie die übrigen Meteore, sondern auch wichtigere Dinge verkündigen. . . . " The statement by Stoddart, 255-6, who regarded Paracelsus as "a keen student of positive astronomy", must be disregarded.

205 Sudhoff, 14, knew but two copies of the 1531 tract on comets, none, except in the collected works, of the 1532 tract, and three of the 1534 tract. The various editions of Paracelsus' collected works, which included many apocryphal treatises, were widely circulated at the time of the comet of 1577 and later.

206 Pingré, I, 230; Rauscher, 273.

207 Johnson (1937), 155, 310.

208 Rauscher, 273-6. Nausea believed that comets could portend either good or evil, depending on their origin, and that the ills could be alleviated through

opinion of the " Pythagorists ", that comets are perpetual stars, with a regular course to run.[209] Melanchthon, in an unpublished address to the students of Wittenberg, announced that he was about to tell about the significance of eclipses and comets.[210] Jacob Milich, an associate of Melanchthon, a professor at Wittenberg, and a physician, in a commentary on book II of Pliny's *Natural History,* finished in 1534 and first printed in 1535, maintained and defended Aristotle's theory of the origin of comets, although he himself realized that many men were treating them as stars.[211] Matthias Brotbeyel or Brotbeyhel was one of the many Germans who wrote *practicas* and tracts on comets in their native tongue. His tracts include one on the comet of 1532, one on that of 1533, and one on that of 1539.[212] Antoine Mizauld or Mizaldus was another sixteenth century astronomer who wrote about comets but added nothing new to the knowledge concerning them. He was well known to his contemporaries, and was mentioned in Cardan's commentary on the *Quadripartitum* and in Squarcialupus' treatise on comets. His *Cometographia, crinitarum stellarum . . .,* to which was added a catalogue of comets to 1540, appeared in Paris in 1549, and his *Meteorologia* appeared in the same city two years earlier.[213] Like so many distinguished scientists in his time, he

prayers. Pingré, I, 78-9, interpreted Nausea's work as attempting to unite the Aristotelian system with that of the theologians, but there never seems to have been any discrepancy and Nausea seems not to have brought up the question. According to Allen, 74 note 63, Nausea said that comets are not all formed as Aristotle said, but are created by God as warnings against His just vengeance.

209 Johnson (1937), 155.

210 Thorndike, V, 401.

211 *Ibid.*, V, 387, 389. See also note 74, chapter I, above, and Allen, 66. Milich's *Oratio de dignitate astrologiae*, delivered sometime between 1524 and 1533 when it was printed, was a conventional defense of astrology (Allen, 66).

212 Günther (1887), 86-8, 91-2; Schottenloher, IV, 376; Struve, II, 550 (the tract on the comet of 1532, only).

213 Nicéron, XL, 203; B. M. catalogue; Thorndike, V, 299-300 and note 59. See also Allen, 252-3.

seems to have been a credulous person [214] and was mentioned as " Mizzaldus ineptus " by Squarcialupus.[215]

In the same century, Giovanni Ferrerio wrote a tract, *De vera cometae significatione contra astrologorum omnium vanitatem, libellus,* which was published in Paris in 1540, but dealt with the comet of 1531.[216] It seems not to have had any bearing on the development of cometary theory, although it denied any natural connection between comets and kings and expressed doubt that comets could affect men at large. Five books on the elements by Gasparo Contarini (1483-1542) were published posthumously in Paris in 1548.[217] Although in a sense favorable to astrology, Contarini believed that human affairs could not be determined from celestial causes alone. He believed that comets were generated in the sphere of fire rather than in the upper region of the air; that this generation depended on a constipation of parts of that element, so large as not to be easily dissolved; and that the earthly exhalation merely provided fuel. Peter Haschard or Haschaert, a physician of Brussels, in an astrological work printed in 1552, gave examples of the effects produced by comets.[218] In 1564 Cyprian Leowitz (Leovitius) published a work [219] intended to prove the veracity

214 Nicéron, XL, 201.

215 Item 37 (3) of appendix, below, E_3v.

216 Thorndike, V, 293-5; B. N. catalogue; Houzeau, 5569; Riccardi, 452; L'Art Ancien catalogues, 22, 25. Another tract, or a translation of this one, by the same author, on comets and their significations, in French, was also published in Paris in 1540 (Houzeau, 5570). A translation into Italian was published in Florence in 1577 (Houzeau, 5569; Riccardi, 452). This has led to confusion because the treatise has been listed with those on the comet of 1577 (Tiraboschi, VII, part I, 433).

217 Thorndike, V, 552-3.

218 *Ibid.*, V, 329. See note 244 in this chapter. See also Hellmann (1924), 31.

219 *De coniunctionibus magnis insignioribus superiorum planetarum, solis defectionibus, et cometis . . . cum eorundem effectuum historica expositione* (Langingae ad Danubium, 1564). Allen has given the date of this work variously as 1554 (pp. 73 and 74) and 1544 (p. 262), but the earliest edition listed in either the B. M. or B. N. catalogue is 1564, which is consistent with the fact that the work mentions the comet of 1558 (Allen, 74).

of astrological prediction. In this he recounted many events which he considered due to comets.[220]

The comet of 1556 attracted much attention which resulted in many tracts. These were not very important in formulating new theory, except that the data concerning the observations were fairly accurate and enabled later astronomers to compute the comet's orbit. Several of the tracts are illustrative of a growing tendency to write in the vulgar tongue. The mere number of tracts is indicative of an increased interest in the subject, the ability of more people to write and of more people to read about it, and of the increased use of printing.

Probably the best known observations of the comet of 1556 were made by Paul Fabricius (1529-1588).[221] Subsequent calculations of the orbit of that comet were made chiefly from his observations.[222] He was an experienced astronomer who leaned toward the theories of Seneca rather than toward those of Aristotle, attributing to Seneca the belief that comets are stars created by God in the beginning and made to appear in order to show His power and to announce future events, a theory also expressed by Dudith.[223]

220 Allen, 74. Leovitius was a Bohemian mathematician in the Palatinate, who died in 1574. He is mentioned below, in this chapter, as an observer of the nova of 1572.

221 Wolf (1877), 407-8 note 5, gave the date of Fabricius' birth as 1529 and of his death as 1588. Houzeau, II, 130, Jöcher, Fortsetzung, II, 990, and Poggendorff, I, 711-2, gave the date of his birth as 1529 or possibily 1519. Dreyer, editor, VIII, 456, gave the date of his death as 1589. He received the degree of doctor at Vienna in 1557 (Jöcher, Fortsetzung, II, 990) and later was mathematician to Charles V. (Pingré, I, 72). His tract on the comet of 1556 was published in Amsterdam in 1557 with the title *Le cours et signification du Comete qui a este veu l'année precedente, dans le discours du quel il dispute doctement de son opinion touchant la fin du monde* (Wolf (1877), 408, note 7, citing Libri's catalogue). Jöcher, Fortsetzung, II, 990, gave the title in German as mentioned by Suevus in the latter's 1578 work on comets.

222 Pingré, I, 72. These were the earlier calculations, prior to Pingré's writing in 1783.

223 Pingré, I, 72. For Dudith's work see items [34] and [35] in the appendix, below. Fabricius repeated these sentiments in his tract on the comet of 1577, item 39, appendix, below, A_3v.

Fabricius described the comet of 1556 in a leaflet dated March 14th of that year, which contained a map of the comet's path.[224] In the leaflet, he said that in his *Pratika* he had predicted a comet, and that the comet appeared in March and rose night after night, and he referred his readers to his Latin *Judicium* for further particulars.[225] He began the *Judicium,* like the pamphlet, with the prediction of the comet, and then gave its nightly positions from March 5th to 15th, guessing at the position of the comet on March 4th, and saying that on the 16th the comet was so small as to be barely visible.[226] He gave the hours of the observations in round figures only, and mentioned that the comet passed over Boötes and Ursa Major.[227] He also wrote a *Prognosticon* for 1565,[228] wrote on the comet of 1558,[229] and, as will be shown below, observed the nova of 1572 and the comet of 1577.

Another important observer of the comet of 1556 was Joachim Heller (1518-1590).[230] For about thirty years he issued *Practicas,*[231] and in a cautious way predicted comets for 1556 and 1557. Indeed, he made predictions of comets for nearly every year, and consequently his predictions were usually not fulfilled.[232] Of course, he had no knowledge of actual periodicity. He described the comet of 1556 in a book which was

224 Littrow, 634. According to Littrow, this leaflet was entitled *The Comet Seen in March, in the year LVI. in Vienna.* It and its title were probably in German.

225 Littrow, 634. This *Judicium* is not the one on the comet of 1577.

226 Littrow, 634-5. Pingré, I, 502, said that Fabricius observed the comet from the 4th to the 15th of March.

227 Littrow, 635.

228 Hellmann (1924), 27. Jöcher, Fortsetzung, II, 990, said that it was for 1567 and was printed in 1566. Jöcher, Fortsetzung, II, 989-991, listed many other works by Fabricius.

229 Thorndike, VI, 184 note 17.

230 Richel, 67; Thorndike, V, 337, 394-6.

231 Hellmann (1924), 27; Thorndike, V, 396. For a discussion of Heller as a defender of astrology, see Allen, 65-6.

232 Richel, 67.

printed in Nuremberg, probably in 1557.[233] He also wrote on the comet of 1577 which he observed in Nuremberg in October of that year and mentioned as well the comet seen in Milan and Lyons in May of that year and a comet seen in the spring of 1558.[234] Heller, in the fourth chapter of his *Practica,* spoke of having predicted the appearance of the comet of 1556. He told how and in what position he first saw it on February 27th, while on a journey. However, not seeing a tail, he had been doubtful of the nature of the phenomenon until he arrived in Nuremberg, when he learned that a comet with a little tail had been seen there on March 3rd.[235] In Nuremberg, Heller observed the comet from March 6th to April 19th, when it reached the Tropic of Cancer, and glimpsed it once more before the 22nd. He expressed the hope that he had given sufficient particulars to enable the learned to calculate the comet's real course.[236] Indeed, his many observations have proved most valuable. The observations of both Fabricius and Heller were used in 1857 in making a new computation of the orbit of the comet of 1556.[237]

233 The title of the tract is *Practica / auf das M. DLVII. Jar / sampt Anzeygung vnnd erclerung / Was die erscheinung / vnnd bewegung / des vergangenen vnnd zuuor angezeygten Cometen / Im sechs vnd funfftzigstem Jar gewesen / vnd bedeutet habe. Auss warem grundt der Astronomey von newen Practicirt vnd gestellet durch M. Joachim Heller verordenten Astronomum zu Nůrmberg. Regirende Planeten dises Jars. Saturnus. Mohn. Sonne.* Beneath the title is a woodcut showing the comet's path on a celestial hemisphere. At the close of the book are the words "Gedruckt zu Nürmberg, | bey Joachim Heller, Mit | Kayserlicher vnd Chur- | fürstlicher zu Sach- | sen Freyheit nit | nachzudru | cken." See Richel, 70.

234 Richel. 71. Richel said that the appearance of a comet in the spring of 1558 is uncertain, but Zinner (1934), 66 ff., noted Heller's, Flock's and B. Herzog's observations of the comet of that year. According to Pingré, I, 507, the comet was first seen on July 14th.

235 Littrow, 635-6.

236 *Ibid.*, 636-8.

237 See Hoek, a thesis presented in Leyden. He found that the comet of 1556 was not identical with the comets of 975 and 1264, as was supposed by some of his contemporaries. In the middle of the nineteenth century, the

Cornelius Gemma[238] was one of the best known men to observe the comet of 1556. He stated that its tail faced east, that is, directly opposite the sun. And he expressed his disagreement with Fabricius as to its apparent size, saying that in the beginning he found it at least as large as Jupiter.[239] He spoke of it again in his tract on the comet of 1577.

One of the astronomers who observed the comet of 1556 from Nuremberg, and who wrote a book about it, was Erasmus Flock (1514-1568). He observed the comet from the 18th to the 20th of August, and was prevented by bad weather from observing it the next day.[240] This comet was also observed and described by Johann Hebenstreit (d. 1569),[241] a physician of Erfurt, who said that it would not be unlikely that another comet would follow. He differed from the Polish astrologer, Peter Prosuossczwice, and others who said that the comet seen

comet of 1556 was thought identical with those of 975 and 1264 and was generally expected to return in the middle of the nineteenth century (Littrow, 633). Therefore, search for all possible records was made, resulting in the finding, by Karl v. Littrow, of particulars concerning the observations of Fabricius and Heller.

238 See chapter III, below, for a fuller discussion of Cornelius Gemma.

239 Pingré, I, 502.

240 Richel, 71-2. See note 234, above. Thorndike, V, 342, says that Flock's book on the comet of 1556 appeared in German in 1557, and that in the following year his review of recent comets was published in the same language. Richel, 71-2, said that this was printed in Nuremberg by Valentin Neuber in 1558 and has the title, *Von dem jüngsten vnnd | achten Cometen, deren, so von dem Jar | M. D. XXXI an, biss auff das yetzig | lauffend M. D. LVIII. Jar, er- | schinen sein, im Augstmonat | gesehen. || Christus Luce am 21. Cap. | Auch werden schrecknuss vnd grosse | zeychen vom Himmel | geschehen. || Eras. Flock Doctor. || Nürnberg. || M. D. LVIII.*

241 Richel, 68-9. Hebenstreit's treatise on the comet of 1556 was printed in Wittenberg that year and has the title, *Des Cometen / so dieses 1556. Jars von dem 5. tag Marcij an / bis auff den 20. Aprilis zu Wittemberg erschienen / bedeutung. Darinne auch derer meinung / zu zween Cometen gesatzt / gründlich refutirt wird / durch M. Johannem Hebenstreit juniorem Erphordensem.* On the title-page, beneath the title, is a woodcut of a celestial sphere showing the path of the comet, the whole in a border of figures pointing to a comet in the sky. See also Hellmann (1924), 27; Thorndike, VI, 106.

in April was different from the one observed earlier because no comet was seen between the 16th and 23rd of March. Hebenstreit believed that they both moved in the same path and were of the same color. He related many natural events for the year 1556. Hebenstreit wrote several *Practicas* and calendars.

There are other tracts on the comet of 1556, including several written at the time of the comet's appearance, and several written in the nineteenth century.[242] A tract on comets appeared in Lyons in 1556. This was by the Italian author, Gabriello Simeoni (1509-1575).[243] The same year another tract in French, by Peter Haschard or Haschaert was printed in Louvain.[244] Also in that year a *Prognosticatio von dem Cometen 1556* was printed in German in Erfurt.[245] It was by Adam Ursinus, the author of several *Prognostications.*[246]

The comet of 1556 was also the occasion of the publishing of one of the earliest catalogues of comets. This catalogue was by Benedict Marti von Bätterkinden (1505-1574), better known as Aretius.[247] Another catalogue of comets was by Lud-

242 Some of these were listed by Carl, 50-1, and by Scheibel, 23-4. A book intended for the layman, Hind, appeared in 1857. Hagecius wrote about the comet of 1556 in the Czech language (see below, chapter IV). Observers of the comet of 1556 who did not publish a tract on it include Johannes Homelius (Thorndike, V, 397). See chapter III, note 118, below.

243 Hoefer, XLIII, 1020-2; Jöcher, IV, 594. It had the title, *De la génération, nature, lieu, figures, cours et significations des cometes. A monsieur le Seneschal de Lyon, plus un sonnet et une élégie au roy.* (Nourry catalogue 53 (1933), item 706. There is a copy in the B. M.)

244 *De l'horrible comete, qui sest apparu en ces regions, environ le premier iour de mars l'an 1556, au quel est adiouste un petit traicté de la preservation contre la peste.* (Houzeau, 5575; see also Thorndike, V, 330). See the discussion of Peter Haschaert, an upholder of astrology, above.

245 Houzeau, 5576.

246 Two of them are listed in the Crawford library catalogue, 452.

247 Wolf, I, 21 note 20. It was included in his *Brevis cometarum explicatio, physicum ordinem et exempla historiarum praecipua complectens,* which was published in Berne in 1556. According to Wolf, Berchtold Saxer seems to have relied heavily on Aretius' catalogue for material for his *Comet Sternen,* which was printed in Berne in 1578. For the full title, see Weller (1857-8), 360, 215.

wig Lavater (1527-1586), a pupil of Dasypodius, Bucer, and Ramus, and appeared in Latin in Zurich in 1556, was reprinted in 1587, and was translated into German, augmented and re-edited in Zurich in 1681.[248] As we have seen above, Flock published a catalogue of comets from 1531-1558.[249]

The comet of 1558, mentioned above and observed by the Landgrave William[250] and Cornelius Gemma[251] as well as by Heller, Flock and Herzog,[252] and the comets of 1560 and 1569 led to no valuable additions to the relevant literature.

The next astronomical event important in this connection was the new star of 1572, a phenomenon which presented many problems similar to those of a comet, and which was called a comet by many of its observers. Many of them had previously observed comets, many were to observe the important comet of 1577. To all, the physical problem of distance or parallax and the philosophical problem of change in the supposedly immutable heavens presented themselves.

The influence of the new star of 1572 in moulding the astronomical thought of the period cannot be overestimated. Many more tracts were written about it than had been written about any comet before then. Many more men made observations of it than had observed any one celestial phenomenon in the past. As is evident from the preceding pages, the number of tracts relevant to any one comet increased with the appearance of each new comet. Of course, the increase in the number of tracts is partly due to the use of printing and partly to the fact that fewer of the later tracts were lost with passing time. Certainly,

248 Wolf, III, 106-7 and notes 29, 30, and 31. The catalogue had the title *Cometarum omnium fere catalogus, qui ab Augusto ... usque ad hunc 1556 annum apparuerunt, ex variis historicis collectus* ... Tiguri, per A. et J. Gesnerum fratres (B. N. catalogue, where, however, the year 1566 appears in place of 1556).

249 See note 240, above.

250 Wolf (1877), 408.

251 *Idem*; Carl, 52.

252 Zinner (1934), 66 ff.

there seems to have been a growing body of data. Partly because of its long duration, but also because of the interesting problems it presented, nearly every astronomer in Europe directed his attention to the nova.

In England, it both marked a step in the gradual acceptance of the Copernican theory and a deterrent to that acceptance. The deterrent was the failure of the observers of the nova to detect any annual parallax for that star. But far more important, the step in advance was due to the positive determination of the position of the star outside the lunar orbit. In England, the two most prominent observers of the nova were John Dee and his pupil Thomas Digges.[253] Dee published, in March, 1573, a book of trigonometric theorems for determining stellar parallax.[254] He abandoned "Aristotle's" idea of solid orbs, and, in order to explain the gradual disappearance of the nova, even suggested that it receded from the earth in a straight line.[255] However, this argument could not account for the star's sudden appearance. In 1573, Digges also presented a work on the nova. This contained a record of observations of the nova which surpassed in accuracy those of all other astronomers with the exception of Tycho.[256] It contained no astrological matter,[257] and emphasized the importance of a large body of observations of the new star and other heavenly bodies, in order to determine, experimentally, a true system of the universe, or to verify or correct the Copernican theory.[258] Thus

253 Johnson (1937), 155.

254 *Ibid.*, 156. The work was called *Parallaticae Commentationis Praxeosque Nucleus quidam.* Dee left, among his unpublished manuscripts, a work entitled *De stella admiranda in Cassiopeiae Asterismo, coelitus demissa ad orbem usque Veneris, iterumque in Coeli penetralia perpendicularitèr retracta.* This was in three books and was written in 1573.

255 Johnson (1937), 155. This was also suggested by Elias Camerarius and Gemma (Pingré, I, 83; Dreyer (1890), 63).

256 Johnson (1936), 390-1. Digges' book bore the title *Alae seu Scalae Mathematicae.*

257 Johnson (1934), 110 note 2.

258 Johnson (1936), 391.

Digges' book was important both as a treatise on the nova and as a plea for the use of the experimental method in astronomy.[259] He hoped to find, in his observations of the new star, positive proof of the truth of the Copernican system, and in order to demonstrate, geometrically, the earth's revolution around the sun, he suggested careful determination of the annual parallax of the star of 1572.[260] But stellar parallax, the lack of the determination of which led Tycho to reject Copernicus' theory, was not to be detected until 1838, nor could it have been, with the instruments available to sixteenth century astronomers.

On the continent, the greatest astronomer of the period, Tycho Brahe, recorded most accurately his observations of the phenomenon,[261] and proved conclusively, by his failure to detect any parallax, that it was in the region of the fixed stars. However, he used that proof to show that the new star could not be a comet or meteor, because these were generated below the moon.[262] He hoped in the future to discuss the position of comets.[263] His first book on the nova was printed in Copenhagen in 1573.[264] After presenting the observations and the mathematical deductions from them, Tycho, in keeping with

259 *Ibid.*, 399. Johnson explains the term "experimental method in astronomy" by adding: "Digges makes it clear that he fully appreciates the essential value of new and brilliant hypotheses for furthering scientific research, but he is uncompromising in his insistence that such hypotheses must be grounded upon observations, and accepted only as a guide for future investigations." He stresses Digges' plea for a larger body of accurate observations.

260 Johnson (1937), 158-9, 215; Johnson (1934), 112-3.

261 Dreyer (1890), 38-69. Tycho measured the angular distance of the new star from known fixed stars, especially from Shedir at upper and lower culmination. See also chapter III, note 116, below.

262 Dreyer (1890), 48.

263 *Idem.*

264 *Ibid.*, 44 note 1. The title is *Tychonis Brahe, Dani, De Nova et Nullius Aevi Memoria Prius Visa Stella iam pridem Anno a nato Christo 1572 mense Nouembrj primum Conspecta, Contemplatio Mathematica....* Tycho later wrote about the new star in volume I of his proposed trilogy. See items [17 and 18] of appendix, below.

the tenor of his times, explored the astrological possibilities of the phenomenon. His work was held in high esteem by his contemporaries, and sixty years after the appearance of the nova, a partial English translation of his book appeared in London.[265]

Although Tycho's observations of the phenomenon far surpassed all others, his are not the only ones which merit recognition. No solitary genius could have accomplished for cometary theory what the combined efforts of the cream of the scientific intellectual circle was able to do. It was the background of the observations of the new star of 1572 which enabled the astronomers in 1577 to rise to the level they achieved.

Michael Maestlin's conversion to Copernicanism, as will be shown below, was largely due to his observations of the nova, which he described in his book on that phenomenon in 1573,[266] and which Tycho later discussed. Maestlin, like Digges, so observed the star as to show that it did not move relative to four fixed stars.[267] He concluded that the star had no parallax and was among the fixed stars. He did not discuss its signification.[268]

Munosius' observations of the nova, made in Spain, and Hagecius', made in Bohemia, have been considered among the best after those by Tycho and Digges.[269] Indeed, Hagecius [270]

265 Johnson (1937), 330. Hazlitt (1876-1903), 4th series, 389, gave the full title and a description of the work. Bruun, II, 67, also cited a London 1632 edition but with a slightly different title.

266 The title as given by Hoefer, XXXII, 649, is *Beobachtungen des neuen Sterns in der Cassiopea,* but it has been variously given by other sources, such as Scheibel, 74, and Chalmers, XXI, 99. Possibly Maestlin's work was never published separately, but only as it was discussed in Tycho's *Progymnasmata* (Brahe, III, 58-62). Galileo, II, 524, gave no title for Maestlin's work on the nova and seems to have been acquainted with it through the *Progymnasmata.* Hagecius, in his second work on the comet of 1577, which was written before the appearance of the *Progymnasmata,* spoke of Maestlin's work on the nova.

267 Dreyer (1890), 60. As was shown above, Geoffrey of Meaux used a similar method to observe the comet of 1337.

268 Pingré, I, 84.

269 Johnson (1934), 108.

270 See Hagecius (1574).

not only presented his own observations, with diagrams and tables, but also those of Paul Fabricius and Cornelius Gemma [271] and a letter from Munosius to Bartholomew Reisacher, professor of mathematics at Vienna.[272] Fabricius could find no perceptible parallax for the star, which he first saw at the end of October.[273] Munosius was certain that the star was not visible on November 2nd.[274] The Landgrave of Hesse, who likened the star to that of Bethlehem,[275] was sure that the parallax of the new star did not exceed three minutes.[276] Clavius, the Jesuit astronomer, in his *In Sphaeram Ioannis de Sacro Bosco Commentarius* (1593 but with a 1581 dedication) expressed belief that the nova was in the firmament, whether it was a comet or a new creation; he quoted works written about it in 1572 by Paulinus Pridianus of Antwerp and by Maurolycus.[277] Paul Hainzel, an able astronomer, and friend of Tycho, using the quadrant at Augsburg, was among those who found no perceptible parallax for the star.[278] Nevertheless he and Caspar Peucer, who also found no parallax, thought that the star was beneath the moon.[279] Many of the ablest astronomers

271 In his *De Naturae Divinis Characterismis, seu raris & admirandis Spectaculis,* Gemma further discussed his observations of the star. He, with William Postel, and again with Leovitius, published tracts on it.

272 Reisacher also wrote in Latin about the nova. See Reisacher. Weller (1857-8), 322, listed a work in German on the same subject by the same author, printed by the same printer in the same year.

273 Dreyer (1890), 60-1, 60 note 1, 61 note 4. Fabricius, in a letter written in 1573, seemed to consider comets as meteors, but this may be due to a broad use of the term comet (Pingré, I, 72).

274 Dreyer (1890), 61; Galileo, II, 524.

275 Pingré, I, 82.

276 *Ibid.*, I, 80; Dreyer (1890), 57-8, 65.

277 Thorndike, V, 74.

278 Dreyer (1890), 60 and 60 note 1.

279 Pingré, I, 81. On this point Dreyer (1890), 58, wrote: "Peucer and Wolfgang Schuler at Wittenberg found a parallax of 19′, which Tycho believed was a consequence of their having used an old wooden quadrant; and, in fact, when he learned that the Landgrave had found little or no parallax,

would not altogether deny the existence of a parallax, but were sure that it was at least so small as to place the star above the moon.[280]

Many astronomers, however, truly thought that they distinguished a large parallax for the new star. Among these was Andreas Nolthius, whose tract was printed in Erfurt. He called the star a comet and, trying to find its distance trigonometrically, found it in the elementary circle of air and consequently concluded that it was composed of elementary matter.[281] His observations attracted Tycho's attention, for he had used the hour angle, the azimuth, and the latitude of the observation station, but chose a bad time, when the altitude was very great.[282]

Other observers of the new star, who were later to observe the comet of 1577, included the following: Theodorus Graminaeus,[283] professor at Cologne, George Busch of Erfurt,[284] Johannes Praetorius,[285] Hannibal Raimondus,[286] and David Chytraeus.[287] Other astronomers of good repute who observed the new star included Cyprianus Leovitius, who thought he observed a movement in the star, and likened it to the stars or "comets" in 945 and 1264.[288] Theodore Beza (1519-1605),

Schuler had a large triquetrum constructed, and also found that the star had no parallax, or at most a very small one." Thus, Dreyer disagreed with Pingré's statement that Peucer found no parallax for the nova. See Brahe, III, 121.

280 Pingré, I, 80-1.

281 Nolthius (1572), B_{11} v-B_{111}r.

282 Dreyer (1890), 60 and 60 note 2.

283 *Ibid.*, 68-9; Pingré, I, 81; Schottenloher, IV, 377.

284 Busch found a parallax of 22° 40′ for the star (Hagecius (1574), 74-5). See also chapter IV, note 130, below.

285 Pingré, I, 81; Dreyer (1890), 60 note 1.

286 Dreyer (1890), 61; Pingré, I, 82.

287 See chapter V, below.

288 Leovitius, A_2v-A_3r; Pingré, I, 81; Dreyer (1890), 63 note 2, 65, and 65 note 2. See note 220, in this chapter.

the theological reformer, deriving his idea by analogy from the star of Bethlehem, announced Christ's last coming and the end of the world in a Latin poem included in the work on the nova by Gemma and Leovitius.[289] Adam Ursinus, who had observed the comet of 1556, described the new star in his *Prognosticatio.*[290] Valesius of Covarruvias, physician to Philip II of Spain, like Reisacher, thought the star was an old and faint one which had suddenly become brighter because of a change of air between it and the earth or a condensation in one of the spheres through which its light passed.[291] Other observers of the nova were Philip Apian,[292] Maurolycus, Bernhard Lindauer, Frangipani,[293] and Aegidius Misner [294] and authors of anonymous treatises.[295]

From this short sketch it is possible to see how great was the interest in the astronomical problems presented by the nova, problems which also are presented by comets. Tycho Brahe, in his later book on the nova, carefully analyzed the findings of other observers and coordinated them. But even before this was done, men were eager to test further their new ideas and newly discovered facts by applying them to another phenomenon. The opportunity for which they so eagerly waited was soon afforded them by the comet of 1577, and they watched intelligently and carefully. That is one reason why its appearance was so important in the history of comets.

289 Dreyer (1890), 68 and 68 note 1; Bouché-Leclercq, 613 note 1. This short poem is given by Lubienski, 365.

290 This work, which was printed in Erfurt in 1574, has the title, *Prognosticatio. Auff das Jhar / ... M. D. LXXIIII. Beyneben einer kurtzen Beschreibunge des erschienenen Cometens / 1572. vnd 1573. Jhare.* Crawford library catalogue, 452; Schottenloher, IV, 377.

291 Dreyer (1890), 60 note 1, 63-4; Pingré, I, 83; Gassendi, 93.

292 Lubienski, 366; Günther (1882), 118-9. See above.

293 Dreyer (1890), 62 and 62 notes 3 and 4.

294 Crawford library catalogue, 309.

295 Such as the one cited by Dreyer (1890), 63 note 1. Other authors were cited by Lubienski, 364-6.

CHAPTER III

THE COMET OF 1577: BELIEVERS IN ITS SUPRA-LUNAR POSITION

TYCHO BRAHE.—MAESTLIN.—ROESLIN.—WILLIAM IV, LANDGRAVE OF HESSE CASSEL.—CORNELIUS GEMMA

WHEN the comet of 1577 appeared, it was observed by most of the astronomers who had observed the nova of 1572 and by many others. It was the focal point of astronomical thought in the last quarter of the sixteenth century. Because the comet was visible for some time, the observers were able to communicate with each other; and because some of the observers delayed in publishing their works, additional letters passed between them before final decisions were reached. There were no scientific societies to act as clearing houses of information in the sixteenth century, but several of the well known and highly esteemed astronomers acted as centers to which information was sent and whence issued criticism, sometimes constructive, sometimes scathing, sometimes in the form of praise. Tycho Brahe, who at the time of the comet's appearance was living on the island of Hveen, was the best known of these astronomers. Hagecius, who observed from Prague, was another. Writings by both show a grasp of the current astronomical problems and a comprehension of what was being done about them. Correspondence which has survived shows that Tycho received records of observations from the Landgrave of Hesse Cassel. Tycho also corresponded with Paul Hainzel, Johannes Major, Scultetus, Brucaeus and Hagecius.[1] Hagecius published a work by Cornelius Gemma, and wrote to Martin Mylius of Görlitz, discussing the beliefs of Maestlin and Roeslin concerning the comet of 1577. Hagecius also was acquainted with the work of Raimondus. Tycho and Hagecius discussed

[1] Dreyer (1890), 131. Hainzel, because he lacked instruments, did not observe the comet of 1577. This is stated in a letter from Hainzel to Tycho, March 25, 1579 (Dreyer, editor, VII, 49-50).

the parallax of the comet of 1577, which Hagecius believed to be below the moon. Tycho, not publishing his great work on the comet of 1577 until 1588, was able to compare the observations of that phenomenon made by the best known European astronomers. For example, he realized that a comparison of his own observations at Hveen, of distance measures of the comet from stars, and those of Hagecius at Prague, which showed a difference of only one or two minutes, was an added proof of the comet's position outside the lunar orbit.[2]

Astronomers in 1577 were an active group, eager to keep abreast of new astronomical theories, interested in the works of their fellow astronomers, and, on the whole, frank in the exposition of their observations. All this will become apparent from the examination of their tracts on the comet of that year. To understand the significance of this comet, it is not necessary to survey all the works and authors listed in the appendix. A representative cross section should suffice. Although many of the volumes not discussed are not available in any form, it seems unlikely that any valuable information will be overlooked by thus limiting the discussion. The works chosen to be summarized would have cited any important books or data. Even if there were a startlingly new theory advanced in these undiscussed works, that theory can have had little effect on the contemporaries of the man who presented it or on his immediate successors, if it was not repeated in current works on comets. The works to be summarized have been chosen primarily because of the importance or influence of their authors, but also with an eye to covering the different types of tracts and the different sections of Europe, and always with the restriction that copies were available for summarizing.

Although the outstanding astronomer of the second half of the sixteenth century and the most important observer of the comet of 1577 was undoubtedly Tycho Brahe,[3] he will not be

2 Dreyer (1890), 165.

3 Tycho was the most accurate observer since Hipparchus.

dealt with here as fully as he deserves because a great deal of material concerning him is readily available, accumulated by a succession of learned scholars, the most important of whom was the late J. L. E. Dreyer, who wrote the definitive biography of Tycho and began publication of the fifteen volume edition of his works. It is impossible to include in these pages an adequate treatment of Tycho and it is not worthwhile to outline his career and sketch his contributions to cometary theory alone.

Summaries of Tycho's most important work on the comet of 1577, the *De Mundi Aetherei Recentioribus Phaenomenis,*[4] are numerous.[5] However, little space is devoted to the very lengthy tenth chapter, which concerns observations of the comet by others, and to which reference is often made in the present study. That chapter is not very important to a study of Tycho as an astronomer, although it is a monument to his powers of analysis and judgment.[6] It is important in a study of the contemporary writings on the comet of 1577, and in many instances gives the only reference to those books which has hitherto appeared. In most cases, the statements by Tycho concerning them cannot be improved. Besides, there is value in his comparison of the observations of others with his own. In this tenth chapter Tycho first discussed the observations of the comet by the four men who considered it supra-lunar, namely the Landgrave of Hesse Cassel, Maestlin, Gemma and Roeslin. Then Tycho sketched the works on the comet by other observers, namely Hagecius, Scultetus, Nolthius, Wincklerus, Johannes Praetorius, Squarcialupus and Erastus and Simon

4 Item 20 of appendix, below.

5 Chapter VII of Dreyer (1890) is devoted to that work, and Delambre (1821), I, 207 ff., analyses it. Dreyer's edition of Tycho's works makes the original text available to all who wish to consult it and, in addition, gives excellent notes to clarify questionable points in the book itself and concerning its production.

6 Tycho treated the writings on the nova of 1572 in a similar manner in chapters VIII, IX, and X of the *Progymnasmata.*

Grynaeus, Dasypodius, Henischius, Bazelius, Steinmetz, Huernius, Graminaeus, Busch, and finally, Chytraeus.

Tycho's observations of the comet of 1577 are the most accurate which were made. In fact, they were the ones used in the nineteenth century to compute the orbit.[7] Even in his own time Tycho's accuracy was accepted, and thus, by means of his book, he was able to shake the time-worn belief in the immutability of the heavens and pave the way for a new astronomy which his pupil and assistant, Kepler, was to supply, not as Tycho envisaged it, but based upon Tycho's observations as well as on the work of Copernicus. There were but two outstanding dissenters from Tycho's conclusions about the comet, Craig and Claramontius.[8]

The comet of 1577, coming, as it did, so soon after the nova of 1572, enabled Tycho to test the conclusions which he had drawn from his observations of that phenomenon. Although he had said that the nova of 1572 could not be a comet because it was in the region of fixed stars, whereas comets are generated below the moon,[9] nevertheless, he started out to look for an astronomical body, not an atmospheric phenomenon; and this approach to the subject is illustrative of a change in cometary theory which the comet of 1577 ushered in. Even the comet observations of Toscanelli and Regiomontanus and the observations of comets' tails by Fracastoro, Cardan, Gemma, and, above all, Apian, did not do this. Furthermore, Tycho attempted to compute an orbit for the comet, a departure in the treatment of comets.[10] Tycho made no change in the adherence to

7 By Woldstedt. See item 110 of appendix.

8 See chapter VII, below.

9 See chapter II, above.

10 Chapter VI of item 20 of appendix. See also Dreyer (1906), 366. The calculations of comets before that of 1577 had never established an orbit for a particular comet, although the detailed measurements by Toscanelli for the comet of 1472 fell short of this by so little that the concept of a fixed orbit could have been added without alteration of the previous arguments. Roeslin's sphere of comets with poles and axis was applicable to comets as a whole

the postulate of circular motion, and his orbit for the comet was circular.[11]

Although Tycho mentioned the astrological discussions of the comet which his contemporaries wrote, he did not devote space in his *De Mvndi Aetherei Recentioribvs Phaenomenis* to the astrological implications of the comet. Nevertheless, he was not different from other sixteenth century men in this respect, and considered astrology a proper science if kept within bounds, as is shown by his German work [12] on the comet. That work exhibits Tycho's astrological interests and shows that his work, while astronomically and mathematically far more accurate than that of his contemporaries, followed the same lines as theirs.

The German work, probably written immediately after the disappearance of the comet in 1578 but first printed in 1922, is not very well known now and seems to have had little influence when it was written. It often refers to the Latin work on the same subject, much of which was written in 1578, but the two were considered by their author to be different types of treatises. However, the German work, like the Latin, emphasizes the absence of parallax and consequently the untenability of the so-called Aristotelian concept of crystalline spheres and immutable heavens. Tycho's conception, bolstered up by his excellent observations, dealt the death blow to ancient cosmogonies and paved the way for the Keplerian system.

and not only to the comet of 1577, so that here, too, the concept of the orbit was missed. However, Maestlin did find an orbit for the comet of 1577, based on the theory of circular motion for celestial bodies. Scultetus, although he believed that the comet was sublunar, likewise assumed a circular orbit for it.

11 However, he suggested that it might not be exactly circular but an oval. See Dreyer (1906), 366.

12 Item 20a of appendix. In this work Tycho showed that he did not consider astrology a pseudoscience but a science on which one could build only as much as the original premises permitted. According to Thorndike, VI, 70-1, the fact that the German work was not published during Tycho's life pointed to an absorption by scientifically minded persons in positive astronomical activity which tended to preclude their former intense interest in astrology.

Tycho's German book affords interesting reading. In it he showed respect for the Copernican system, with which he disagreed. Although he did not yet put forth a coherent mathematical system of his own, he hinted at his new system, which he introduced in the *De Mvndi Aetherei . . . Phaenomenis.* His jealousy of his system is shown by the fact that he was distressed to hear from Rothmann, after the printing of that book, that there had been a similar system proposed to the Landgrave. The German book shows Tycho in the historical position which he felt he held and which is not so clearly stated in the Latin work; and makes it evident that he realized what he was doing when he cast aside the doctrines of his predecessors. It shows that Tycho had an understanding, not only of the problems confronting the observers of the comet of 1577, but also of the cometary theories extant at the beginning of the century and of what the nova of 1572 had established. Because of the book's late publication, and because it does not go into the mathematics of the subject, it has not received the attention which was meted out to the Latin work. It will, therefore, be summarized here.

The first [13] of the ten sections, on the origin of comets and on what ancient philosophers thought about them, deals with different theories about comets. First Tycho described the universe, placing the earth at the center, and mentioning the great speed with which the outer spheres must travel. Then he introduced the subject of comets, saying that they stand out above all wonders visible in the heavens, and that they have always been the object of inquiry by philosophers because they are visible only at certain times. Tycho appreciated that Pythagoras, Anaxagoras, and Democritus thought that comets originated in the heavens and were special stars only sometimes seen on earth, having their existence and place in the heaven. He noted that Aristotle refuted their argument on the grounds that there can be no change in the heavens, saying that comets are engen-

13 The first section has the title, "Vonn Der Cometten Uhrsprung Was Die Alten Vnnd Neuen Philosophi Inn Denselben Vermaint Vnnd Dauon Zuhalten Sei."

dered, not in the heavens, but in the highest portions of the atmosphere and are made of dry, dense exhalations from the earth which burn until they are consumed. Tycho pointed out that philosophers had accepted Aristotle's arguments until the appearance of the new star in Cassiopeia. Because this star had no parallax and belonged in the sphere of the fixed stars, the philosophers began to doubt Aristotle and to believe that changes can take place in the heavens, and that it is possible that other comets originate there and are not formed from earthly vapors. Tycho also reported that the followers of Paracelsus believed that the heavens were of the fourth element, fire, and that "generation" and "corruption" could take place there. They considered it not impossible that comets were born there as monsters are among animals. Paracelsus held that the guardian deities above ("penates superi") occasionally make stars and comets out of celestial material and make them visible to men as a sign of some future event, which does not have its origin in planets, but is shown and produced out of a pseudoplanet, as a comet is called.

Tycho said that the nova of 1572, which he called "the star of four years previous", was shown to be in the heavens and that he, through hasty observations and demonstrations of the present [1577] comet, discovered that it likewise had its position and pathway above the moon in the heavens. He argued that consequently the opinion of Aristotle, that comets are pulled aloft from the earth and cannot be generated in the heavens, is false, because it was established by cogitation, not by mathematical observation or demonstration. Tycho thought that, because they are formed in the heavens, which are made of the most subtile transparent material, there was even more reason that comets be considered wonder-signs, than if they were made according to the decree of God by the "penates superi", who are unknown to man. He did not wish to dispute, at this point, that God, unaided, put a new sign of warning in the heavens, for, as he said, we on earth can have no more understanding of the material of comets and their generation than we can about the sun and moon and their motion. Furthermore, philosophers should not fight over things they do not know how to determine, but should inform men that comets are a wonderwork of God, coming from a hidden natural cause.

The second section,[14] on the first appearance and the duration of the comet, begins by stating that shortly after sundown, on November 11, 1577, this "newborn" showed itself in the heavens: a comet with a very long tail and a head of white light, not like that of a fixed star but somewhat darkish, much like the "star" Saturn,[15] which was not very distant at the time. The tail was very long, somewhat bent in the middle, of a burning dark red color, like a flame penetrating through smoke. In Tycho's opinion, this comet had its true beginning with the new moon, which occurred shortly before the 10th of November, one hour after midnight. He pointed out that several seafaring people reported that they saw it in the Northwendic [16] sea in the evening of November 9th. Tycho first saw it with his instrument [17] on November 13th, because before then the sky was not clear long enough for such an observation. He said that it lasted more than two months, until January 26th, although it decreased as time went on, so that on January 13th he could scarcely observe it with his instrument, and on the 26th, when he saw it for the last time, it was all but unrecognizable.

In Section III,[18] on the course of the comet, Tycho said that when he first observed the comet at 5 P. M. on November 13th

14 The title of the second section is, "Wen Diser Comett Erstlich Gesehen Vnnd Wie Lanng Er Geschinen Hatt."

15 Tycho, like others at this time, sometimes used the term "star" when speaking of planets.

16 Possibly "North" or "Baltic".

17 Tycho's observations of the comet were made with a radius and a sextant and occasionally with a quadrant which had an azimuth circle. He computed the comet's latitude and longitude from measurements of its distance from certain fixed stars.

18 Section III has the title, "Von Dess Cometten Lauff Vnnd Seinem Orth Vnder Dem Firmament."

it was in the 7¼th degree of Capricorn and had a declination north of the ecliptic of 8° 20′,[19] for it was 26° 50′ from the bright star in Aquila and 21° 40′ from the lowest star in the horn of Capricorn, where its tail ended. He found the position of the comet as given above by trigonometry.[20] On the 14th the comet was 23° 45′ from "lucida Vulturis"[21] and 18° 30′ from the above mentioned star in Capricorn.[22] From this Tycho concluded that the comet traveled 3½° in its path in 24 hours, and because its motion was quickest in the beginning, he believed that the day before he observed it, it had moved 4°. Therefore, he said, because it was first seen on the 9th and at the time of the new moon, it must have begun near the ecliptic under the 25th degree of Sagittarius in the edge of the Milky Way from which, he said, most comets have their origin. Tycho thought that the comet had its beginning there, near the ecliptic, not far from the winter solstice and the "tropical circle"[23] and that it continued and ended somewhat north, through a succession of signs in the manner of planets and stars from setting to rising, contrary to the motion of the heavens, until it arrived in the tropic of Cancer at the star in the chest of Pegasus, half way between two small stars and one large one called "Scheat." He found it at that spot for the last time on January 26th. It was so small then that one could hardly see it; and he believed it to have disappeared shortly thereafter. This comet described

19 This last measurement would now be taken from the equator, and the observation would be recorded in right ascension and declination. The right ascension of a star is now defined as "the arc of the celestial equator intercepted between the vernal equinox and the point where the star's hour-circle cuts the equator", and the declination of a body as "its distance in degrees north or south of the celestial equator, + if north and — if south". See Russell, Dugan and Stewart, 17, 15.

20 "durch die *scientia triangulorum*".

21 The bright star in Aquila ?

22 These measurements are given as 23° 23′ and 18° 26′ in the *De Mvndi Aetherei Recentioribvs Phaenomenis,* where the observations have been corrected and apply to the head of the comet.

23 The tropic of Capricorn.

a fourth of a great circle on the sphere, beginning at the 25th degree of Sagittarius in the ecliptic and intersecting the equator 300° 40′ from the vernal equinox, at an angle of 34°. When it disappeared it was 30° north of the ecliptic in the longitude of the 25th degree of Pisces. It traversed a fourth of the heavens, not only in its own orbit but, as Tycho said, also reckoning with the ecliptic.[24] Its apparent motion was not uniform; for in the beginning, as stated above, it had a motion of its own of 4° in a day, and later went more slowly, so that by the 15th it moved 3° in a day; on the 20th, 2¼°; on the 23rd, 2°; on the last day of November, 1½°; on December 5th, 1°; on the 16th [?], 50′; on the 31st, 35′; on January 10th, 25′; and in the end, as it was growing fainter, scarcely 20′ in one day, for from January 13th to the 26th it moved only 4⅓°. From this Tycho concluded that in the beginning it moved as much in one day as in the end it moved in ten days, its progress decreasing like its size, but that in the end the change in motion from day to day was not as great as in the beginning.

In Section IV,[25] on the tail of the comet, one reads that in the beginning the tail was long, stretching 22°, but that it became smaller and shorter so that by the end of January it was barely visible. The comet always turned its tail directly away from the sun, like all other comets previously observed by Regiomontanus, Apian, Gemma Frisius and Fracastoro. Tycho believed that from this fact it follows that the tail of a comet is nothing other than the rays of the sun shining through the comet's body, which, not being diaphanous like that of other stars, cannot transmit the beams invisibly, and which, not being thick and opaque like the moon, cannot reflect them. He thought that, since the body of a comet is neither rare nor dense, it partly holds the sunshine, the light of the head remaining in accordance with the diversity of the celestial ma-

24 According to the *De Mvndi Aetherei . . . Phaenomenis* it had a longitude of 20° 55′ in Pisces when it disappeared and one of 20° 55′ of Sagittarius when it appeared.

25 Section IV has the title, "Von Dess Cometten Schwantz."

terial out of which the head is made. Partly, however, because of rarity and porousness, the comet's body allows the sunbeams to pass through and they are seen by us as a long tail hanging from the comet's head. Tycho cited all comets observed at various times by mathematicians as evidences of this and believed that there could be no further doubt, regardless of Aristotle and those who followed him. They thought that the tail on a comet was a flame of dry fatness burning in the atmosphere. Tycho argued that in that event this flame would have nothing to do with the sun, away from which it is always turned. Tycho thought it difficult to tell in words the way this comet traversed the circle of the heavens and the way the tail was always turned from the sun, although it evidently traveled through the declinations, as he said he showed by figures.[26]

The fifth section,[27] on the position of the comet, says that those who consider Aristotelian philosophy the best believe that all comets are generated and move far beneath the sphere of the moon and that it is impossible for any change to take place or anything new to be generated in the upper air or among the heavenly bodies. They obtained such knowledge and opinions, not from experience or from mathematical observations of industrious masters, but from subtle argument by reasoning alone. Thought, however, in such things can rise no higher toward the truth than what apparent observation with correct instruments interpreted by trigonometry shows should be believed. Moreover, however subtle the argument, it is after all only human and can be refuted by other arguments. Many philosophers, Tycho added, both before and after Aristotle had disagreed with him and recognized comets as heavenly and not elementary bodies. The only way to determine the distance of the comet from the earth is by the parallax; for if the comet

26 These figures can be found in the Latin work on the comet (item 20 of appendix, below).

27 Section V has the title, "Von Dem Ortt Dises Cometten Wo Der Gestanden Sei *In Mundi Diametro* Vnnd Wie Weit Er Von Vnns Ist Erhoben Gewesen."

had a larger parallax than the moon, which is next to us, it would follow that it was closer to us than the heaven in which the moon moves.

Tycho's reason for having taken great pains to investigate the parallax was that therein lay the entire knowledge of the place and characteristics of the comet. He saw from many observations with appropriate instruments, and thereafter found through the scientific demonstration of spherical triangles, that this comet was so far from us that its greatest horizontal parallax could not exceed 15′ and was more likely to be less. This he thought he had thoroughly shown from observations in his Latin work on this comet which he considered understandable by the masters. From this it followed, according to Tycho, by "geometrical distribution" that this comet was at least 230 semidiameters of the earth from the earth. Because one semidiameter equals 860 German miles,[28] the comet was 200,000 German miles away. Since the moon is 52 semidiameters of the earth or less than 50,000 German miles away from us, it is easily understood that this comet was far above the moon in the heaven of the planet Venus. For the sphere of Venus, which astronomers place next beneath the sun, begins 164 semidiam-

28 Dreyer (1890), 167, said that this value was probably taken from Fernel's *Cosmotheoria*, Paris, 1528. However, I disagree. According to Renouard, 117, the 1528 edition of Fernel's work, which Dreyer cited, is the same as the 1527 edition (Fernel), which I have seen. (The N.Y.P.L. copy of the 1528 edition will not be available until after the war.) In the 1527 edition ($B_{ii}v$), using 22/7 as the value for π, Fernel gave 3900 miliaria, apparently Italian miles, as the semidiameter of the earth with the information that 1000 passus [or feet] equalled an Italian mile and 4000 passus equalled a German mile, which would lead to the conclusion that the earth's semidiameter was 975 German miles. Elsewhere (Brahe, II, 383; Dreyer, editor, II, 457) Tycho used the value 5400 German miles for the earth's circumference from which it follows (using $\pi = 22/7$) that the earth's radius is 859 German miles. With the great variations in the sixteenth century in the value of the German mile it is difficult to find the corresponding value in English miles. An English mile equalled approximately 4½ German miles. Dreyer, editor, II, 457, also used the ratio 4 to 1 for the values of the German and Italian miles, saying, on the basis of Tycho's figures, that 1° [of the earth's meridian circumference] equalled 15 German miles or 60 Italian miles.

eters of the earth away from the earth and extends to the sphere of the sun which is 1104 semidiameters way. However, one might prefer not to follow the usual division of the heavens but to accept the idea of some ancient philosophers or the opinion of Copernicus that the sphere of Mercury is nearest to the sun and the sphere of Venus outside that of Mercury. This reckoning, according to Tycho, is not in complete disagreement with the truth, even if the sun is not constrained, as it is by the Copernican hypothesis, to be motionless in the center of the universe. Then, it follows that this comet was generated between the sphere of the moon and the sphere of Venus, for in accordance with this opinion Venus could not come nearer the earth than 296 semidiameters of the earth, and the moon when furthest from us is 68 semidiameters away, so that between the moon and Venus is a space of 228 semidiameters of the earth which should be empty. Tycho believed that the comet was generated in this space and that it was 230 semidiameters above the earth.

Once more Tycho said that the Aristotelian theories of the immutability of the heavens and of the generation of comets were untenable because of what he himself had discovered about the comet of 1577 and what he and other mathematicians had established through careful observation in regard to the new star of four years previous. This was seen for a whole year in Cassiopeia and gave sufficient evidence that something new can be generated in the heavens, because it was not in the lowest heaven but in the very highest eighth sphere and had neither parallax nor proper motion. He considered this new fact less unbelievable because the comet of 1577 had a real head or body like a star and observation showed it to have its position in the heavenly orb. He repeated the opinion that sunshine passing through the comet caused its tail, and added that this comet could have its place in the celestial region of the sky just as well as the new star. Tycho stated that this had now been sufficiently explained and would be demonstrated and proved geometrically in Latin.

The sixth section,[29] on the size of the comet, begins by saying that it was largest in the beginning but gradually decreased in size just as its daily motion decreased. Tycho observed it with a good instrument on November 13th and found its diameter to be 8′ and the length of its tail to be 21°40′. The tail at that time stretched from the edge of the Milky Way, where the head was, to the horn of Capricorn. This Tycho called its apparent size, saying that it was so far away in the heaven of the planet Venus that it had in itself a much greater size than we here below could recognize. For by geometrical division and demonstration, the head was 230 semidiameters of the earth away or 200,000 German miles, and its apparent diameter was 8′. Thus the comet itself must have had a breadth of 465 German miles so that its diameter was nearly a fourth of the earth's and the circumference of its head was 1460 German miles.[30] From this it followed that the body of the comet was as large as 1/30th of the earth and almost as large as the morning star, Venus; and that its tail, 22° in length, calculated geometrically, came to 70,000 German miles, the distance that the sun's rays could be seen through the comet. The greatest width of the tail was 2½°, corresponding to 5,000 German miles, from which Tycho concluded that the comet itself was a tremendous thing even if it looked so small.

Tycho started section VII,[31] on the astrological influence and significance of the comet, by asserting the impossibility of the opinion that a comet is generated with a constitution similar to that of some particular star [planet] with which its influence too is in accord. He also asserted the untenability of the opinions that the eclipse of the moon in Aries on September 27th, observed as a fiery sign, was the cause of the comet and that the comet's signification should be in agreement with that

29 Section VI has the title, "Von Dises Cometten Grosse."

30 In this calculation $\pi = 3.139+$.

31 Section VII has the title, "*Judicium Astrologum* Von Dises Cometten Effect Vnnd Bedeuttung."

phenomenon. His reason was that comets have neither origin nor meaning from any natural motion of the stars nor from any eclipse of the sun or moon, but are new and supernatural works of God. Their meaning and influence, he stated, not only have nothing in common with those of the planets but oppose and interfere with them. They overcome the natural indications of the stars with much greater strength and replace them by their own. Because they are a great wonderwork of God and a miracle of nature, they must cause wonder more than do all other natural motions of the heaven. Men do not know exactly what they presage, but this is disclosed by God. Although men have sought the explanation of God's natural works from the earliest times, they do not understand them. Even less do they understand the unnatural works of God by which He signifies something other than He signifies by His natural works. But when men see such a sign in the heavens they become eager to know what the effects and meaning will be. Even if the truth is hidden from all men, some information about what such a phenomenon could signify can be discovered from old astrological writings without recourse to superstition and without transgressing the bounds of knowledge.

Tycho was more specific in the eighth section,[32] on the significance of the comet, although its indulgence in astrological argument may seem somewhat inconsistent with what he had said in section VII. He thought that all historians would testify that comets have always had some great task to perform in the world, but that usually they have aroused dryness and heat in the air, strong and destructive winds, and in some places uncontrollable floods and in others horrible earthquakes and the spoiling of the grain and fruits of the earth. Comets are followed by plagues, fevers, pestilences, and poisoning of the air from which men and beasts perish; and they point to great disunity among rulers, war, bloodshed, and the deaths of chiefs. This comet would have no less effects than the previous ones, especially because of its size and its saturnine aspect. Because

32 Section VIII has the title, "Volget Nun Was Diser Comett Bedeuttet."

its color was similar to that of Saturn, it was regarded as of the nature of Saturn, toward which it first moved on November 14th in the tenth degree of Capricorn. Other reasons advanced for considering it to be of the nature of that planet were that it passed over Saturn in a conjunction near the beginning and also the same evening after sunset, when Saturn was seen in the eighth heavenly house to which astrologers ascribe death. From this, and because the comet's tail had a dark red martial aspect showing Mars' influence, and because it occurred in a "human" constellation, Tycho reasoned that the comet signified many deaths both by pestilence and by wars. He wanted to make it clear that, because of the unfortunate resemblance of the comet's head to Saturn and of its tail to Mars and because of its origin in the tropic of Capricorn and its approach to the ninth house to which astrologers ascribe an influence on religion, this comet would cause great changes in religious matters. The changes would be greater than hitherto, especially because the comet stood in the fourth house with Saturn at the time of the new moon which Tycho believed to have occurred on November 10th, 1 hour and 20 minutes after the previous midnight. This, according to the ancients, signified the rise of new sects. Earthquakes in the south and other usual misfortunes such as extreme heat and cold were also predicted.

The ninth section,[33] on what regions and peoples the comet will most affect, says that because the comet was first seen at sunset its effects would be felt more in the west than in the orient. The greatest influence was to be over Spain and its possessions, because the comet was seen in Sagittarius which rules over them, toward sunset, where those lands lie. Tycho thought that one could conclude that there would be great evil from the Spaniards in Germany and particularly in the Netherlands where they ruled, because the comet traveled northeast. But he thought that the greatest misfortunes would befall the Spaniards themselves, and perhaps they would lose their chief

33 Section IX has the title, "An Welchen Orthen Der Welt Vnnd Bei Was Volckern Diser Comett Am Maisten Seine Wirckung Volbringen Wirt."

and their cattle [34] and many of their best people. He saw a possibility that there would be war and bloodshed among them because of religious disunion, and that the Spanish would receive just retribution for their oppression of many true Christians. Perhaps not only the Netherlands but also Germany, which includes Saxony, would have enough to do, because Capricorn, into which the comet moved shortly after its beginning, has its influence over the northern part of Germany.

The comet would have its influence over rulers of Spanish extraction, especially a ruler whose birth and crowning were under Sagittarius and Capricorn where this comet was first seen and where its evil effects were being felt. It would affect the eastern lands whither its tail turned. Its influence was considered bound up with its position in Sagittarius, and it was deemed possible that the then ruling Muscovite would be punished for his tyranny. The Jews would suffer persecutions because they were under the influence of Saturn and not only they but all who in the guise of true religion were seeking their own honor and set themselves up as pseudo-prophets would be affected, because the comet allowed itself to be seen as a pseudo-planet.[35]

The tenth and last section,[36] on when the influence of the comet would begin to be felt and how long it would last, says that this influence would begin in 1578 but would have its greatest effects in 1579 and 1580 and would continue until

34 The German reads: "... vnnd villeicht werden si ir Haubt neben vil von irem f[ie] vnnd auch vil von iren besten leuthen verlieren, ...". The word "f[ie]" has been translated as equivalent to "vieh".

35 "Vnnd nicht || allein si sonndern vil anndere, die im schein der waren Religion ire aigne ehr vnnd nutz suechen, vnnd alss Phseudoprophetten, die von himmel vnnd gestirn nicht auss dem gottlichen liecht geporen sein, vnd sich selbst in den weingarten vnberuefen einstellen, von disem Phseudo Planetten gestrafft vnnd zam gemacht werden, dann der Comett hat sich darumb alss ein Phseudo Planetta sehen lassen, das er die kinder der Planetten, baide gaistlich vnnd weltlich, die allzu hoch in irem vbermuech gestigen sein vnnd in gottlicher weisshait nicht wandlen, straffen wirt."

36 Section X has the title, "Vonn Der Zeit Wann Die Bedeuttungen Des Cometten Anfangen Vnnd Wie Lanng Sie Sich Erstrecken Werden."

1583. Thereafter the constellations and their meanings would interrupt. Then the new star of 1572 together with the "greatest" conjunction of planets in the beginning of Aries, which can happen only once in 800 years,[37] would begin their powerful

37 Considerable confusion seems to have arisen over the time between greatest conjunctions, which, though astronomical data, really have meaning only to the astrologer and consequently are not treated in astronomy text books. The printed editions of Peter of Abano's *Conciliator*, cited by Thorndike, II, 898, have given the time as 960 years. Professor Thorndike did not agree with that value (II, 898), although he had used it previously (I, 648) when citing the first use of the theory of conjunctions in Arabian astrology by Alkindi, but Professor Thorndike also thought that 800 years was too short. The Enciclopedia vniversal, XIV, 1296, has given the time as 800 *or* 900 years. Zedler, VI, 980, has given 800. The period given by Tycho, 800 years, seems to be the most accurate. The disagreement seems to have been occasioned by the failure of some writers to account for the precession of the equinoxes.

A great conjunction takes place when Saturn and Jupiter are in conjunction. A conjunction of these two planets occurs first in Aries and repeats itself in approximately twenty years, or more accurately 19.86 years. But this second conjunction takes place in Sagittarius. Twenty years later it occurs in Leo. The fourth time it is again in Aries, and the cycle repeats itself. This happens four times in the three constellations which make up the fiery triplicate, requiring 3 x 20 x 4 or 240 years. Then the cycle moves on into the earthly triplicate, requiring 240 years there. Similarly it occurs in the airy and watery triplicates so that after 4 x 240 or 960 years the conjunction, having passed through the four triplicates, again takes place in Aries. When this occurs a "greatest conjunction" is said to take place, and the time between greatest conjunctions appears to be 960 years. However, making allowance for precession, the period of 240 years becomes approximately 196.6 years and the 960 years become 786.4, which accounts for Brahe, I, 31, giving the period of the complete cycle as "barely" 800 years. Thus we can safely say that a greatest conjunction, or conjunctio maxima, takes place only once in about 800 years. See Loth, 267-9, and more especially Drecker, 164-5, for a detailed discussion of this problem. Drecker said that astrologers usually use 200 years and 795 years for the two periods of time. They believe that the first "greatest conjunction" took place in 3980 B. C., when the world was created, and place the eighth in the year 1583. This is the year when Tycho also placed a greatest conjunction. According to Loth's summary of an Arabic astrological text by Alkindi (Loth, 268), who was interested in the influence of conjunctions, the conjunctions which occurred every 20 years were considered the lesser conjunctions, those which occurred every 240 years the middling conjunctions, and those which were 960 years apart were the great conjunctions. Thorndike, I, 648, referred to this passage

operation.[38] Tycho thought that they would, in the following years, cause great changes which might be best for Christendom. He drew further conclusions from the great conjunction and closed the treatise by saying that one cannot predict the end of the world from heavenly constellations. The eclipses which Christ spoke of as preceding Doomsday are not natural events nor astronomically predictable, said Tycho. Neither did he think that the end of the world could be predicted from the comet, because comets have been seen since the birth of Christ and also since the beginning of the world. The end of the world cannot be predicted by natural events, and is known to God alone and to no creature. Let Him permit us to achieve our short life on earth so that as angels we may praise Him forever, said Tycho.

The detailed summary given above shows how Tycho saw and met some of the astronomical problems of his day. His German work is the very best of all those written on the comet of 1577 in a vulgar tongue and far better than almost any work written in Latin on the same subject. It is surpassed by Tycho's own *De Mvndi Aetherei . . . Phaenomenis,* where details of the observations and mathematical calculations are given, and perhaps by the recording of observations in the works of Maestlin, Roeslin, Hagecius and Cornelius Gemma. The observations indicated in the German work and given in detail in the *De Mvndi Aetherei . . . Phaenomenis* achieved the greatest possible accuracy for their time, although Tycho's observations of later

from Loth. Arabian astrologers began the complete cycle with a conjunction in the watery triplicate at the time of Mohammed's birth, ushering in a new world-period (Loth, 268, 269). Much of Alkindi's theory of conjunctions, which was concerned with conjunctions other than those of Saturn and Jupiter, was taken over by Albumasar (Loth, 271-2).

38 For a discussion of the period of influence of the new star and the "greatest" conjunction see Brahe, I, 31, III, 164, 232, 311-2; Dreyer, editor, I, 310 (the note to Brahe, I, 31); Dreyer (1890), 194-5, 195 note 1, 49, and 49 note 1. There is a discrepancy between statements regarding the conjunction in Tycho's German work on the comet and in his other writings. In his Latin works Tycho located the conjunction in Pisces. Dreyer made no note of this discrepancy.

comets are better, due to the use of improved instruments and the collaboration of a staff of observers.

The second most important astronomer of the second half of the sixteenth century was Michael Maestlin (or Moestlin).[39]

39 Bassaeus, I, 403, 524, II, 313.—B. M. catalogue.—B. N. catalogue.—Bök, 90-1.—Brahe, IV, XV (index). Further information concerning Maestlin can be found in the other volumes.—Brewster, 10-3, 201.—Cantor (1892), 676.—Carl (Ms.), section 20.—Cat. Belg., 389.—Chalmers, XXI, 99.—Doppelmayr, 8, 89, 91, 94.—Dreyer (1890), 59, 171-2, 181, 289, 297.—Dreyer (1906), 348-350, 365-7, 372.—Dreyer, editor, VIII, 458-9, VII, 82, 406.—Favaro, editor, X, 428-9, Maestlin and the *Sidereus Nuncius*, XII, 64, Letter from Maestlin to Kepler, saying that Maestlin's telescope was inadequate for observing the Medicean planets, and that he was unable to read Galileo's "lettere" on sun spots. (These references have been but partially examined.)—Favaro (1876), 8.—Freher, IV, 1489 (Maestlin's picture is opposite page 1486.)—Frisch, editor, I, 44-5, 188, 190, VII, 280 and passim.—Galileo, II, 527-8, Maestlin's observations of the nova of 1572.—Gesner (1583), 607.—Günther (Maestlin).—Hoefer, XXXII, 649-650.—Houzeau, 2746-2751.—Janssen, V, 345.—Jöcher, III, 579-580.—Kästner, 446-451.—Kepler (1596).—Montucla, I, part III, book IV, section VI.—Müller, A.—Peignot, II, 382.—Pingré, I, 79-104.—Poggendorff, II, 170.—Prandtl.—Riccioli, I, xli, II, 152-3, 13, 26, 28, 87, 134. Riccioli, II, 152-3, discussed Maestlin's observations for parallax in the nova of 1572; II, 87, his work on the comet of 1577.—Scheibel, 74, 105, 119; and I, 321.—Stimson, 36, 53-5.—Strauss, 328-355.—Thorndike, VI, 46-7, 76-83.—Vossius, 192 (chap. XXXVI, § 21), and 401 (chap. LXVIII, § 13-4).—Weidler, 394, 396.—Wohlwill, I, 18-9, 28-30, 190, II, 2, 3, 398. Maestlin's observations of moonlight are discussed in I, 255, 260, 261, 281, 297.—Wolf, I, 69, II, 37 ff., 42.—Wolf (1877), 179, 238, 249, 266, 282, 283-4, 286, 290, 291-2, 307-8, 332, 351, 408, 433, 582.—Zedler, XXI, 809-810.

For further information concerning Maestlin see the following references: *Geschichte der Astronomie*, v. 1, Chemnitz, 1792, p. 301 (author not ascertained).

Apelt, Ernst Friedrich: *Johann Keppler's astronomische Weltansicht.* Leipzig, 1849.

Barettus.

Berti, Domenico: *Copernico e le vicende del sistema copernicano in Italia nella seconda metà del secolo XVI e nella prima del XVII con documenti inediti intorno a Giordano Bruno e Galileo Galilei. Discorso letto nella R. Università di Roma in occasione della riconenza del IV Centenario di Niccolò Copernico.* Roma. Tip. G. B. Paravia e C. 1876.

Bök, August Friedrich: *Abhandlung von den Gelehrten Würtembergs, welche sich um die Mathematik verdient gemacht haben.* Tübingen, 1767.

Breitschwert, J.L.C.: *Johann Keppler's Leben und Wirken*... Stuttgart, Löflund, 1831.

Crusius: In *Annalibus Sueviae.*

His importance lies partly in his accurate observations of the nova of 1572 and the comets of 1577, 1580 and 1618, in his ephemerides and in his *Epitome Astronomiae*,[40] but more par-

Delambre (1821).

Der Mathematiker Michael Mästlin, Professor zu Tübingen, der Lehrer Keplers. (*Diözesan-Archiv von Schwaben*, v. 9: 26 (1892). This reference is from Schottenloher, II, 1.

Favaro, editor, II, III, X, XII, XVIII.

Frisch, editor. The entire work is an excellent source for material concerning Maestlin.

Frisch, Ch.: editor of *Maestlin's Briefwechsel mit Kepler.*

Gassendi.

Günther: *Beiträge zur Geschichte des neueren Mathematik.* Ansbach, 1881.

Hansch, Michael Gottlieb (editor): *Epistolae ad Joannem Kepplerem... scriptae; insertis ad easdem responsionibus Kepplerianis,... opus novum... nunc primum cum praefatione de meritis Germanorvm in Mathesin, introductione in Historiam Literariam saeculorum XVI. et XVII. et Jo. Keppleri Vita... ex manuscriptis editum.* [Lipsiae] cIↄ Iↄ cc XIIX.

Hoffmann, Joh. Jac.: *Lexicon universale...* Leyden, 1698.

Mädler, Johann Heinrich: *Untersuchungen über die Fixsternsysteme*, Part II, Mittau, 1848, 36.

Reitlinger (Edmund)=Neuman=Grunner : *Johannes Kepler...* Part I, Stuttgart, 1868, 89 ff.

According to Houzeau, 2751, and Bök, 91, Maestlin's correspondence and manuscripts are in the Imperial Library in Vienna. Houzeau said that Maestlin's works were on the *Index* of prohibited books.

Students of English astronomy may be inclined to rate Thomas Digges ahead of Maestlin, but Digges' observational astronomy was secondary to his translation of the *De revolutionibus* and his popularization of the Copernican doctrine.

40 The *Epitome Astronomiae* did not follow the Copernican doctrine, possibly because of Maestlin's university position. It was written in the usual manner of the fifteenth and sixteenth centuries. For example, on page 73 of both the editions which I have seen (Maestlin (1588) and Maestlin (1610)) one finds the "Argumentum" reading as follows: "Terra vndique à coelo aequaliter abest. Ergo est in medio mundi, & sic centrum Mundi." However, in the later editions of his work, Maestlin added a passage to the already existing appendix to the first book. This passage is in favor of the Copernican theories. Both the pagination and the running head over the pages make this appendix an integral part of the first book. The appendix can be found on pages 82 to 90 of the 1588 edition and on pages 82 to 95 of both the 1610 and 1624 editions. Information concerning the latter was kindly furnished by Dr. A. Pogo who used the H.C.L. copy. The 1610 and 1624 editions seem to be identical from pages 82 to 95. The appendix in all three editions has the title "Appendix Tertiae Partis Libri Primi. De dimensione globi

ticularly in his teaching, his defense of the Copernican ideas, and his great influence over Johann Kepler. This influence was particularly noticeable in the latter's *Mysterium Cosmographicum,* which Maestlin saw through the press and to which he added an edition of Rheticus' *Narratio Prima.*[41] This book appeared in 1596 and attracted the attention of Tycho. The long corespondence between Maestlin and Kepler and the latter's repeated references to the former, as in the *Mysterium Cosmographicum,* testify that Maestlin's influence over Kepler extended into the seventeenth century. In 1600 when Kepler was forced to leave the Austrian provinces because of his refusal to become a Roman Catholic, Maestlin was one of those to whom he applied for advice.[42] Maestlin had taught him the Copernican system, and at that time he was loath to ally himself with Tycho who did not believe in it.

Michael Maestlin was born in Göppingen on September 30, 1550 and died in Tübingen December 20, 1631.[43] As a young

Terreni." The appendix to the 1588 edition deals with terrestrial measurements. It can be found on pages 82 to 90 of all three editions, and seems to be the same in the 1588 and 1610 editions with but minor additions to the latter. The added passage might be called the "celestial" section. It is not present in the 1588 edition, but occupies pages 91 to 95 of the 1610 and 1624 editions. It begins as follows: "Lubet hic (exemplo aliorum quorundam) dimensiones altitudinum Sphaerarum Coelestium, iuxta coniecturam & calculum Aphragani, in gratiam Tyronum, subiungere." On pages 94 to 95 is the statement: "Inter caeteras rationes, quae Copernico de alijs hypothesibus, aliaqúe Sphaerarum Mundi dispositione, quae cum Ratione, cum Natura & Obseruationibus melius corresponderent, cogitandi occasionem praebuerunt, haec incomprehensibilis & incredibilis in celeritate rapiditas, haud dubiè non postrema, si modò non prima, fuit." Dreyer (1906), 349-350, referred to this passage.

41 Maestlin's name does not appear on the title-page of Kepler's book (Kepler (1596)), but does appear at the beginning of the preface to the reader at the beginning of the *Narratio Prima.* This preface was dated October 1, 1596. Maestlin was mentioned throughout Kepler's preface, pages 6 to 10. On pages 161 to the end, page 181, there is an appendix, by Maestlin, entitled "De Dimensionibvs Orbivm Et Sphaerarvm Coelestivm Ivxta Tabulas Prutenicas, ex sententia Nicolai Copernici."

42 Dreyer (1890), 297.

43 Favaro, editor, XX, 478. Frisch, editor, I, 188, said October 16, 1631 and Zedler gave the date as either 1631 or 1635.

man he went to Italy, where he is said to have been won over to the Copernican theory and to have influenced Galileo.[44]

44 Günther said that it was not certain that Maestlin ever made such a trip, and that if he did, it was between 1571 and 1576. It is interesting to note (Favaro, editor, XII, 64) that Maestlin was not proficient in the Italian language. Maestlin is even supposed to have made a public speech in Italy in favor of the Copernican system. Weidler said that Galileo was impressed by Maestlin's arguments and won over to Copernicanism by them, although previously he had been a follower of Aristotle and Ptolemy. Chalmers' dictionary, not a very reliable source, gave this same information, as did also Doppelmayr, who cited Vossius. Although Weidler as a rule seems to have been highly accurate, his opinion must be questioned in this instance. What seems to be the earliest statement to the effect that Maestlin won Galileo over to Copernicanism was made by Vossius. Wohlwill, too, referred to this statement by Vossius, but did not give it much weight. He discussed it from the point of view of Maestlin's development, doubting that at the time of his visit to Italy he could have been a sufficiently staunch supporter of Copernicus. Perhaps Vossius' statement was the source of the others, particularly since Weidler gave Vossius as a reference and since the statements of the two men are very similar. The following sentence is from Vossius, 192, chap. XXXVI, § 21: "Iunior in Italia egit; ubi cùm pro Copernicana sententiâ publicè in Lyceo orationem habuisset; Galilaeus Galilaeïus, perpensis ejus argumentis, etsi antea Aristoteli, & Ptolomaeo, penitùs addictus, postea pedibus, sive animo potiùs, in ejus ivit sententiam." The following sentence is from Weidler, 396: "Natus ille est Goeppingae, in ducata [sic] Wirtenbergico, iuuenis egit in italia, ubi cum *pro copernicana sententia* publice in lyceo orationem habuisset, Galileus perpensis eius argumentis etsi antea Aristoteli & Ptolemaeo penitus addictus, postea in eius iuit sententiam." One must conclude that the latter sentence was taken bodily from the former. The fact that neither the town in which the Lyceo was situated nor the year of the speech is given might be considered suspicious, and certainly makes checking impossible. Günther considered Vossius the originator of the tale. Perhaps a more thorough examination of Frisch's edition of Kepler's *Opera Omnia* or of Favaro's edition of Galileo's writings and Favaro's and others' works about Galileo would clear up the point, but probably not. The story of Maestlin's journey to Italy is obscure, but the question can well be left to some future treatise devoted to Maestlin. Such a piece of work would be worth while. Galileo, not born until 1564, must have been a mere child and not interested in scientific cosmogonies when Maestlin was in Italy. Günther, as well as Wolf, pointed to the extreme youth of Galileo at the time of Maestlin's supposed trip; and in criticising the statement by Vossius, Günther quoted Favaro as saying that Maestlin and Galileo do not seem to have been personally acquainted. Even during a large part of his lectureship at Padua, Galileo still taught the Ptolemaic cosmogony. It is possible that, when he grew up, he heard of a speech by Maestlin, and it is probable that he heard

However, Maestlin's observations of the nova of 1572 and the comet of 1577 really made him believe in the new system,[45] and furthermore, with him it was no longer a hypothesis. Maestlin studied mathematics and theology at Tübingen. In 1576 [46] he took the office of deacon at Backnang in Würtemberg. Wolf [47] said that Maestlin was a pupil of Philip Apian at Tübingen and later succeeded him.[48] In 1580 Maestlin became professor of mathematics at Heidelberg and in 1584 [49] at Tübingen. There Kepler was one of his pupils. Maestlin's influence through his teaching and his writings [50] must have been enor-

and read about Maestlin, as did all the astronomers of the time. Wolf, too, (Wolf, II, 37 ff., 42 and Wolf (1877), 249), doubted Vossius and made the point that there is little reason to suppose that Galileo would refer in his *Dialogue on the Two Principal Systems* to a comparatively unknown man, like Christian Wursteisen or Vurstisius, as the man who early lectured on Copernicanism in Italy if he himself had really been influenced by Michael Maestlin. For additional information concerning Wursteisen see Wolf (1852). This article gives further references for Wursteisen and discusses Galileo's conversion to Copernicanism by either Maestlin or Wursteisen. Stimson, 53-5, said that the dialogue on the two systems gives the only source of information concerning Galileo's conversion to Copernicanism. Brewster, 10-13, too, cited Vossius as the source of the story of Maestlin's speech converting Galileo, but he did not go so far as to say that the dialogue gives the true account of Galileo's conversion. Galileo's indebtedness to Wursteisen seems taken for granted, although information concerning it rests on a dialogue in which the speaker, Sagredo, does not necessarily speak for Galileo himself. It is often mentioned casually in general writings on Galileo, such as Müller, A., 9. It seems unlikely that Maestlin's journey to Italy can be credited with Galileo's conversion to Copernicanism. However, it is somewhat surprising to read Favaro's statement (Favaro (1876), 8) that Maestlin was but slightly acquainted with the Copernican theory.

45 This was stated by Wohlwill, I, 18, and also, which is more important, by Maestlin, himself, in his book on the comet of 1580, where this later phenomenon was on the list of those which showed Maestlin the untenability of the Peripatetic doctrine. See Thorndike, VI, 80-1.

46 Jöcher and Zedler said 1570.

47 Wolf (1877), 266.

48 Apian died in 1589.

49 Jöcher said 1583.

50 In addition to his book on the comet of 1577, Maestlin's works include the following:

Alterum examen novi Pontificialis Gregoriani Kalendarii, quo ex ipsis fontibus demonstratur, quod novum kalendarium omnibus suis partibus, quibus quam rectissime reformatum vel est, vel esse putatur, multis modis mendosum et in ipsis fundamentis vitiosum sit . . . Tubingae apud G. Gruppenbachium, 1586.

Aussführlicher Bericht von dem allgemeynen Kalender . . . *Sampt Erklärung der newlichen aussgegangenen Reformation, von Bapst Gregorio XIII. und was darvon zu halten sey, etc.* . . . in *Notwendige und gründtliche Bedennckhen, Von dem allgemeinen, uhralten* . . . *Römischen Kalender* . . . *Sampt Erklärung und Widerlegung dess ungegründten unnd unnohtwendigen neuwen Bäpstischen Kalenders* . . . *durch etliche hochverstendige Theologen unnd Mathematicos* . . . *beschrieben unnd erkläret.* Heydelberg, 1584. (This title may be merely a variation of the title directly below.)

Aussfůhrlicher vnd Gründtlicher Bericht Von der allgemainen / vnd nunmehr bey sechtzehen Hundert Jaren / von dem ersten Keyser Julio / biss auff jetzige vnsere Zeit / im gantzen H. Rŏmischen Reich gebrauchter Jarrechnung oder Kalender / In was Gestalt er anfänglich gwesst / vnd was durch långe der Zeit fůr Irthumb dareyn seyen eyngeschlichen. Item ob / vnd wie er widerumb ohn merckliche verwůrzung zu verbesseren were. Sambt erklårung der newen Reformation / welche jetziger Bapst zu Rom Gregorivs XIII. in demselben Kalender hat angestellet / vnd an vilen Orten eyngefůhret / Vnd was darvon zuhalten seye. Gestellt durch M. Michaelem Maestlinvm Goeppingensem, Matheseos Professorem zu Heydelberg. (at end: Getruckt in der Churfůrstlichen Statt Heydelberg / durch Jacob Můller / im Jar M. D. LXXXIII.) (This title is taken from a copy of the book in the N.Y.P.L.)

Beobachtungen des neuen Sterns in der Cassiopea, 1573.

Bericht von der Allgemeinen vnd nun mehr bey 1600. Jahren / von dem ersten Kayser Julio biss auff jetzige Zeit im gantzen H. Röm. Reich gebrauchter Jarsrechnung oder Calender / in was Gestalt er anfänglich gewest vnd was durch lenge der Zeit für Irrthumb darinn sind eingeschlichen / Item ob vnd wie er wiederumb ohn merckliche verwirrung zu verbessern were. Heydelberg bey Johan Spiess. 1583. (This item may be the same as the second item on this list.)

Chronologicae theses et tabulae breves contractaeque, ad investiganda tempora historiarum et epocharum potissimarum, praesertim sacrarum . . . *editae studio et cura Samuelis Hafenrefferi,* . . . *cum exegesi quaestionum chronologicarum* . . .Tubingae, typis P. Brunii, 1646. (First edition, Tübingen, 1641.)

Consideratio et observatio cometae aetherei astronomica, qui anno 1580, mensibus octobri, novembri et decembri, in alto aethere apparuit. Item, descriptio terribilium aliquot et portensorum chasmatum, quae his annis 1580 et 1581 conspecta sunt. . . . Heidelbergae, excudebat J. Mylius, 1581.

De astronomiae principalibus et primis fundamentis. Heidelbergae, 1582.

De cometa anni 1618. Tübingen, 1619.

De Dimensionibvs Orbivm Et Sphaerarvm Coelestivm Ivxta Tabulas Pruntenicas, ex sententia Nicolai Copernici, printed as an appendix, pp. 161-181, in Kepler's *Prodromus Dissertationvm Cosmographicarvm* (1596) (See below, Rheticus' *Narratio Prima.*)

Defensio alterius sui examinis, quo ex ipsis fundamentis demonstraverat, quod Gregorianum Novum Kalendarium... totum sit vitiosum, adversus cujusdam Antonii Possevini, jesuitae, ineptissimas elusiones, . . . Tubingae, apud G. Gruppenbachium, 1588.

Demonstratio astronomica Loci Stellae nouae, tum respectu Centri Mundi, tum respectu Signiferi et Aequinoctialis.

Dialexis Germanica, 1583.

Disputationes tres astronomicae et geographicae. Tübingen, 1592 (According to Houzeau, 2748, there was a German translation of this work printed in 1619 under the title "Problema astronomicum, die Situs der Sternen, Planetarum oder Cometarum zu observirn (par Begern).")

Divino Rectoris Astrorvm Favente Nvmine, De Astronomiae Principalibvs Et Primis Fvndamentis Dispvtatio ad discutiendum proposita, à M. Michaele Maestlino Goeppingensi, Matheseos in antiquissima & inclyta Academia Heidelbergensi professore, in Auditorio philosophico, ad diem 20. Ianuarij. Respondente Hieremia Iacobo Vlmensi. Heidelbergae Excudebat Iacobus Mylius. M D LXXXII. (This title-page was copied from the copy in the N.Y.P.L.)

Disputatio de eclipsibus Solis et Lunae. Tübingen, 1596.

Disputatio de multivariis motuum planetarum apparentibus irregularitatibus. Tübingen, 1606.

Ephemerides novae, ab anno salutiferae incarnationis 1577. ad annum 1590. Supputatae ex tabulis Prutenicis. Tübingen, 1580.

Epitome Astronomiae, Qva Brevi Explicatione Omnia, Tam Ad Sphaericam quàm Theoricam eius partem pertinentia, ex ipsius scientiae fontibus deducta, perspicuè per quaestiones traduntur Conscripta per M. Michaelem Maestlinvm Goeppingensem, Matheseos in Academia Tubingensi Professorem. Iam nunc ab ipso Autore diligenter recognita. Cum Priuilegio Caesareae Maiestatis. Tvbingae, Excudebat Georgius Gruppenbachius. Anno 1588. (Dreyer (1906), 350, gave the Tübingen 1588 edition as the first. Houzeau, 2747, listed octavo editions for Heidelberg 1582, Tübingen 1588 and 1593 and also later editions. Hoefer gave the dates for the Heidelberg editions as 1582 and 1588 and said that the Tübingen editions were 1593 and later. Delambre (1821), I, 312-3, spoke of the Tübingen 1588 edition. Cat. Belg., 389, listed a Heidelberg 1582 octavo edition. Gesner (1583), 607, listed the Heidelberg 1582 edition. The title for the work as given above is taken from the copy in the reserve room of the N.Y.P.L. The words "Iam nunc ab ipso Autore diligenter recognita" show that the 1588 edition was not the first.)

Horologiorum solarium sciatericorum in superficiebus planis descriptionis universalis informatio. (According to Doppelmayr, this is a manuscript found among those of Praetorius. However, Poggendorff listed it just as though it had been printed, although he gave no date of publication.)

Judicium M. Moestlini de opere astronomico D. Frischlini. (dated from Tübingen, January 18, 1586, but not printed).

Perpetuae dilucidationes Tabularum Prutenicarum coelestium motuum. Tübingen, 1652.

mous. He is said to have been the first to explain correctly the ashen color of moonlight after new moon,[51] and, indeed, made many observations of the moon.

Maestlin's observations of the nova of 1572, although made without elaborate instruments, are exceedingly accurate. He picked out four stars such that the nova was at the intersection of two lines, each drawn through two of the stars. He held a thread before his eyes so that it passed through the new star and two of the others and thus assured himself that the nova did not move relatively to the four stars during the daily revolution of the heavens. From this he concluded that it had no parallax and was among the fixed stars, which according to Copernicus were extremely distant.[52] Maestlin also observed the nova of 1604.

Regiomontanus' *Ephemeriden,* (a commentary). Tübingen, 1582. Second edition, 1610.

Rheticus' *Narratio Prima,* new edition in Kepler's *Prodromus Dissertationvm Cosmographicarvm, Continens Mysterivm Cosmographicvm, De Admirabili Proportione Orbivm Coelestivm, Deqve Cavsis coelorum numeri, magnitudinis, motuumque periodicorum genuinis & proprijs, Demonstratvm Per Qvinqve regularia corpora Geometrica, A. M. Ioanne Keplero, . . .* Tvbingae Excudebat Georgius Gruppenbachius, Anno M. D. XCVI. (This title was copied from the title-page of the copy in the reserve room of the N.Y.P.L. For the complete title see the bibliography of references for this dissertation.)

Synopsis chronologiae sacrae. Lunebourg, 1642.

Theses de Eclipsibus, 1606.

Tractatus brevis de dimensione Triangulorum rectilineorum & sphaericorum. (According to Doppelmayr, this is a manuscript found among those of Praetorius. However, Poggendorff listed it along with Maestlin's other works, but without a date of publication.)

51 According to Houzeau, 2749, and Wohlwill, I, 260, this was in Maestlin's *Disputatio de ecclipsibus Solis et Lunae* (Tübingen, 1596). Leonardo da Vinci's explanation of earthshine on the moon had lain buried in his notebooks.

52 See Brahe, III, 58-62, where Maestlin is quoted at length and a diagram is given. Maestlin also described this method of observation in his tract on the comet of 1577 (item 70 of appendix, 21-2). See the summary of chapter V of that work, given below, and Dreyer (1890), 59. Geoffrey of Meaux, as was said in chapter I, above, in observing the comet of 1337 used a method similar to Maestlin's. Geoffrey observed the fixed stars nearest the comet, drawing circles to them from the poles.

Maestlin's work on the calendar, a subject which has absorbed the interests of many able scientists, was considerable. He corresponded with Kepler on the subject.[53] Tycho's library contained at least one book by Maestlin on the calendar, and also Clavius' answer to it.[54]

Tycho spoke at length of Maestlin.[55] His book on the nova of 1572 speaks of Maestlin's observations, and the *De Mvndi Aetherei... Phaenomenis,* book II, chapter 10, discusses Maestlin with three other astronomers who had acknowledged that the comet of 1577 was beyond the lunar orbit, praises him in glowing terms, and deals at length with his observations of that comet. Tycho considered Maestlin's book chapter by chapter, and analysed his observations, comparing them with his own. Tycho particularly commented on Maestlin's description of the comet's tail and referred to the elder Gemma's translation of Apian on the subject of comets' tails. At the same time Tycho spoke of other observations by Maestlin, such as those of the comet of 1580. However, as Riccioli said and as we shall see below, Maestlin believed that some comets were above the moon, others below it.

Maestlin was in contact with other astronomers of his time, as is shown by his controversy with Hagecius,[56] in which Tycho sided with Maestlin. Controversies of this type clarified the opinions of the more able astronomers of the time and aided materially in advancing the knowledge of cometary theory. It must be remembered that advances in science are accepted gradually, not instantaneously.

The importance of Maestlin's treatise on the comet of 1577 was recognized by his contemporaries, especially by Tycho. It

53 See Janssen, V, 345, which cited Frisch, editor, IV, 6 ff.

54 Prandtl. A copy of Clavius' answer can be found in the N.Y.P.L. See the bibliography of references below.

55 See Delambre (1821), I, 225 ff. as well as Brahe, especially IV which contains Tycho's *De Mvndi Aetherei... Phaenomenis.*

56 See item 49 of appendix and the summary of that item in chapter IV, below. Dreyer, editor, gives information concerning Maestlin's correspondence.

was frequently cited by later writers on comets.[57] The work,[58] published the year after the comet's appearance, when Maestlin was in Backnang, was a scholarly treatise, intended for scholars, and written in the Latin language. It emphasized the position of the comet "in the sphere of Venus," that is, further from the earth than the moon.

In the dedication of his book, Maestlin talked of immortality achieved through deeds. Those, he said, who helped humanity, such as Alphonso of Spain, Alexander the Great, and Aristotle and those who brought glory to the Church have kept their names alive. Maestlin dedicated his book to Duke Louis in recognition of the liberal attitude of that ruler and in gratitude for the assistance he himself had received in pursuing his mathematical and astronomical studies. He listed, with respect, the great astronomers, Hipparchus, Ptolemy, Albategni, Peurbach, Regiomontanus, and Copernicus, whose observations, he said, can be calculated and reduced. Furthermore, he was not forgetful of his religious affiliations,[59] despite his scientific interests.

Maestlin spoke of the comet as furnishing an opportunity to show the glory of God rather than as an evil omen. In the first chapter of his treatise on the comet, not relying on past observations of such bodies, he announced that the comet would be shown to be, not in the elementary region of the world, but in the sky. Although he knew of past accounts of comets with irregular motions, he found that this one had a definite motion. He believed that the comet and its streamers were the largest ever heard of, since they extended 30° and the comet of 1531 covered only 20°. He described the comet of 1577 as a terrible sight which caused men to stand in rapt admiration and to turn to prayer. He said that it was whitish in color and later turned to a leaden hue and that the streamers or tail were turned away from the sun, and were merely sunlight broken up by the body

57 For example, Lubienski, II, 374-5, and Riccioli, II, 87.

58 Item 70 of appendix.

59 He was a Swabian Protestant by birth. See White, 184.

of the comet. Maestlin observed the comet on November 12th, when it was not far from the bow of Sagittarius in the direction of the ecliptic in the first degree of Capricorn, and he told the positions of the planets at that time. The last day on which Maestlin saw the comet, it was near two little stars in the chest of Pegasus. When he first saw the coma,[60] it extended from the first degree of Capricorn past the stars in the clothing of Sagittarius toward the horns of Capricorn. He said that it was narrow at the beginning and wider at the end; that at its origin it was not as large as the comet's head; and that on November 12th its beams extended to the head of Sagittarius. He stated the position of the tail on November 17th, when the comet's head was near the star in the knees of Antinous, on December 2nd, on December 6th and on December 31st.

The second chapter was given over to a discussion of the time of the comet's first appearance and its duration, both of which were uncertain. The uncertainty was partly due to inclement weather. Maestlin observed the comet for the first time on November 12th and for the last time on January 8th, although he looked for it on the 14th. He thought that it had probably lasted from November 5th to January 10th, and that Thurneysser could not have been correct in saying that he saw it on October 19th.

Maestlin devoted the third chapter to a proof that the comet was not sublunar. He said that although the motions of heavenly bodies had long been observed, nobody had thought that comets followed the same natural laws. Yet it seemed unlikely that God would have hidden this knowledge, since it is by His Divine will that man is able to determine the future positions of stars, just as he does when foretelling eclipses. Maestlin stated that in the previous century, Regiomontanus, whom he called "another Ptolemy," had taught the method of parallaxes for determining the distance of comets. Thus, said Maestlin, we learn whether or not comets transcend the elementary world. He wanted to measure the comet's distance but thought

60 Maestlin used the words "coma" and "cauda" interchangeably.

that the calculations of Regiomontanus could not be applied to the comet of 1577 because he himself had found no parallax in any part of the comet's daily revolution. Then Maestlin gave observations to show whether the comet was in the superior or inferior world. For, since no difference of parallax had been found, Maestlin concluded that the comet must be far above the moon.[61] He took into consideration only one observation, that of December 2nd at 6 P. M., when the comet was near the little stars in the nose of Equiculus,[62] and gave the exact position of the comet with reference to the fixed stars at 6 and at 9 P. M., and an entire day later. The comet's own motion was such that on December 2nd the comet was north of the stars, and on the 3rd, south of them. Considering this motion with regard to the observed positions of the comet at 6 and 9 P. M., he concluded that the comet by no means turned in the sublunary orb. He made a diagram with the earth at the center and two quadrants representing the sphere of the stars and a sphere at the supposed distance of the comet. He drew lines of sight from the center of the earth and from a point on its surface through points on the inner sphere and extended them to the sphere of the stars. He pointed out that the lines of sight from the two points of origin were identical in the case of a comet in the zenith, but that they diverged more and more the closer the observed body was to the horizon. He called the point where the line from the earth's center cut the sphere of the stars the " true " position of the body; the other point he called the " apparent " position, which, he said, is always nearer the horizon. Citing Ptolemy as his authority, Maestlin said that if the true place of the star is known and the apparent place is noted from

61 " Interim verò ex illis obseruationibus colligi dabatur, vtrum in inferiori vel superiori mundo Cometa hic versatus sit. Etenim cum omnem parallaxeos differentiam excludat, omnino necesse est, eius distantiam à terra tantam fuisse, vt terrae crassicies ad eam vix comparabilis sit, sed supra Lunam multis partibus exaltatus fuerit."

62 " Equiculus " is a constellation just east of the Dolphin, near Pegasus. It is also called Equuleus, or Equulus, Equus minor, Equus prior, Sectio Equi, Hinnulus, and the little horse. (Zedler, VIII, 1458).

the observation, the distance of the observed body from the earth's center can be found, and that, even if the true place is not known, that distance can be computed by using the difference of parallax or the difference between the angular separation of the two lines of sight for two observations at different heights above the horizon, just as was done by Regiomontanus. Maestlin thought the altitude of the star great enough to make the size of the semidiameter of the earth inconsequential. In order to determine the difference in parallax, Maestlin used the observed distances between the comet and a fixed star at 6 and at 9 P. M. He found no sensible difference and concluded that the comet was beyond the lunar orbit.

In Maestlin's fourth chapter the opinions of some who considered the comet elementary were reviewed and refuted. Maestlin did not believe that these men had made the necessary observations or that they could have made the necessary computations in the short time which elapsed between their observations of the comet and the publication of their works. He said that many of them wanted the comet to be elementary and made every effort to follow Aristotle. Maestlin thought those men worthy of indulgence who did not fight with reasons but followed this ancient authority, but he censured those who supported their arguments by geometrical demonstrations. He, however, admitted that he himself, with all his careful work, had not easily been persuaded that the comet was aethereal.

Maestlin mentioned a certain man who boasted of having been Reinhold's disciple and who contended that the nature of this comet in no way disagreed with Aristotle's ideas. This man, according to Maestlin, tried to show that the comet was sublunar, using Aristotle's reasoning that all phenomena which, like the comet in question, cannot be moved so perfectly and rapidly by diurnal motion as the moon, are below the sphere of the moon. Maestlin reasoned from the motion itself and the parallax that the reverse was true, and cited Aristotle's *De caelo et mundo,* book II, chapter X, as bolstering his argument. Maestlin showed that if the reasoning given above to prove the comet sublunar were followed, Saturn would be shown to be below the moon. He turned the argument to show that the comet was above the moon.

Maestlin said that others believed they could observe the parallax of the comet from its meridian altitude. He told that an observer [63] found that the altitude of the comet was 46° and that of Aquila 36° at 5 on the afternoon of December 6th. Maestlin added that this observer found from the Prutenic tables that the latitude of Aquila was 29° 10′,[64] which he added to the difference in altitude, obtaining the value 39° 10′, and that he called those degrees rejected from the observed altitude the inclination of the equator or the parallax, namely 6° 50′.[65] Thus this observer was cited as claiming the comet to be 8 semidiameters of the earth from the earth, or 6872 miles and some paces from the center of the earth.

Maestlin wished to mention a few points before leaving the above demonstration for the reader to examine. He did not think that the observations used were accurate, and showed why Aquila could not have been on the meridian at the prescribed hour. He called the subtraction and the derived parallax incongruous, the one latitude being measured from the ecliptic, the other from the equator; and he disagreed with other meridian observations of the comet recorded by the same observer. He even suggested that the demonstration was imitated from the ninth chapter of Hagecius' book on the new star of 1572.

Next Maestlin discussed Nolthius' two observations of the comet on December 7th, one hour apart, the first giving the

63 According to Brahe, IV, 213, this was Winckler.

64 " Sc " and " Scr ", abbreviations for " scrupuli ", have been translated as " minutes " throughout the summary of Maestlin's book, on page 14 of which was written " quadrante gradus, vel. 15. scr." Tycho used this meaning when discussing his own data (i. e. IV, 90) and when discussing the data of others (i. e. IV, 254, 283, 338). See chapter IV, note 39, below, for a discussion of the use of "Aquila " as a point of reference.

65 Maestlin's sentence (p. 13) is: " Hos gradus ex altitudine obseruata reiectos, appellat inclinationem aequatoris, quae sit ipsa parallaxis, scil. 6, gr. 50. scr." Probably Maestlin was purposely quoting from one of the poorer tracts in order to prove his point concerning their poor quality.

altitude of the comet as 41° 8′ and the azimuth as 44° 25′; the second giving the altitude as 33° 15′ and the azimuth as 27° 30′. Maestlin said that from these values, by the doctrine of triangles, following Regiomontanus, Nolthius computed that the parallax in the first observation was 4° 59′ and in the second 5° 32′ and thence concluded that the comet was eight and two thirds semidiameters of the earth from the earth. Maestlin regretted these observations by an otherwise learned man, and set forth his own observations for the same day. At 6 P. M., on December 7th Maestlin had found the comet very near the straight line between the beak of Cygnus and the jaws of Pegasus except that the line from the beak of Cygnus through the comet passed a semidiameter of the moon or 15′ east of the jaws of Pegasus and that therefore the comet was west of the straight line through the stars. Its apparent place was 22° 53′ in Aquarius with a north latitude of 26° 2′. The same night at 9:15 Maestlin saw that the comet approached the line through the stars and was one third of the diameter of the moon east of its first place, the line from the beak of Cygnus through the comet missing the jaws of Pegasus at the east by scarcely one twelfth of a degree. Hence its position was 23° 2′ in Aquarius, with a latitude of 26° 4′. Maestlin thought it evident from a careful consideration of these two observations that the comet had no parallax, as had been shown from the observation of December 2nd. He said that the comet was further east, not further west, by the amount of its daily motion, and that therefore its altitude was greater than eight or nine semidiameters of the earth. Maestlin repeated the diagram of the previous chapter and applied it to the particular case of the comet of 1577 to refute Nolthius' observations. Maestlin pointed out that at the time when Nolthius made his observations the comet was at such an altitude that a slight error in observation would be many times multiplied in computation. Maestlin believed that the fault of Nolthius' work on the star of 1572 arose from the same cause. Of this, said Maestlin, Regiomontanus had warned when he observed the comet of 1475 (sic),[66] relating it to the position of Spica.

Maestlin realized that he was breaking with the Aristotelian tradition which placed comets in the air, so he said that Aristotle wrote about comets in his own time and that then parallax was unknown.[67] Maestlin repeated the Aristotelian theory of comet generation, which he said could be applied only to comets in the elementary not in the aethereal regions. He thought that Aristotle would have changed it had he known of heavenly bodies found by parallaxes.[68] Maestlin seemed to think that some comets were elementary, others aethereal. He rejected Aristotle's theory and concluded that the generation of comets was a mystery,[69] the key to which was held by God.

Chapter V concerns the method of observing the true places of stars, without instruments, by arithmetical calculation. Maestlin enumerated the instruments invented for observing the motion of the stars: the plane astrolabe, the astrolabe or armillary sphere, the quadrant, and so forth, but said that he used none of them except a great quadrant, because he did not trust them. He said that his custom was to observe the body, whose position he sought, with regard to four fixed stars, the longitudes and latitudes of which were known. He made the condition that any two of the known stars be in the same great circle with the observed body and the other two be in the same circle with it, and moreover, with care that the second circle be as much higher than the other as possible.[70] Since one should

66 This seems to have been Jacob Ziegler's error. See chapter II, above.

67 "... quanquam etiam tum parallaxeos indagandae ratio ignota fuit, ..."

68 "Dubium non est, si Aristoteles de aethereis per parallaxes inuentis certus fuisset ... profectò sententiam conceptam mutasset."

69 Tycho differed from Maestlin in regard to the generation of comets; and, discussing the latter's chapter 4, said (Brahe, IV, 213), "Ego materiam omnium Cometarum prorsus Coelestem esse iudico, siquidem etiam omnes in ipso Coelo generantur."

70 The body under observation would be at the intersection of straight lines between two pairs of stars.

not have faith in naked eye examination alone, said Maestlin, this affair can be examined by means of a straight rule as Gemma Frisius advised in chapter twenty-one of his book on the astronomical radius, because in this kind of observation it is useful to stretch a thread in the line. This enables observation of the motion of the body in question relative to the known stars. From the longitude and latitude of those four stars, said Maestlin, the doctrine of triangles discloses the mutual section of their circles which is the true place of the celestial body. The remainder of the chapter is given over to two problems to illustrate the computations necessary for the above described method. In the first problem there are given two right spherical triangles, with one angle of one, in addition to the right angle, equal to one angle of the other. The sides opposite the equal angles also are given, as well as either the difference or the sum of the sides beneath those right and equal angles. It is required to calculate both the sides and the equal angles. In the second problem a spherical triangle, not a right triangle, is given with two angles and the side between them known. From the third angle a perpendicular arc is drawn to the given side. It is required to determine the size of that perpendicular and the two sections it cuts on the given side [or that side extended].

In the sixth chapter Maestlin recorded his observations of the comet's motion and showed from them under what circle in the firmament the comet proceeded throughout its appearance. He said that he did not care to write about the comet's meaning, concerning which much had been written, but he was interested in its position and motion. He felt that he could not rely on predictions made by men who erred in the comet's position and motion. In this connection he cited chapter II of the tract by Dasypodius.[71] According to Maestlin, some astronomers said that the comet was in the sixth degree of Capricorn on November 12th, others that it was in the middle of that sign, and some that it was not far from the beginning of Aquarius. He added that a certain person claimed that on November 23rd

71 See, below, the summary of Dasypodius' tract.

the comet was as much as 2° from the bright star in Aquila, a distance, which, according to Maestlin, was always more than 10°. Maestlin noted differences in the recorded width of the streamers. He then gave his observations of the comet's position at six in the evening of November 12th with reference to certain stars for which he gave the longitude from the first star in Aries and the northern latitude. From these values, allowing for equinoctial precession, he found the comet's place, 3° 43′ in Capricorn, with a latitude of 7° 5′. He did likewise for his observations of November 17th, placing the comet 20° 50′ in Capricorn, with a latitude of 15° 26′. From similiar information and by similar calculation he found the comet on December 2nd at 6 o'clock to be 17° 17′ in Aquarius with a latitude of 24° 46′. The same evening at 9 o'clock he noticed that it changed its position, following the signs; for it receded from the above mentioned line toward the east by almost eight parts of a degree. From this he found it clearly shown that in both the observations those two stars formed a nearly right triangle with the comet. At 6 o'clock the right angle stood by the third star of Equiculus but at 9 it was carried down to the comet. He found the comet to have a longitude of 289° 33′, or 17° 25′ in Aquarius with a northern latitude of 24° 47′. On December 15th at 6 P. M. he discovered the comet in the same circle as the second star of Antinous and the eleventh of Pegasus, likewise in the same great circle as the second of Pegasus and the eleventh of Cygnus. He added that he numbered the stars of Pegasus in the manner of Copernicus and Reinhold and he gave the longitudes and latitudes of the stars he mentioned. Maestlin said that up to that date he was uncertain whether or not the comet's motion was regular. Therefore he began more thorough observations. He found that on December 15th the comet's longitude was 301° 48′, its place 29° 40′ in Aquarius, and its northern latitude 27° 20′. Moreover, on that day it was in the circle which led through the second star of Antinous and the eleventh of Pegasus. On November 24th at 6 P. M. Maestlin found it in the circle which passed through the second of Antinous and the twelfth of Pegasus; likewise the circle from the fourth of Aquarius through the comet divided the space between the bright star and the fifth of Aquila nearly in half; whence the position was shown to be 5° 47′ in Aquarius with a latitude of 21° 18′. Hence, as in the first problem above, it was concluded that the intersection of the circle of the comet was 44° 48′ distant from the first observation, and 68° 41′ from the last, which fell nearly on 21° 0′ of Sagittarius. The angle of obliquity in the same place was shown to be 28° 58′. He used the same calculation for the other observations, none of which deviated from the path of the circle.

On December 31st at 6 P. M., at which time the comet had not yet receded from this circle, the line from the ninth star of Pegasus through the comet divided the space between the sixth and seventh stars of Pegasus nearly in half, being a little nearer the sixth than the seventh. He gave the longitude and latitude for the three stars of Pegasus and for the place between the stars in which was the circle of the comet. This gave the longitude of the comet as 311° 38′, and the position as 9° 30′ in Pisces with a northern latitude of 28° 32′. On January 8, 1578 at 6 o'clock, when Maestlin last saw the comet, it was in that circle which from the sixth star of Pegasus turns away for a bit from the right shoulder of Cepheus into the south. For this star he gave a longitude of 340° 0′, and a latitude of 69° 0′. He found the longitude of the comet to be 314° 40′, its place 12° 32′ in Pisces and its latitude 28° 40′. On December 7th at 6 o'clock the straight line from the first star of Cygnus through the comet was a semidiameter of the moon west of the first of Pegasus. But at 9:15 those stars were almost in a straight line with the comet, the comet being slightly west, perhaps not beyond a twelfth of a degree. He gave the longitude and latitude of the stars mentioned and the position of the straight line on which the comet was at 6 and 9:15 P. M. Combining these with the circle of the comet he found the comet's longitude at 6 P. M. to be 295° 1′, its place 22° 25′ in Aquarius and its latitude 26° 2′; and at 9:15 he found that the longitude was 295° 10′, the place 53° 2′ in Aquarius and the latitude 26° 4′. From these observations the positions of the comet were found for the prescribed days. From them it was shown that the comet had followed one circle of the first and highest heaven, the obliquity of which was 28° 58′, and which cut Sagittarius in the 21st degree. This circle he chose to call the circle of the comet, which advanced from November 12th to January 8th from the 3rd or 4th degree of Capricorn to nearly the 13th degree of Pisces, increasing the northern latitude from 7° to nearly 29°, with a motion, apparently irregular, although in itself regular.[72]

Next,[73] Maestlin explained how the sphere and circle of the heavens or of the upper world in which the comet moves is found, and what it is. He said that he had already shown, by the doctrine of parallaxes, that this comet by no means took its place in the elementary region of the world. The circle in which all the observations fit emphasizes this. This chapter and the next, he added, will prove it by a third way, and will show the comet's position and size. He gave a table showing, in the first column, ninety divisions of the above defined circle of the comet. These divisions may be considered positions of the comet in its circle. The second column of the table gives the comet's longitude along the ecliptic; and the third, its latitude from the ecliptic. The table is preceded by a diagram showing an arc of the circle of the comet and its intersection with an arc of the ecliptic. This intersection takes place at 21° 0′ of Sagittarius. In the diagram, a perpendicular dropped from the circle of the comet onto the ecliptic represents the comet's latitude measured from the ecliptic. In the right triangle thus formed, one side, which represents the comet's distance from the ecliptic along the comet's circle, is known; and one angle is known, namely the obliquity of the circle of the comet to the ecliptic, or 28° 58′. Maestlin also gave a table of the motion of the

72 "... motu tamen (quoad apparentiam) non vniformi, sed admodum inordinato : quanquam is in seipso regularissimus fuerit, sicut sequitur." The path was even but the rate was not uniform.

73 Chapter VII.

comet in its circle, thus showing the unequal rate of the motion. Using book VI, chapter II, of Copernicus' *De revolutionibus,* where the boundaries of Venus are set forth, he came to the conclusion that the comet was in the sphere of that planet.

Unwillingly, indeed, as he put it, did he depart from the general opinion concerning the distribution of the spheres of the world. He stated the fundamentals of Copernicus' hypothesis. With a diagram he showed the sphere of Venus between the earth and the sun and said that the comet was in that sphere. He made allowance for libration. In his diagram the earth had a circular orbit about the sun.

The eighth chapter describes the motion of the comet in its circle, and the apparent divergence from the equal and regular motion of a circle, and its distance from the earth. First Maestlin called attention to the small velocity of the comet in November. He said that the comet must be in the sphere of Venus, and by diagram and table showed its distance from the sun on different days. He also gave a table of the inequality of the comet's motion. He had several diagrams for different dates, with the sun at the center, showing the comet's motion. From Copernicus he borrowed the use of a center for the comet's circle other than the center of the earth's. He showed the points of the comet's "apogee" and "perigee" [74] and the apparent positions of the sun. Furthermore, the irregularity of the comet's motion was accounted for by having the comet move on a small epicycle within the sphere of Venus.[75] For his diagrams he supplied the relative values of the distances, and finally added that one semidiameter of the earth contained 860 German miles.

Chapter IX sums up, and supposedly clears up whatever was left unexplained in chapters II and VI. Desiring to see what effect turbid air had on his observations. Maestlin re-examined

74 The terms apogee and perigee are now exclusively applied to the orbit of the moon, and the more general terms, aphelion and perihelion, which can apply to any planet, would be used to describe the points on a comet's path which lie farthest from and nearest to the sun.

75 See note 10 in this chapter.

the figure for November 24th in the preceding chapter, in the light of observations on the 12th of November. Maestlin guessed that the comet was kindled at about 4 A. M. on November 5th. He deferred to Johannes Praetorius in regard to the ultimate consumption of the comet because Praetorius had been favored by better weather following the appearance of the comet. The latter said that after the 10th of January 1578 no shadow of the comet could be seen. Maestlin came to the conclusion that the comet shone for 66 days and 12 hours; that it covered 118° in its circle, but in the zodiac, three twelfths of the great circle and 23°, from 20° in Scorpio to 13° in Pisces. He mentioned the constellations through which it passed. A table shows the mean motion of the sun, the distance of the comet from the mean sun, the motion of the comet in its circle, the longitude of the comet in the ecliptic, the latitude of the comet from the ecliptic and the distance of the comet from the center of the earth, for the morning and evening of November 5th and for the evening of each day thereafter through January 10th. At the close of the chapter Maestlin acknowledged his indebtedness to Corpernicus' way of thinking and also expressed his admiration for the immense power and wisdom of the Lord.

The last chapter has no direct connection with the preceding nine. It deals with conjectures concerning the significance of this comet,[76] is not important, and was barely mentioned by Tycho. Maestlin wished to steer a middle course between the types of astrologers, following those who have learned from trials over many centuries that eclipses of luminaries and great conjunctions of the superior planets are unlucky and that general changes have followed. These men, he said, discovered from history what dire events have followed the appearance of other comets. He repeated that comets were created by God, and he enumerated comets of the past and told what events they

76 Thorndike, VI, 77, suggested that "Prediction of the future from the stars is slighted, not for the negative reason that Maestlin considers it superstitious... but because astronomical observation and measurement make a greater positive appeal to him." See also Thorndike, VI, 78.

were supposed to have signified, whence he drew very general conclusions for the comet of 1577.

A third member of the group who considered the comet of 1577 further from the earth than the moon was Helisaeus Roeslin,[77] but unlike the other pathfinders who likewise published books to uphold that contention, his main occupation was not astronomy. Like the illustrious Roeslin of the early sixteenth century, with whom he must not be confused,[78] Helisaeus Roeslin was a doctor by profession.

In estimating what Roeslin added to the development of cometary theory by his book on the comet of 1577, it is necessary to have some idea of his standing and influence in his community. There has been surprisingly little written about him.[79] Born in Pleiningen (or Plieningen) in Würtemberg in

77 Bassaeus, I, 476.—Brahe, VIII, 206.—Dreyer, editor, I, xlii, IV, 251-8, V, 115, 323, VIII, 462.—Dreyer (1890), 171, 274.—Frisch, editor, I, 224, 228-9, 497-9, II, 399, 809, IV, 5, 169, 170.—Janssen, VI, 439-440.—Jöcher, III, 2175.—Kepler, I, 64, 215-287, 501-542, II, 15, 740, IV, 201-269, VIII, 316.—Kestner, 716-7.—Le Long (1719), 800.—Pingré, I, 85-6.—Riccioli, II, 13, 28, 87.—Scheibel, 106-7.—Schenck, 216-7.—Scheuchzer, 53.—Thorndike, VI, 74-6, 79-80.—Zedler, XXXII, 465.

78 Hellmann (1883), 620, confused the two men.

79 The only recital of the details of Roeslin's life which is, at this writing, available was quoted by Frisch, editor, I, 497-8, from Roeslin's own *Historischer, Politischer und Astronomischer naturlicher Discurs von heutiger Zeit Beschaffenheit, Wesen und Standt der Christenheit...*, which was published in Strasburg in 1609. Janssen dealt only with the astrological aspects of Roeslin's work, that is with the predictions from comets, which Roeslin considered signs placed in the heavens by God. He quoted at length from the *Discurs*, but from that work alone, and he drew his conclusions from it. He pointed out that Roeslin believed that both good and evil could be foretold from comets, and that he distinguished between comets and stars, because the latter were concerned only with generalities whereas the former dealt with particular events and did not always have immediate effect but sometimes not until the seventh year. Pingré was interested only in Roeslin's sphere of meteors; Dreyer, in his *Tycho Brahe*, only in Roeslin's observations of the comet of 1577 and the system of the world, and in his edition of Tycho's *Opera Omnia* only in the points of contact between Tycho and Roeslin. Frisch similarly emphasized the controversy between Kepler and Roeslin, the discussions of the latter being, however, but incidental to the comprehension of Kepler's works. Riccioli was interested in the comet observations.

the Filder region of Stuttgart in 1544, Roeslin began his studies and teaching in that capital (with ducal subsidy) and continued them at the university of Tübingen where he received his doctorate. In 1569, before entering into practice, he acted as a medical assistant at Durlach and Carlsbad, where Samuel Eisemenger [80] was his teacher in astronomy and where he also learned some alchemy. He considered those two studies very useful in the practice of medicine. He began his own practice in Pforzheim. There the pharmacist Gröninger gave him good practical guidance in exchange for theoretical instruction in chemistry, and they worked together and distilled medicaments which became widely known throughout Germany.

By that time Roeslin was already doubting Aristotelian philosophy and Galenic medicine. He was summoned by Prince G. Johann, Count Palatine of the Rhine,[81] and became his physician in ordinary. Roeslin gave up his studies to devote himself to his medical practice. However, the star of 1572 awoke in him the desire to find out its meaning. Therefore, he studied history and chronology. He mentioned having written a book, *Speculum Mundi* or *Weltspiegel,* in 1579, which was printed in 1605. Because of uncertainty due to war, Roeslin left the mountains and went into the country, first to Zabern in Alsace and then Hagenau, but he remained the physician to Count Johann until the latter's death in 1592. In the meanwhile, in 1584, he went into the service of the Count of Hanau.[82]

He said that Roeslin used the rude observations of Gemma and that his conclusions, like Gemma's, did not follow from the demonstration. Riccioli cited Tycho's discussion and added nothing to it.

80 Frisch, editor, I, 497, citing Roeslin's *Discurs.* In spite of the difference in spelling the name, this may be the physician and mathematician, Samuel Eisenmenger, known also as Siderocrates. See Adam (1705), 114b-115a; Allen, 253; Hellmann (1883), 593; Hellmann (1924), 29; Jöcher, II, 301-2; Smith (1917), 135; Thorndike, VI, 123; Zedler, VIII, 635.

81 This count was among the powerful princes of Germany and had the electoral vote. Therefore, he could rightly be given the title "Elector", although Roeslin called him "Fürst".

82 Philip IV, who died in 1590, was count of Hanau in 1584.

Roeslin also wrote, as he said, several theses concerning physical things, *De Opere Dei Creationis,* which appeared in 1597. These were opposed by Ursus (or Reymers) and were again upheld by Roeslin in his *Discurs.*

Prince Johann Augustus, Count Palatine in the Rhenish province, and Johann Reinhard, who became Count in Hanau in 1599, also took Roeslin into their service and admonished him to pursue his studies and publish what he had so far discovered. At the age of fifty-eight he began the study of Hebrew and undertook to read the Cabala. To pursue those studies and better to serve his patrons, Roeslin left the city of Hagenau, where he had worked for twenty-six years, and went to Buchsweiler,[83] with the hope that his princely protectors would aid him.

This much of his story Roeslin himself told in the introduction to his *Discurs.* He died in 1616, probably early in September, for on the twenty-first of that month Maestlin, in a letter to Kepler,[84] informed the latter that Roeslin had recently died and commended his soul to Christ. Maestlin had previously said of Roeslin, in a letter dated July 20, 1613,[85] that the man's intentions were of the best, but he doubted if they were always sufficiently prudent. He considered the man wise but too obstinate in his first conceived opinions so that instead of comparing reasons he merely collected them from all sides.

Roeslin's activity as an astronomer began after the appearance of the comet of 1577. His treatise on that comet,[86] printed in 1578, was his first astronomical work.[87] The observations

83 In 1480 Buchsweiler, or Bouxviller, became subject to the Count of Hanau-Lichtenberg. See Brockhaus, III, 467.

84 Frisch, editor, I, 498.

85 *Idem.*

86 Item 93 of appendix.

87 Roeslin's works, other than the tract on the comet of 1577, are:
Disputatio de his, quae pertinent ad definitionem medicinae propositam a Galeno, etc. (Praes. I. Schegkio.) Túbingen, 1569. Quarto. (B. M. catalogue, where the letters "Resp." stand before the title and might mean "respondit". The work was printed the year Roeslin left Tübingen, where he received his

doctorate. The item is probably his doctoral dissertation, which he presented to Schegk, his examiner. See A.D.B. V, 21, article "Jacob Degen" by A. Richter; Jöcher, IV, 235-6, article "Schegk (Jacob)"; Hellmann (1883), 424. The work was reprinted in 1611 with a work by J. R. Camerarius.)

Kurtz Bedencken von der Emendation dess Jars, durch Babst Gregorium den XIII. fürgenomen, vñ von seinem Kalender, nach ihm Kalendarium Gregorianum perpetuum intituliert, ob solcher den Protestierenden Ständen anzůnemen seie oder nicht. Mit angehencktem Prognostico inn was zeiten wir seien ... und was wir zůgewarten haben. Strasburg, J. Rihel, [1583?]. Quarto. (Frisch, editor, IV, 5, and the B. M. catalogue, which also gives an edition of 1584, with a slightly different title, and lists them under the name of Lambertus Floridus Plieninger [see text below].)

Des Elsäss und gegen Lotringen grentzenden wassgawischen gebirgs gelegenheit und comoditeten inn victualien und mineralien, etc. Strasburg, B. Jobin, 1593. Octavo. (H.C.L.; LeLong (1719), 800; Jöcher; Scheuchzer, 53; Zedler.)

Von dem warmen Bade zu Niederbrun in der Grafschaft Hanau. Strasburg, Jobin, 1595. (Jöcher, Kestner; Schenck.)

Tractatus meteorastrologiphysicus, das ist auszrichtigem Lauff des Cometes. Strasburg, 1597. Quarto. (Cat. Belg., 2501, describing a copy in the Catholic University of Louvain; Houzeau, 2852; Hellmann (1883), 620; Frisch, editor, I, 499, II, 809.)

De Opere Dei Creationis Sev De Mvndo Hypotheses ... Frankfort, "Apud haeredes Andreae Wecheli, Claudium Marnium, & Joannem Aubrium", 1597. (N.Y.P.L. Reserve. This is a quarto with signatures A_1—G_4. The pages are numbered 3 to 55 from A_2 r to G_4 r inclusive. The preface is dated from Hagenau, August 24, 1595. On pages 53 (marked 51) to 55, there are five diagrams of the systems of the world by the following five men: Ptolemy, Copernicus, Ursus, Roeslin and Tycho. The work seems to have been reprinted in 1619, three years after Roeslin's death. In that year there appeared in Geneva an edition of Nicolaus Hill's *Philosophia Epicvrea, Democritiana, Theophrastica proposita simpliciter, non edocta* which includes a work entitled *Panepistemon* by Angelo Politian and a work called *Conclusiones* by Pico della Mirandola. These are all paged continuously and the signatures are consecutive. The volume is a duodecimo. Bound with it, in the C.U.L., is a copy of the *De Opere Dei Creationis* with new signatures and new pagination, but the type of the two sets of signatures is similar, and, although there is a bastard title-page, no date nor place of publication for Roeslin's work is given. It is also a duodecimo. The preface is not dated and there are no diagrams of the different systems.)

Vermuthungen von Veränderung des Regiments bis 1604. Strasburg, 1597. Quarto. (Jöcher, Zedler.)

Judicium oder Bedencken vom Newen Stern, welcher den 2. Oct. erschienen vnd zum erstenmal gesehen worden. Strasburg, 1605. (Frisch, editor, I, 497.)

Historischer, Politischer und Astronomischer naturlicher Discurs von heutiger Zeit Beschaffenheit, Wesen und Standt der Christenheit und wie

recorded in it attracted the attention of Tycho, who, in 1588, discussed Roeslin fourth on his list of observers of that comet.[88] Tycho considered Roeslin's work intelligent, but was not particularly complimentary in his analysis of it. He did not consider possible the sphere of celestial meteors which Roeslin had invented as a compromise between ancient and modern opinions in an effort to account for the position of the comet with respect to the rest of the universe.[89] Roeslin's treatise on the comet was followed in 1579 by his *Speculum Mundi,* the existence of which is substantiated solely by his own statement. In 1580 he observed the comet then visible.[90] In 1583 there appeared a work on the calendar which was reprinted in 1584. It was signed "Plieninger," and, considering Roeslin's birthplace,

es ins künfftig in derselben ergehen werde, aus Anleitung dero von anno 1600 her am hohen Himmel erschienenen grossen Wunderzeichen, sonderlichen dess Cometens anno 1607 genommen &c. Allen Gelehrten in allerley Faculteten zu lesen, sowol lustig und nützlich, als menniglichen zur Warnung und Auffmunterung, gestelt durch H. Roeslin, Med. D. &c. Strasburg, 1609. (Crawford library catalogue, 113, 388; Frisch editor, I, 228, 498-9, where the work is discussed.)

Von der Mitnächtigen Schiffart, Strasburg, 1610. (Frisch, editor, IV, 169.)

Mitternächtige Schiffarth, von den Herrn Staden, inn Niderlanden vor XV. Jaren vergebenlich fürgenommen, etc. Oppenheim, H. Gallart, 1611. Octavo (B. M. catalogue; H.C.L.; N.Y.P.L.; Jöcher.)

Praematurae Solis apparitionis in Nova Zembla causa vera : et de magnete nonnulla J. G. Brenggeri : cum accurata instructione navigationis septentrionalis ad Indias Orientales dextrae instituendae. Strasburg, C. Kieffer, 1612. Quarto. (B. M. catalogue; Frisch, editor, I, 498, where it is discussed, and IV, 169; Goldschmidt, catalogue 51; Jöcher. The N.Y.P.L. copy has been stored until after the war.)

Prodromus dissertationum Chronologicarum, das ist der Zeitrechnung halben ein aussführlicher und gründtlicher Teutscher Bericht an Vnsern allergnädigsten Herrn, Matthiam den I. erwählten Römischen Kaysern : das nemblich den Jahren vnd dem Alter vnsers Herrn vnd Heylandts Jesu Christi nicht 5 Jahr zuzusetzen seyen, wie Irer Kays. Maj. Mathematicus Johan Keplerus haben wil, sonder mehr nit als fünff viertheyl Jahr &c. Frankfort-on-the-Main, 1612. (Frisch, editor, IV, 169, 170, 201, ff; Jöcher.)

88 Brahe, IV, 251-8.

89 See Pingré, I, 85-6 and chapter III of item 93, summarized below.

90 Pingré, I, 86.

may be safely attributed to him.[91] In 1593 he wrote a geographical description of Alsace and two years later a work on the warm springs in Hanau. In 1597 he wrote a *Tractatus meteorastrologiphysicus,* dealing with the paths of comets, and the *De Opere Dei Creationis* mentioned above.

By the latter work Roeslin again attracted Tycho's notice, as well as that of the other important astronomers of the day. Roeslin claimed to have invented independently a system of the world similar to the one which Tycho had put forth in 1588. Tycho, who was always angered by what he considered plagiarisms, resented Roeslin's presentation of the new system. He pointed out that nine years had elapsed between the publication of his own book and Roeslin's,[92] and he repeatedly referred to Roeslin's work as taken from his own.[93] Tycho did not vent his anger solely against Roeslin. He was equally vehement toward Ursus, who, he thought, had taken the idea from him while on a visit to Uraniborg.[94] In 1605 Roeslin wrote about the nova of the previous year and in 1609 he published his *Discurs,* in the third chapter of which he said that he preferred Tycho's system to Ptolemy's old one and to Copernicus' new one, because it agreed best with scripture and physics and because he had thought of it before he had read about it elsewhere. He admitted, however, that he had seen Ursus' work, which he considered to have been taken from Tycho's. Roeslin did not like the changes of Tycho's system by Ursus, who permitted the rotation of the earth.[95]

91 See the B. M. catalogue and Frisch, editor, IV, 5.

92 Frisch, editor, I, 224.

93 Brahe, I, xlii, V, 115, 323, VII, 206.

94 There is little to substantiate this latter assertion of Tycho's and even Kepler's defense of his master [1600] (Kepler, I, 215-287) is only half hearted. Kepler (I, 64) mentioned the systems of Ursus, Tycho and Roeslin in a letter written in April, 1598. Ursus' book, which put forth his new system, appeared the same year as Tycho's *De Mvndi Aetherei . . . Phaenomenis.*

95 Frisch, editor, I, 228.

By 1609 Roeslin was launched in a battle against Kepler and also against Maestlin. To the latter he wrote that Copernicus had opposed physical principles and scripture.[96] Kepler wrote several answers to Roeslin's work, especially one answering the *Discurs.*[97] Roeslin replied to him,[98] and spoke [99] of five different hypotheses of the world, by Ptolemy, Copernicus, Tycho, Ursus, and himself, and added that more people were attracted to the Tychonic than to the Copernican hypothesis.[100] Roeslin had already made this distinction between the five systems in the appendix to his *De Opere Dei Creationis.* Roeslin's controversy with Kepler involved much more than the question of priority in formulating the Tychonic system. Among other points, the date of the birth of Christ was in question. Also, Roeslin objected to some of Kepler's conclusions concerning the comet of 1577. Roeslin resented Kepler's correcting him in regard to a phenomenon which he had observed, and Kepler had not. The two also clashed over the interpretation of the nova of 1604. Fundamentally the dissension was due to Roeslin's inability to accept the heliocentric system of Copernicus (or more properly of Kepler) and to concede the motion of the earth. The battle with Kepler was entirely a friendly one. Kepler even said that Roeslin had been his teacher through some of his publications.[101] Each man had respect for the learning of the other. Roeslin called Kepler his good friend and compatriot [102] and praised him in the *Prodromus* for his work concerning Mars.[103] Roeslin's difficulty lay in accepting radical changes

96 *Ibid.*, I, 229.

97 Kepler, I, 495-542.

98 *Mitternächtige Schiffarth* (Strasburg, 1610, Oppenheim, 1611); *Praematurae Solis apparitionis in Nova Zembla causa vera* (Strasburg, 1612); *Prodromus dissertationum Chronologicarum* (Frankfort-on-the-Main, 1612).

99 In the *Prodromus dissertationum Chronologicarum.*

100 Frisch, editor, IV, 170.

101 Kepler, I, 505.

102 Frisch, editor, I, 499.

103 *Ibid.*, IV, 170.

in astronomical thought which the late sixteenth and early seventeenth centuries were ushering in, and, like many others, he sought a compromise. He said of himself that he was no astronomer,[104] but that he desired to know how things work on earth and in the spheres.[105] He was well aware of the astronomical problems facing his contemporaries and he played an important part in their solution. In evaluating Roeslin it should be borne in mind that mustering the facts in support of an opinion which later proves to be wrong, is frequently an important step in establishing a true theory.[106]

The preface of Roeslin's tract on the comet of 1577 [107] deals with omens of " our " times, especially of the year 1578. Roeslin said that some believed that the comet of 1577 was sublunar and others that it was exactly like past comets. However, he singled out Gemma as having written a good book on the subject. This is significant in determining Roeslin's keen powers of judgment, regardless of the fact that Gemma's observations have been considered too inaccurate to build upon.[108] Of the books treating the comet as supra-lunar, Tycho's, containing his own observations as well as those of the Landgrave and others, did not appear until 1588, but Gemma's, Maestlin's and Roeslin's books appeared in 1578, and Maeslin's work may have appeared after Roeslin's.[109] Roeslin, in his preface, spoke

104 Frisch, editor, I, 228, mentioned the distinction between "astrologer" and "astronomer" with regard to Roeslin, whom he classed as the first, and whom he called "medicus, etsi non astronomus".

105 Frisch, editor, I, 663, IV, 170.

106 Günther (1901) wrote on the compromise systems of the world without mentioning Roeslin, except as his name appeared once in a foot-note giving the title of a work by Kepler in answer to Roeslin. Roeslin does not deserve that oblivion.

107 Item 93 of appendix.

108 Dreyer (1890), 165-6.

109 Roeslin's book was finished after May 16, 1578, that date having been mentioned in the preface. Maestlin's work certainly appeared after February 1, 1578, since, on the recto of E_3, he mentioned Dasypodius, whose work on the comet of 1577 contains a preface dated February 1, 1578.

of the star of 1572 and of the two "chasms" in 1575, which were described by Gemma and were probably manifestations of the aurora borealis. Roeslin described an unidentified celestial phenomenon, similar to a comet, seen on December 5th, 1577. He also said that he had received a letter from Samuel Siderocrates [110] about a phenomenon which occurred after sunset, at 9 o'clock, on May 16th, 1578. Siderocrates described it as a tailed comet with two smaller comets following it. It soon disappeared. Others also observed this phenomenon, said Roeslin. He compared it to the star of 393 A. D. described by Nicephorus, which, singularly, according to Roeslin, did not have its tail turned away from the sun but rather from the moon. Roeslin was especially interested in the position of the comet of 1577 between the "tropics," [111] and its beginning in Cassiopeia where the new star had likewise been seen. He announced that he would show in his book that the comet was not elementary nor sublunar nor vulgar and natural but metaphysical and an exhibit of Providence, a sign from God.

Chapter I of Roeslin's tract deals with the fundamentals of the time and position of the comet of 1577, its motion and boundaries. According to Roeslin, when the comet was first observed after sunset on November 12th, its head was in the longitude of the sixth degree of Capricorn, with a latitude of 4° from the ecliptic, toward the equinox. Roeslin first saw the comet on the 14th, situated above Saturn, about in the second degree of Capricorn, with a latitude of 7½°. The tail was turned toward the "wintry place" of the sun, and extended as far as

Maestlin was fifteen years younger than Gemma, and, in 1578, not nearly so well known. His work, even if it had just appeared, might not yet have been known to Roeslin. On October 18, 1578, after reading Maestlin's treatise, Roeslin wrote to him exhibiting great admiration for him and accepting his corrections. Roeslin even suggested that Hagecius might realize his error in giving the comet a parallax of 5°. Roeslin's letter was printed in 1580 with Maestlin's ephemerides for 1577 to 1590. See Thorndike, VI, 79-80.

110 See above, chapter III, note 80.

111 The comet moved between those two circles.

the horns of Capricorn, in the extremities of that sign. The comet stretched nearly 20°, with a curve toward the meridian. Roeslin concluded that its daily motion was nearly 3° in longitude and nearly 2° in latitude, and that at its beginning, about the 9th of November, when it was first visible at new moon, it was in the winter solstice on the ecliptic. Roeslin said that the material of the comet was collected in October in the twentieth degree of Capricorn near the ecliptic. Thence retrograde motion drove it through the head of starry Sagittarius as far as the galaxy and the solsticial colure. In contact with this it was nearly stationary, or at least had a slow motion in latitude, in the beginning of November at new moon. On the 9th day of that month it moved toward the equinox through the other degrees of the sign and on the 14th passed Saturn in the direction of Antinous, and was beneath Aquila, so that it went out of Capricorn at the greatest solar declination, 23°, and arrived in the equinoctial circle, which it passed about the 21st of November, thus having traversed 30° of longitude from the beginning of Capricorn. Then it was 20° in latitude from the ecliptic, extending its tail close behind the longitude of the equinox towards the equinoctial position of the sun and towards the moon, then situated in Aries. Roeslin believed that from there the comet's motion was in a straight line from the sun's position at the beginning of winter toward the summer solstice, which he called the line of the comet's motion or the place where the middle of the tail stretched. After reaching the middle of Capricorn it began to decline and turned its course toward Cassiopeia. Roeslin remarked on the direction of the comets' tail, "toward the summer rising of the sun". He continued to record his observations and once remarked on his agreement with Gemma. He told how many days the comet was visible in Capricorn, in Aquarius, and in Pisces. Then he gave a table of the motion of the comet, including its longitude from the beginning of Capricorn, its latitude from the ecliptic, and its declination from the tropic of Capricorn,[112] partly from his own and

112 See note 19 in this chapter.

partly from Gemma's observations. The table covers the time from November 9th to January 14th, and also tells in which constellation the comet was situated on given dates.

Chapter II is concerned with the "proportional motion" or rate of the comet. Roeslin said that in the beginning the longitude was one and a half times the latitude or declination, which were equal. Then the motion in longitude and in latitude became unlike by an "imperfect" proportion, but they soon returned to a "perfect" proportion, having a common multiple. For this information Roeslin gave what he called a "table of the proportions of the true motion of the comet according to the longitude and latitude and declination," giving the values for the motions in longitude, latitude, and declination, and the ratio of each to the others. Then he considered the motion with respect to the straight line between the "winter and summer settings," obtaining ratios which he called more nearly perfect, and he gave a table which he called a "table of the proportions of the mean motion of the comet according to the longitude and latitude and declination".[113] However, what he was really trying to find was the rate of the comet's motion in longitude and latitude. Roeslin concluded that the rate of change of the comet's motion was constant and uniform, and began to wonder what caused that motion. He made use of Gemma's work on the nova of 1572. Roeslin decided that the comet was moving toward the place where the nova had been. He noticed that in the beginning the comet moved rapidly, in the middle more gently, and at the end most slowly. He made a table to show the number of days the comet required to cover the 30°, each, of the signs Capricorn, Aquarius, and Pisces and the number of days needed by the comet to move 12° in those signs, basing his figures for Pisces on the fact that in twenty-one days less twelve minutes the comet covered the first 12° of that sign.[114]

113 These proportions were "dupla", "sesquialtera", and "sesquitertia".

114 In the table Roeslin said that the comet required twenty and four-fifths days for 12° of Pisces, and in the text he gave, once twenty-one days less twelve minutes, and once, twenty days and forty-eight minutes, as the required time.

In the text Roeslin took into consideration Gemma's observations of the twenty-six days required by the comet to traverse 15° of Pisces, which covered the time up to January 13th. Roeslin gave two more tables representing the proportional motion of the comet.

The third, and most unusual, chapter of the book concerns the sphere and circles of comets and the poles and axis.[115] Roeslin thought it essential, since the comet had its own regular motion from west to east in addition to the diurnal motion of the heavens, that it have its own circle, an arc of which it must describe. Therefore, he suggested a sphere of comets to be determined by using the center of this arc and three fixed points. Judging from his observations of the comet of 1577, he concluded that the first point would of necessity be Shedir, the bright star in the breast of Cassiopeia;[116] the second point would be the two stars in the chest of Pegasus which the comet reached and where he believed it to have been extinguished; the third point would be the position of the head of the comet on November 14th, when Roeslin first observed it above the head of Sagittarius about 2° in Capricorn, at the northern latitude of approximately 8°. The circle determined from these three points had its center in the solsticial colure, like the center of the zodiac and the world, exactly opposite the pole of the world and 23½° from the pole of the zodiac. The circle cut the solsticial colure 8° above the point of the winter solstice. To allow the comet to stretch out in length and width it was necessary to give this circle a width of 8° on either side, thus presupposing a band, like a zodiac, 16° wide. However, this band was not to surround the whole universe like the zodiac or the equator but was to be like the small tropic circles, having in the middle, a

115 Chapter III has the title, "De Sphęra & Circulis Cometarum, de Polis item & Axi."

116 "... primum ergò punctum sit Scheder seu pectus Cassiopeiae, ...". Scheder, Seder, Schedir, (or Shedir) is a star of the third magnitude (see Zedler, XXXVI, 985).

semidiameter of 60°.[117] The width of the band allows for the comet's motion in latitude and longitude and the comet was at all times within the band. However, there is a vertical motion of the comet, which Roeslin said Gemma had shown by stating the variation in the comet's distance from the earth. In order to save this motion, Roeslin proposed an additional circle or epicycle, with a maximum parallax of 40′. He added that in the past comets had appeared in this sphere of meteors and cited Joachimus Camerarius on the comet described by Synesius, Nicolaus " Bruckner " on the comet of 1533, Vögelin on that of 1532, and Johannes Homelius [118] on that of 1556. The fact that Roeslin's circle of comets cut the ecliptic in the equinoctial points, through which the Milky Way does not pass, allowed for the motion of the comets of 1556 and "1475". Roeslin cited the description by Regiomontanus of the earlier comet. In order to be still more explicit concerning the sphere of comets, Roeslin wanted to fix the positions of its poles and axis. Its pole was on the axis of the zodiac, 30° from the pole of that circle, and 60° from the autumnal equinox. Roeslin felt that he could not describe the axis of the comets better than by the circle which extends from its poles through the two equinoctial points. However, he made three further computations to fix the axis and finally concluded that he had sufficiently defined the sphere.

Then [119] Roeslin dealt with the similarity, which he described as symmetry, between the comet and the star [of 1572] and

117 Roeslin's phrase is " sed erit instar parvorum circulorum Tropicorum, cum Semidiametro 60 graduum exactè in medio ". Pingré, I, 86, interpreted Roeslin as imagining a sphere of meteors " déterminée par un cercle tracé autour du pôle du monde à la distance de 60 degrés . . . ".

118 Homelius (1518-1562) became professor of mathematics at Leipzig in 1551. His observation of the comet of 1556 was also mentioned by Tycho and Praetorius. The name can also be spelled " Hommelius ", as Roeslin did, " Homilius ", as Dreyer (1890) did, or " Hommel " or " Hummel ". He was the mathemàtician under whom Scultetus studied (see below, under the sketch of Scultetus' life). See Thorndike, V, 397, VI, 411.

119 Chapter IV, which has the title, " De Symmetria, quam Cometa & Stella ad se invicem, & praecipua Coeli loca habuerunt, ratione Globi coelestis."

their positions in the sky with respect to the celestial globe. This discussion has little value for the present dissertation, except perhaps in that it illustrates one trend of thought and what might be called the persistence of Pythagorean influence. Roeslin attempted to express the comet's motion in geometric proportion, in the manner of a musician. He maintained [120] the tenor by dealing with the relationship which the star and comet have one with the other and with the poles of the world, with respect to the celestial globe.

The sixth chapter [121] aims to show that the comet of 1577 is not an ordinary comet but belongs in the class of portents and unusual signs, created by God miraculously rather than naturally. Roeslin cited Anaxagoras and Democritus and "others" who, he said, thought that comets were celestial and generated from celestial matter. This idea, said Roeslin, was opposed by Aristotle. However, there were those who argued from parallax and demonstrated that comets belong to the aethereal regions. Hagecius demonstrated that the nova belonged there, and Gemma placed the comet of 1577 in the sphere of Mercury. Roeslin concluded that Aristotle's doctrine did not apply to all comets, and told why he thought that the comet of 1577 was aethereal and belonged in the aethereal region. For his first reason he cited the comet's motion, in accordance with geometric progression,[122] as discussed in his second chapter. Such motion he considered impossible in the elementary region. Roeslin's second argument was based on the regular distribution and purity of the comet's light, which showed the body's material to be celestial rather than elementary. His third argument was based on the comet's long duration and regular dimin-

120 Chapter V, which has the title, "De Harmonia, quam Stella & Cometa ad se invicem & ad Mundi Cardines hahent [sic] respectu globi terrestris."

121 Chapter VI has the title, "Hunc Cometam non esse ex vulgarium Cometarum numero, sed ex portentis & insolitis Signis."

122 "... secundum Geometricam proportionem scil. duplam, sesquialteram & sesquitertiam, ..."

ution. His fourth was based on the comet's "perfect" motion, since sublunar phenomena were not moved more perfectly or more rapidly than the moon. In the fifth place, he argued from the analogy of the comet's motion and that of Mercury. It is characteristic of Roeslin that he was unwilling to break entirely with tradition and that he did not say that all comets were celestial, but only that such was the case with the comet of 1577.

The seventh and last chapter is a long one, entirely devoted to "significations." These were based on Roeslin's belief that the comet was the work of God. Roeslin quoted from ancient and modern authors, from sermons, and from the Bible, and included twenty-five "propositions" by which to judge the comet's meaning. The book ends with an appendix dealing with the critical years before the nova, from which prognostications were sought.

With a background and many interests sharply contrasted with those of Roeslin, William IV, Landgrave of Hesse Cassel,[123] (1532-1592), occupies a unique position in this dis-

123 A.D.B., XLIII, 32-9, article by Walther Ribbeck (good sketch of the Landgrave's life and work, but with little space devoted to astronomy).—Arago, 198-9.—Archives. (valuable in determining the Landgrave's political activity).—Bailly, I, 372-5. (The nova of 1572, but not the comet of 1577, is mentioned.) — Brahe, IV, (*De Mundi Aetherei... Phaenomenis*), VI, (*Epistolae Astronomicae*, Liber Primus [1596]), VII, *Epistolae Astronomicae*, hitherto unedited).—Delambre (1821), I, 223, 261.—Doppelmayr, 85, (how Praetorius almost came to Cassel, and references for William IV), 163-4, (Bürgi's connection with the Landgrave).—Dreyer, editor, passim, especially VI, 345-6, giving an excellent account of the Landgrave's astronomical activities.—Dreyer, (1890), passim. This is of particular value for an understanding of the Landgrave's astronomical activities, and places him and his work with respect to the other astronomers and astronomical work of his century.—Dreyer (1906), 350, 359 and note, 362 and note 2, 370 and note 2. This emphasizes Rothmann and is not of great importance for William IV.—Frank, I, 310. This bears on the Landgrave's relation to the "Concordienformel".—Freher, II, 756. A picture of the Landgrave is opposite page 754.—Janssen, VII, 198, 203, 208, 317, 342-5 (the Landgrave as botanist), 350.—Jöcher, IV, 1965.—Mädler, I, 183-7.—Monatliche correspondenz, XII, 267-302 (general discussion of the Landgrave).—Riccioli, II, 12-3 (the comet of 1577), 134, 138, 139, 150-1, 191-2 (all of which are concerned with the nova of 1572), I, xxvii, I, xxxv, (Tycho's presentations

cussion. From one point of view he does not belong here at all; for he never published a tract or treatise on the comet of 1577.[124] But he was one of the foremost astronomers of sixteenth century Europe and is of great importance in a study of the comet of 1577. Tycho preserved his observations of that comet for posterity, and listed him with the believers in the supra-lunar position of the comet.[125]

The Landgrave, although interested in cosmological systems, was devoted to observational astronomy, in which respect he closely resembled Tycho. The Landgrave observed the altitudes and azimuths of the new star and noted the time of his observations. However, this method did not receive the approval of Tycho, who did not trust clocks. Justice cannot be done to the Landgrave in these pages and the details of his life and work cannot be expounded here. They are of great importance in religious, political and cultural history, the latter including

of the Landgrave's observations).—Rommel, I, book 2 ("Hessen-Cassel. Die Zeiten L. Wilhelms des Vierten oder des Weisen. 1567-1592."), section 7, 758-808 ("L. Wilhelm als Gelehrter und Beförderer der Wissenschaften"); I, book 1; II.—Strieder, 69-82 (good general account).—Weidler, 373-4.—Wohlwill, I, 29.—Wolf, I, 58-9.—Wolf (1877), 244, 266-9, 272-6, 332, 381-3, 408.—Zedler, LVI, 1228-1233.

For still further information consult:

Barrettus (observations made at Cassel).

Christianus, Andreas. *Oratio de vita et morte... Gulielmi, Landgravii Hassiae, etc.* Herborn, 1592. (B. M.)

Duncker, Albert. *Landgraf Wilhelm IV. von Hessen, genannt der Weise, und die Begründung der Bibliothek zu Kassel im Jahre 1580.* Kassel, 1881. (B. M.)

Treutlerus, Hieronymus. *Oratio historica de vita et morte... Wilhelmi, Hassiae Landtgravii... publico nomine scripta, & in solenni Academiae Marpurgensis consessu recitata.* Marburg, 1592. (B. M.)

Other works are listed in the B. M. catalogue and its supplement under the name of William IV. See also Schottenloher, III, 199-200, for a bibliography of works about the Landgrave. Probably the short article, *Landgraf Wilhelm IV. von Hessen und Tycho Brahe,* listed there as by "B., E."; which appeared in *Hessenland,* v. 15 (1901), would prove of interest.

124 See appendix, below. The book, which was numbered 108 in the original bibliography, was erroneously included there.

125 Brahe, IV, 182-207.

much more than astronomy.[126] The only part of the Landgrave's activities which will be discussed here is that concerned with the comet of 1577; and the ramifications of this discussion must be omitted. Not even his pertinent observations of the nova of 1572, for which he found little or no parallax, nor those of later comets can be reviewed.

Although the Landgrave observed the great comet of 1577 independently, all extant information concerning those observations is dependent on Tycho Brahe. Despite the fact that Tycho visited Cassel in 1575, he and the Landgrave were not in communication for the next ten years; and it was not until October 1585, upon the occasion of the comet of that year, that the Landgrave, with the assistance of Heinrich Rantzov, established the correspondence between himself and Tycho which has so enriched the annals of astronomy. The observations by the Landgrave of the comet of 1577 first appeared in Tycho's *De Mvndi Aetherei . . . Phaenomenis,* chapter X, where they were discussed at length. Tycho said that they were communicated to him by the Landgrave in a letter. That letter, dated April 14, 1586, was preserved by Tycho in his *Libri Epistolarum Astronomicarum.*[127] It is curiously written in the style which the Landgrave used in most of his letters to Tycho, in German with so many Latin phrases that the Latin translation which Tycho made is scarcely necessary even to one who cannot read German. The Landgrave's observations, quoted in full in the *De*

126 For example, the observations made at Cassel by Rothmann, and, after his strange disappearance, by Bürgi, are of great interest and could be reviewed in a volume devoted solely to the Landgrave. He deserves such a study. There the founding of the observatory at Cassel and the subsequent improvement of the instruments could be discussed and an explanation of the Landgrave's opposition to the Gregorian calendar, which was purely political, could be given. The outlines of the Landgrave's life and bits concerning his activities can be gathered from various of the usual sources, but he has nowhere, save possibly in the work of Rommel, been treated with the patience and care necessary to write a unified history of a man of such diversified interests.

127 Item 19 of appendix, below. The letter is given in full by Dreyer, editor, VI, 48 ff., and has been quoted at length by Wolf, I, 58-9.

Mvndi Aetherei . . . Phaenomenis, have been repeated in several places in the years following the publication of Tycho's book, and they have been discussed at considerable length, but nothing essential has been added to the comments made by Tycho. The important point to stress here is that the Landgrave considered the comet supra-lunar. Also, his observations of the tail are important, because he was one of those astronomers who had read Apian's works, and was in a position to transmit his knowledge of astronomical progress on the subject. In fact, Apian's *Astronomicum Caesareum* had greatly influenced the Landgrave who had had the three dimensional paste-board demonstrations in that book reproduced in copper. He observed the width and length of the tail as well as its latitude and longitude, and Tycho compared the Landgrave's observations with those of Maestlin.

It is in his above-mentioned letter to Tycho that the Landgrave called Bürgi " another Archimedes," but in the same letter he pointed out that at the time of the comet of 1577 he had no court mathematician.[128] Therefore, he said, he was unable to observe that comet for parallax. Also, he stressed the importance of the instruments used in observing and said that his instruments were not as large as Tycho's. He drew some of his conclusions from observations of the comet of 1585, for which he found no parallax. Thus he denied the " philosophical fundamental " that comets are engendered in the sublunar atmosphere. In this letter he also explained how he reversed his sextant to allow for instrumental errors, and how he noted errors due to refraction at the horizon.[129] With his letter, as he stated in his postscript, he enclosed observations of the nova of 1572, for which he had found no parallax, of the comet of 1577, of the comet of 1585, and of the sun. Receipt of these observations was acknowledged by Tycho in May.

128 Rothmann, together with Ioannis Troldenirerus, observed the comet of 1577 on November 10th in Bernburg. See the letter from Rothmann to Tycho written at Cassel in October, 1588 (Dreyer, editor, VI, 160-1).

129 Arago said that the Landgrave knew of atmospheric refraction but did not make use of it.

The Landgrave was the first man dealt with in the tenth chapter of Tycho's book on the comet of 1577. His observations were given in detail, as they were forwarded by him to Tycho. For November 16th, 17th, 20th, 21st, 23rd, 25th, 30th, December 1st, 2nd, 3rd, 6th, and 30th, the times of the observations in hours and minutes, together with the corresponding occidental azimuths [130] and altitudes in degrees and minutes were given. Several observations were given for each day. Where the original observation was of the tail it was so stated, and the part of the tail observed was indicated. The data given by Tycho included the height of the pole at Cassel. The observations by the Landgrave were analyzed by Tycho and compared with his own. He made four investigations of the parallax from them, without being altogether satisfied, but he concluded that the Landgrave's observations show the comet to have been supra-lunar.

One more believer in the supra-lunar position of the comet of 1577 remains to be treated, namely, Cornelius Gemma,[131]

130 "The *azimuth* of a heavenly body . . . may be defined as *the angle formed at the zenith between the meridian and the vertical circle which passes through the object*; or . . . it is *the arc of the horizon measured westward from the south point to the foot of this circle*." Russell, Dugan and Stewart, I, 12.

131 Adam (1705), 33, 106, (*Vitae . . . Medicorum*).—Bib. Belg. Valerii Andreae, 149.—Bassaeus, I, 454.—Biographisches lexicon der hervorragenden ärzte, II, 711. — Cantor (Gemma). — Castellanus, 237-8. — Chalmers, XV, 396.—Degeorge, 88. This mentions the *De Naturae divinis* . . . —Dreyer, editor, VIII, 456.—Dreyer (1890), 38-69, 158-185.—Foppens, I, 200, with a portrait of Cornelius Gemma (missing in the C.U.L. copy) on the opposite page.—Gesner (1583), 175.—Ghilini, II, 220.—Hoefer, XIX, 854.—Jöcher, II, 914.—Kestner, 337.—Michaud, XVI, 137.—Ortroy, (1920), (the definitive work on the two Gemmas).—Pingré, I, 65.—Poggendorff, I, 872.—Quetelet, 83-4, 89-90.—Riccioli, I, xxxiii, II, 13, 28, 87, 122, 134-5, 138, 153, 191.—Scheibel, 69-70, 102-3.—Smith (1917), 133-4, 136.—Struve, I, 230, 549, 550, 739.—Thorndike, VI, 406-9.—Weidler, 394-5.—Zedler, X, 809.

Other sources for material concerning Gemma might be:

Ekama, C. *Verhandeling over Gemma Frisius, den eersten Grondlegger tot het bepalen van de Lengte op Zee*. Amsterdam, 1825.

Favaro, Ant. *Nuovi studi intorno ai mezzi usato dagli antichi per attenuare le disastrose consequenze dei terremoti*. Venice, 1875, p. 29.

who was born in Louvain on February 28, 1535,[132] the son of Gemma Frisius, the well known doctor and astronomer,[133] and who was destined for a similar career. He received his first instruction from Bernhardus, director of a school in Malines, in which city Gemma spent at least two months in either 1546 or 1547. He began his studies at the university of Louvain in 1549. In 1561 he called himself " medicus " on the title-page of his *Ephemerides,* but he really became a doctor of medicine on May 23, 1570. At the same time he was made a professor of medicine in the university of Louvain.[134] He was much influenced by the errors of astrology, and Quetelet said of him that his reasoning did not measure up to his learning. He nevertheless ranks as one of the foremost astronomers of the sixteenth century,[135] and received the praise of Tycho Brahe, who listed [136] him with those who considered the comet further away than the moon. Gemma was in correspondence with many of the leading scholars of his day. He wrote several works about the nova of 1572, which he had observed and for which he had

Garnier, J. G. and Quetelet, L.A.J. *Correspondance mathématique et physique* . . . 8 v., Gand, 1825-35, v. 1.

Miraeus (Le Mire, Aubert). *Elogia illustrium Belgii scriptorum, qui vel ecclesiam Dei propugnarunt vel disciplinas illustrarunt, centuria decadibus distincta. Ex bibliotheca Auberti Miraei,* . . . Antwerp, successors of Bellerus, 1602.

Sweertius, Franciscus. *Athenae Belgicae, sive nomenclator Infer. Germaniae scriptorum,* . . . Antwerp, 1628.

132 Jöcher gave the date as 1534, but the weight of the other authorities overrules him.

133 See chapter II, above.

134 Van Ortroy (1920), 123, 125, said that Gemma was named professor as early as 1569. Smith (1917), 136, said that Gemma was professor of astronomy, also, at Louvain.

135 A succinct description of Cornelius Gemma's prowess and failings was given by van Ortroy (1920), 127-8. Thorndike, VI, 30, said of him: " The interest and activity in the field of astrological prediction of Cornelius Gemma were even more intense and prolonged than those of his father, Gemma Frisius, while his intellectual ability and scientific aptitude were distinctly inferior."

136 Brahe, IV, 238-251.

found no perceptible parallax.[137] According to Dreyer,[138] he "had a great deal to say about the star, but most of his distance measures are upwards of a degree wrong". He was severely criticized by Squarcialupus for having arranged, because of poor observations, the stars of Cassiopeia in the form of a cross.[139] Wolf said that he observed the comet of 1558.[140] Gemma realized the importance of the direction of a comet's tail and his book on the comet of 1577 pictures the tail extended away from the sun. Of course, this was not a new observation on his part. No doubt it had been drawn to his attention at an early date by his father, the translator of a work by Apian.[141] The younger Gemma edited some of his father's work. He considered the distance of the comet of 1577 from the earth important enough to mention on the title-page and agreed with Tycho in the deductions from this observation. According to Riccioli, he believed that some comets were above the moon but that others were below it. His works [142] are numerous and

137 Gemma in Hagecius (1574), 137-145. Unable to find a parallax in excess of 4′, Gemma, nevertheless, thought the phenomenon must have one. He did not classify the star, saying (145), "Nec stella neque exhalatio dici possit, multò minus cometa". See also Dreyer (1890), 60 note 1.

138 Dreyer (1890), 58-9.

139 Ortroy (1920), 132, citing Squarcialupus. See page 69 of item 37 (3), of appendix, below, for the exact citation.

140 Wolf (1877), 408.

141 See Pogo (1934), 443-4, and Riccioli, II, 122.

142 Gemma's works are cited in many places. Van Ortroy (1920), 367-396, has given an almost complete list of them. Some of them, with information concerning the location of copies, are:

Gemma Frisii,... de Astrolabo catholico liber, quo latissime patentis instrumenti multiplex usus explicatur... (Edente C. Gemma)... Antwerp, J., Steelsius, 1556. (B. M. and B. N. catalogues. The younger Gemma added a preface, a dedication, eighteen chapters and a panegyric in verse to his father's work. The work was reprinted in 1583.)

De arte cyclognomica, tomi III. Doctrinam ordinum universam, unaque philosophiam Hippocratis, Platonis, Galeni et Aristotelis in unius communissimae ac circularis methodi speciem referentes,... Antwerp, Plantin press, 1569. (B. M. and B. N. catalogues. There were two editions of this work printed by Plantin in 1569. The work is preceded by a Latin poem: *Menti rerum architectrici, divini amoris et Psyches Hymeneum Cornelius Gemma, loco hymni, magici consecravit.*)

De natvrae divinis characterismis, sev raris & admirandis spectaculis, causis, indiciis, proprietatibus rerum in partibus singulis uniuersi, libri 2 . . . Antwerp, Plantin press, 1575. (B. M. and B. N. and Crawford library catalogues. This work deals with the nova of 1572, and is followed by these medical treatises: *Casus mirabilis cuiusdam abscessus in puella Louaniensi* . . . ; *De raro genere Epidimicae febris ac pestilentis, quae ad Galeni Hemitrit os accedens proximé, magna contagij vi totum pergrassata est* . . . ; and *De vlteriore transmutatione febris pestilentis in pestilentiam veram* . . . The Crawford library catalogue says: "These volumes treat chiefly of prodigies and portents, amongst which the author includes aurorae, parhelia, comets and the New Star of 1572." Struve, I, 230, made a similar statement and listed, I, 550, the following title for chapters 3 and 4 of book II of the work: "De prodigioso phaenomeno syderis novi, et de memorabilibus quae terrae Belgicae post apparitionem novi syderis contigere. Additamentum exhibens excerpta ex libello hispanico Hieronymi Munnos, quae tractant de nova stella (p. 267 et seq.) . . ." As an alternative title for the work Foppens gave *Cosmocritico,* which seems to have been taken from a subtitle in the second volume.)

Ephemerides meteorologicae . . . institutae per C. G. 1561. (B. M. catalogue. Van Ortroy (120), 367, seems to have thought that these began in 1560 and that there were at least five of them.)

De Peregrina stella quae superiore anno primum apparere coepit, clariss. virorum Corn. Gemmae, . . . et Guliel. Postelli, . . . ex philosophiae naturalis mysticaeque theologiae penetralibus deprompta judicia . . . (The dissertation by Postel is entitled: *De Nova stella quae jam a XII. die novembris anni 1572 ad XXVI. junii. 1573 . . . durat . . . judicium.* B. M., B. N., and Crawford library catalogues; Struve, I, 549. The book is without date but the Crawford library catalogue says that it was printed in 1573. Van Ortroy (1920), 141, seemed uncertain of the existence of this work with an Antwerp imprint, but cited it on the authority of Houzeau.)

De Communi Cometarum natura (A poem; see Ghilini.)

Hymnus ad S. Christi crucem. (This is a fourteen line poem and can be found in the N.Y.P.L. in Gruterus, 458-9 (wrongly marked 495).)

There are several editions of Gemma's work on the nova of 1572, including the following in the Crawford library:

Stellae Perigrinae iam primvm exortae et caelo constanter haerentis φαινόμενον vel observatum, diuinae prouidētiae vim & gloriae maiestatem abunde concelebrans. Louvain, Bogardus, 1573. (This was also listed by Struve, I, 549 and van Ortroy (1920), 378-380, or 380-1, discussing second edition by the same printer in the same year.)

Stellae perigrinae iam primùm exortae . . . (Included in Hagecius (1574).)

De nova stella jvdicia dvorvm praestantivm mathematicorvm. D. Cypriani Leovitii a Leonicia. et D. Cornelii Gemmae, Professoris regij Louaniensis. Addita svnt de eadem epigrammata quaedam lectu dignissima. 1573. (According to the Crawford library catalogue, 277, this tract seems to be a reprint of the treatises by Gemma and Leovitius, and was probably brought out by Laurentius Benedictus in Copenhagen. See also Ortroy (1920), 381-3.

deal with astronomy, astrology, and medicine, and include some Latin verse which has been considered good. It is said that shortly before his death, Gemma was summoned to Nimwegen in consultation by the Duke of Alba,[143] but he died in Louvain, a victim of the plague, on October 12, 1579,[144] at the age of forty-four.[145] He left a son, who was also a doctor of

Both the Crawford library catalogue and van Ortroy gave the title of the original work by Leovitius. The book was also listed by Struve, I, 549.)

Van Ortroy copied a list of Gemma's works from Andreas and Foppens. Foppens gave the list as follows: "Leguntur denique Lovanii in Bibliotheca Academica ejusdem

Poëmata: *De Mundi coelestis cum clementari* [sic] *symmetria.*

Emblemata Philosophica ruinae Belgicae.

Fabula Mulieris redivivae, Comoedia.

Item *Oratio De necessitudine mutua Praxeos & Theoriae in arte seu Facultate qualibet.*

143 Van Ortroy (1920), 131, wrote: "On rapporte que le duc d'Albe le fit appeler à Nimègue, pour le pressentir sur les événements futurs..."

144 Most authorities gave the above date. However, the date of Gemma's death has been given as October 12, 1577, by usually competent authorities including Dreyer, editor, VIII, 456, Cantor (Gemma), Hoefer, and Poggendorff, who, in addition, gave Quetelet as a source for the later date. Cantor even went so far as to discuss the "erroneous" later date. Van Ortroy (1920), 61, wrote the phrase "Corneille Gemma, mort en 1577 seulement, ...", in a sentence, however, where he was interested in the fact that Cornelius had sufficient time to edit his father's translations. Cornelius Gemma cannot have died before 1578 if, as there is no reason to doubt, he is the author of item 43 of the appendix below, which deals with the comet visible and observed by the author from November 1577 to January 1578. Another argument for the later date is that the university of Louvain was asked by Gregory XIII in 1578 to consider calendar reform and Gemma and Pierre Beausard (see Smith (1917), 133-4) were supposed to go to Rome to report for their colleagues on that question. They both died of the plague and the report had to be sent to Rome without them. Van Ortroy (1920), 137-8, and Quetelet gave 1578 as the date of this episode, which Cantor mentioned without giving a date. Castellanus, 238, dated Gemma's death in 1578. The date of Gemma's death was discussed at length by Van Ortroy (1920), 117-8, where the dates 1577, 1578, and 1579 for the event were attributed to Valerius Andreas.

145 Maximilian Vrientius or Laurent Beyerlinck (see Ortroy (1920), 401-2, and Foppens, I, 200) wrote the following inscription for Gemma: "Quis lapis hic? Gemmae. Gemmam Lapis an tegit? inquis. At condi in Gemma debuerat potius. Non ita: nam quaevis minor illâ Gemma fuisset, Et posito à Gemma Gemma sit iste lapis."

medicine, but was not as famous as Cornelius and Gemma Frisius.[146]

Gemma's work on the comet of 1577 [147] begins with a short poem entitled "ΕΥΡΩΠΗ ΝΟΣΗΛΕΥΟΜΕΝΗ EIDYLLION, In apparitionem Cometae anni 1577", which predicted that there would be great turmoil and that the Netherlands would be purified by flames. Figure I shows the path of the comet on the celestial sphere. The first chapter is medical as well as astronomical and astrological, describing the nova in Cassiopeia from November 1572 to April 1574, the "chasms" observed in the Netherlands in February 1575, and the significance of these phenomena.

The most important chapter of the work [148] describes the physical appearance of the comet and its motion and magnitude. It discusses the comet's tail, its form, its red color, its direction away from the sun, its area of visibility, north, south and east, its varying brilliancy and its curvature. Gemma gave day-to-day observations of the comet, speaking of the position, the motion, and the parallax, and of the two tails seen on November 28th. According to Gemma, after December 3rd the brilliance of the comet decreased. After December 13th the tail grew brighter. About the middle of January the comet was almost stationary, then went rapidly through Capricorn and Aquarius and on to the middle of Pisces.

Because of the parallax of not more than 40′, which he had found, Gemma concluded that the comet was further from the earth than the moon, in the sphere of Mercury. He spoke of his disagreement with Hagecius, Munosius,[149] Thomas Digges, John Dee and Tycho.

146 For a discussion of Gemma's wife and family see Ortroy (1920), 119-121. Another son, Raphaël, also survived his father.

147 Item 43 of appendix, below.

148 Chapter II.

149 Munosius is supposed to have copied from Gemma concerning the nova of 1572. See Weidler, 394-5, quoting Tycho.

The third and last chapter adds nothing to astronomical knowledge but discusses methods of predicting the future from comets, and in doing this goes partly into the history of cometary observation, but only superficially. Such names as those of Albategni, Regiomontanus and Vögelin are mentioned. The comet of 1556 is spoken of as foreshadowing disasters in Gaul, and reference is made to the nova of 1572. Quotations in classical Latin referring to comets, such as those from Virgil's *Aeneid,* are given, and there are moral and religious digressions.

The epilogue starts on the recto of E_3, opposite a picture of the weeping Belgica, and is entitled "EIMAPMENH Sive EIDYLLION Fatalis Vicissitvdinis In Belgico Statv". It is a dialogue between Sibylla Erythraea and Belgica Virgo and discusses the comet of 1556, the nova of 1572 and the comet of 1577. It is followed by another short poem. The book as a whole clearly reflects the unrest in the Netherlands, then in conflict with Spain, and soon to be divided.

In the tenth chapter of his own treatise,[150] between his summaries of Maestlin's and Roeslin's works, Tycho summarized Gemma's. He praised it and called it learned, but his discussion was principally confined to the second chapter. He said that Gemma found a straight path for the comet and showed that it moved above the moon.

150 Brahe, IV, 238-251.

CHAPTER IV

THE COMET OF 1577: THOSE WHOSE COMPUTATIONS OF ITS PARALLAX PLACED THE COMET BENEATH THE MOON

HAGECIUS.—SCULTETUS.—NOLTHIUS.—BUSCH

THADDAEUS HAGECIUS AB HAYCK,[1] or Tadeáš Hájek z Hájku, also called Thaddaeus Nemicus, was the foremost astronomer of eastern Europe in the second half of the sixteenth century. The son of Katerina and Simon Hájek, he was born in Prague between 1525 and 1527,[2] and died there on the first of September, 1600.[3] Katerina Hájek was of noble blood. However, she died before 1528, and before 1530 Simon married Dorothy, the sister of Jindřich Jaromiřský of Vlcnoves, who, fortunately for Thaddaeus, proved a good stepmother. Simon Hájek had a bachelor's degree from Prague and was a biblio-

1 Bailly, I, 375-6, 411. — Bassaeus, I, 339, 421, 556. — Crawford library catalogue, 216, 217.—Delambre (1821), I, 195, 226-9.—Dreyer, editor, IV, 507, VIII, 457.—Dreyer (1890), 58, 64, 82-3, 222-3, 269-270, 302-3, 321.—Frisch, editor, VII, 288.—Gesner (1583), 774.—Hagecius (1574).—Jöcher, II, 1315.—Lalande, 109.—Mädler, I, 186.—Poggendorff, I, 991.—Riccioli, I, xlv, II, 13, 28, 40, 89, 134, 138, 139, 151.—Rosen.—Scheibel, 65-6, 70-2, 90, 103-4, 111, 125.—Smith (1917), 136.—Struve, I, 550.—Thorndike, VI, 504-6.—Vetter. This seems to be the fullest account of Hagecius which has been written. A translation for use in the present study was made by Dr. J. Novak.—Vetter (1926).—Vetter (1928).—Vetter (1937).—Weidler, 393-4. —Wolf, J. C., I, 1157.—Zedler, XII, 159.

Further information concerning Hagecius' calculations and observations of the nova and comets can be found in Favaro's edition of Galileo's works, vols. II, VI, and VII. Unfortunately, Pelzel's *Abbildungen böhmischer und mährischer Gelehrten und Künstler*... (Prague, 1773-1782) is not available, although the Library of Congress tried to locate a copy. Nor is there an available copy of Professor Vetter's *Poznámka k astronomické činnosti Tadeáše Hájka z Hájku* (Čas. mus. Král. čes. XCI, 330). Other sources were given by Vetter, 1 note 1.

2 Vetter, 1, and Vetter (1928), 500.

3 Dreyer (1890), 302-3, said that Hagecius died after a prolonged illness, Vetter, 16, that he died suddenly.

phile, especially interested in religious books, and in alchemy. His gorgeous study rooms were described by John Dee.[4] Thaddaeus inherited a love of study and of books from his father, and himself became a connoisseur. During his lifetime he accumulated a valuable library and a collection of astronomical instruments. He was renowned both as a mathematician and as an astronomer, and besides was physician to the emperors Maximilian II and Rudolph II. He has been considered the most remarkable of the representatives of Czech science under the latter ruler.[5] Furthermore, he is an important figure in the his-

4 In 1584 Thaddaeus Hagecius gave the use of the old Hájek house to Dee and a companion, who, in the presence of Thaddaeus Hagecius and another, proceeded to transform mercury into gold.

5 The list of Hagecius' writings is long and, in addition to the *Dialexis de novae... stellae... apparitione* (Hagecius (1574)) and the two books on the comet of 1577 (items 48 and 49 of appendix, below), includes:

Actio medica T. ab Hayck adversus P. Franchelium,... (*I. Exegesis curationis foedae scabii, simul etiam querela in P. Fanchelium* [sic in B.M. catalogue]. *II. P. Fanchelii responsum ad eandem exegesin. III. Eidem responso aliud responsum oppositum sub titulo Antifanchelius.*) 2 pt. Amberg. 1596. (B.M. catalogue; Vetter, 12)

Aphorismorum metoposcopicorum libellus unus. (According to Vetter, 6, the first edition was published in Prague in 1562 by Melantrich. A second unchanged edition appeared in Frankfort in 1584 and copies of it can be found in the B.M. and B.N. This was translated into German. However, considering the reference by Thorndike, VI, 72, to Hagecius' remarks on the nova of 1572 in the 1584 edition of the work, that edition must be different from the 1562 edition. A French translation, printed by G. Chaudiere, appeared in Paris in 1565. It had the title *Nouvelle invention pour incontinent juger du naturel de chacun par l'inspection du front et de ses parties, dicte en grec Métoposcopie, le tout extraict du latin de M. Thaddée Hagèce.* (B.N. catalogue).)

Aphorismorum medicorum libellus unus... Frankfort, 1597. (Vetter, 3)

Apodixis physica et mathematica de cometis tum in genere, tum in primis de eo qui proxime elapso anno LXXX in confinio fere Mercurii & Veneris effulsit & plus minus LXXVI dies duravit. Görlitz, 1581. (Vetter, 9)

Astrologica opuscula antiqua. Fragmentum astrologicum, incerto autore, in quo, praeter caetera, aliquot exemplis ostenditur, quomodo medicatio ad Astrologicam rationem sit accommodanda. Liber Regum de significationibus Planetarum in duodecim domiciliis Coeli... Liber Hermetis centum Aphorismorum, cum commentationibus Thaddaei Hagecii ab Hagek D. Omnia nunc primum in lucem edita. Prague, 1564. (B.M. catalogue; Vetter, 5. According to Vetter, the first part of this work is the publication of an old astronomical manuscript from the library of the college of Charles IV, containing a fragment by an unknown author, which quotes the author of the *Sphaerae* of about 1220, probably Sacrobosco. The fragment also speaks of cures for St. Vitus dance and leprosy. Hagecius added notes to the fragment.)

De cervisia, ejusque conficiendi ratione, natura, viribus & facultatibus, opusculum. Frankfort, 1585. (B.M. and B.N. catalogues; Vetter, 11. According to Vetter this work was written because of an inquiry by Julius Alexandrinus, then foremost physician of the emperor and author of a book on hygiene, who wanted to know how beer was being brewed.)

Diagrammata seu Typi Eclipsium Solis et Lunae futurarum. Anno a Christo natu 1551 una cum eorundem explicationibus in gratiam Venerabilium virorum Joannis Albini & Syxti Beüschl Canonicorum Cenobii Neuburgensis etc. per Thaddaeum Nemicum alias Hagek Pragensem conscripta & aedita. 1551. (Vetter, 5, 7. According to Vetter, an astrological poem by Nicolas Bourgois was added to this work. The work also contains a calculation of a lunar eclipse on February 20 and of a solar eclipse on August 31.)

O některých předešlých znameních a úkazech v povětří a o kometě tohoto r. 1580. Hořelice, 1580. (Vetter, 9. The translation of the title is: *On several foregoing signs and phenomena in the air and on the comet of this year 1580.*)

Beschreibung des Cometen, der im J. 1580 erschien. Prague, 1580. (Poggendorff, I, 991. This work may be a translation of the one above.)

Oratio de laudibus geometriae, scripta et recitata in Academia Pragensi, sub initium lectionis Euclideae, XII Februarii die A M. Thaddaeo Nemico Haykone ab Hagek. Prague, 1557. (Vetter, 11; Poggendorff, I, 991. According to Vetter, this is the address with which Hagecius began his geometrical lectures at the university of Prague. It contains a short history of mathematics of Bohemian origin.)

Responsio Ad virulentem & maledicum Hannibalis Raymundi, Veronae, sub monte Baldo, nati, scriptum: quo iterum confirmare nititur, Stellam, quae Anno LXXII. & LXXIII. supra sesquimillesimum fulsit, non nova, sed veterem fuisse. Prague, George Nigrinus, 1576. (Crawford library catalogue, 217; B.M. and B.N. catalogues; Struve, I, 550; Dreyer (1890), 64 note 1; Vetter, 11.)

Spongia contra rimosas & fatuas Cucurbitulas Hannibalis Raymundi, Veronae sub monte Baldo nati, in larua Zanini Petoloti à monte Tonali.... Prague, 1578. (This was printed with item 48 of appendix, below. It was also cited by Vetter, 11. Struve, I, 550, listed an edition printed in Prague in 1577 and the B.N. catalogue listed an edition without a date.)

Ad secundas insanas cucurbitulas Hannibalis Raymundi-Zani Itali Veronae sub monte Baldo nati Spongia Secunda. Prague, 1579. (Vetter, 11; Lalande, 109; Scheibel, 111.)

Tabule Dlouhosti Dne i Noci, Východu, Poledne i Západu k zpravování Orloje obojího, celého i polovičního, kterak ten srovnán býti má podle svých hodin, a to přes celý Rok, ku položení České Země a k vyvýšení Polum L graduov, Od D. Tadeáše Hájka z Hájku gruntovnie zpravená a nyní znovu vytištěná. Prague, 1574. (Vetter, 10. The translation of the title is: *Tables*

tory of geodesy and cartography. He began a measurement of Bohemia in order to make an accurate map of it, but was forced to stop because of the lack of funds. He was professor of mathematics at the university of Prague and opened his lectures there in 1554 with an address on the history of mathematics in Bohemia, which gave proof of his interest in the history of science.[6]

Hagecius first attended the university of Prague, but in 1548 or there-about he went to the university of Vienna, where he studied mathematics under Perlachius and gave private instruction in geometry. He also studied medicine at the university of Vienna, but in 1550 he was back in Prague and on July 14th of that year received the A.B. degree from the university there. On April 29, 1552, he obtained the degree of master of arts from the same university and in the autumn of that year he went to Bologna to study. From there he made a trip to Milan to visit Cardan. In October 1553 he was back in Prague as prefect of the *Collegium Carolinum*.[7] Hagecius lectured in the uni-

of the length of the day and the night, of the rise, noon and fall for adjusting both clocks, entire and half, how they are to be regulated according to their hours, and this for the whole year, for the situation of the Bohemian country and for the altitude of fifty degrees. By D. Thaddaeus Hajek ab Hayek thoroughly revised and now again printed.)

Výklady na proroctví turecké. 1560. (Vetter, 5. The translation of the title is: *Explications of the Turkish prophecy.*)

Vypsání s vyznamenáním jedné i druhé komety, kteréž vidíny byly března a dubna měsícuov létha tohoto MDLVI. (Vetter, 5, 9. The translation of the title is: *Narration with the description of the one and the other comet, which were seen in March and April of this year 1556.* 1556.)

According to Vetter, 3, 5, Hagecius published, in 1562, a translation of Mattioli's *Herbarium*, and, in 1566, Laurentius Gryll's *De sapore dulci et amaro*, and is said also to have published a book by Paul Alexander entitled *Rudimenta pro natalitiis.*

6 See note 5, above, and Vetter (1937), 243.

7 This college, named after Charles IV, was founded in 1366, although 1353 might be called the date of the actual opening of the university. As a professor of the faculty of arts, Hagecius was treasurer of the faculty in the years 1555 to 1557, and on November 20, 1555 was elected to the board of directors of the college. He confirmed the rector's accounts for the year 1555 to 1556.

versity as late as February 1557 but seems to have left soon after. He gave up his professorship to be married and to practice medicine. The place and date of his receiving his doctor's degree are not known, but he used the title of doctor for the first time in his translation of Mattioli's *Herbarium* in 1562. Previous to that date he styled himself " Master." His reputation as a physician was excellent, and he had a large practice, and also wrote several medical works. An iatromathematician, he practiced medicine in conjunction with astrology. Besides, he was intensely interested in chiromancy and metoposcopy, that is, in divination from the hand and in the art of judging a person's character or telling his fortune from his forehead or face.[8] He even wrote a work on metoposcopy which was published in Prague in 1562.[9] In the preface to the second edition of it, Hagecius complained that many pseudo-prophets were misusing human credulity; and he severely criticized those who added false prophecies to calendars. He said that with advancing age he was losing interest in all kinds of prophesying. This is an interesting observation on the part of a man who, although not entirely free from belief in strange influences, was a believer in modern exact scientific methods. On the one hand he was influenced by the atmosphere of credulity and on the other he was thinking for himself and growing less credulous. In his youth he wrote many astrological calendars, but he ceased to do so in 1555 because of other activities. However, he yielded to pressure and began again with one for 1558, because at that time there was no one else in Prague who was qualified to produce one. In the calendar for 1558 he replaced the discussion of " aspects " with one of " suitable times for agriculture." However, the " aspects " appeared in the calendars for 1561, 1564,

8 When in Milan, Hagecius asked Cardan to show him his work on metoposcopy but the request was refused. Cardan's work first appeared in 1658. According to Sotheran, catalogue 861, item 2425, a translation into French was printed in that year. According to Davis and Orioli, catalogue 91, item 119, the Latin edition of 1658 was printed in Paris. See Thorndike VI, 505-6.

9 See note 5 above. Rheticus suggested that Hagecius expand this pamphlet, but the latter did not have sufficient free time.

1567, 1568 and 1570. In his later books Hagecius showed less belief in " aspects " and considered " celestial signs " as general warnings of God to sinful mankind. Already in 1574, in his *Dialexis,* he blamed astrologers for frightening people with nonsensical stories.

Hagecius openly attacked opinions which he considered incorrect. In return, he himself was attacked. His leading opponents were Raymundus,[10] Graminaeus,[11] and Bishop William Lindanus. These opponents insulted Hagecius personally and attacked his religion. They were answered by Hagecius in several works, both printed and unprinted.[12] Tycho tried to dissuade Hagecius from answering Raymundus' assertions concerning the new star, but his pleas had no effect. Hagecius was also involved in a medical polemic against Philip Franchelius.

Hagecius was one of the observers of the comet of 1556, but his tract on that comet is not so important for the history of cometary theory as his tract[13] on the nova of 1572. The tract on the comet of 1556 was written in the Czech language, and therefore its circulation was limited. The first three chapters gave a description of comets based on the writings of ancient authors, and the fifth to seventh chapters discussed the astrological effects of comets. Only in the fourth chapter did Hagecius give his observations of the comet of 1556.[14]

Hagecius first communicated the results of his observations of the nova of 1572 in a letter to Bartholomew Reisacher, which was published as an appendix to Reisacher's work on the

10 See appendix, below.

11 See appendix, below, and Thorndike, VI, 72 note 8.

12 The printed works are cited above in note 5. A pamphlet, probably by Hagecius, entitled *Aquilo historicus,* was sent to his opponents in manuscript. (See Vetter, 11.)

13 See chapter II, above.

14 According to Vetter, 9, Hagecius mentioned the comet of 1557 at the end of this work. However, from the title given by Vetter, 5, it seems that the work was written in 1556. It is possible that an appendix was added and that the book was printed in 1557.

nova.[15] Hagecius' instruments were accurate only to 10′ or at best 5′. Nevertheless, his results were nearer to Tycho's than were those of any other observer of the phenomenon.[16] Hagecius treated the new star in some detail in 1574 in his *Dialexis*.[17] There he gave a number of observed distances between the nova and neighboring stars, some of the distances being as much as 7′ to 12′ in error. However, some years after the publication of his book, Hagecius sent Tycho a copy with manuscript additions and corrections, in which the most erroneous measurements had been struck out; and this emendation Tycho quoted extensively in the *Progymnasmata*.[18] The most important fact about Hagecius' book on the new star is that in it he showed that he was ready to admit the possibility of something new in the heavens. Furthermore, he had observed the transit of the nova over the meridian and had noted its altitude and the time. Credit for being the first to find the positions of heavenly bodies by observing azimuths [19] and altitudes and time, or meridian altitudes and time, has generally been accorded the Landgrave of Hesse Cassel, who thus observed the nova. However, he did not publish the results of his observations and they are now

15 The letter has the title " De Investigatione loci novae stellae in zodiaco secundum longitudinem ex unica ipsius meridiana altitudine & observationis tempore, geometrica deductio ". See chapter II, above, and Vetter, 8. Reisacher's work appeared in Vienna in 1573 (see Reisacher), and the appendix to it is undoubtedly the work to which Poggendorff, I, 991, and Scheibel, 70, referred.

16 Vetter, 8.

17 Hagecius (1574). See chapter II, above.

18 Dreyer (1890), 58 and 58 note 2. The *Dialexis de novae ... stellae ... apparitione* was dedicated to Maximilian II and contains Fabricius' and Gemma's works on the new star, Regiomontanus' *De cometa anni 1475* [sic], and Vögelin's tract on the comet of 1532. It also contains a letter from Johannes Crato of Crafftheim, an excerpt of a letter from Munosius to Reisacher, and Hagecius' own *Historia Stellae Novae, Apparentis In Asterismo Cassiopeae*, which is in fifteen chapters and has an appendix opposing Raymundus and Graminaeus.

19 See note 130 in chapter III, above.

chiefly known through his correspondence with Tycho.[20] Tycho probably knew William's method when he visited him at Cassel in 1575, but it is unlikely that Hagecius knew it in 1572. Nevertheless, Hagecius, by observing only at the time of meridian transit, employed a method similar to and simpler than the Landgrave's. Tycho did not approve of the method because it necessitated reliance on clocks, and nobody at the time had erected an instrument permanently in the meridian.

Although Hagecius refused to accept offers of positions in foreign countries, he traveled a great deal and was friendly with many scholars of other nationalities. He and Tycho met in Ratisbon in 1575, at which time Hagecius gave Tycho a copy of a letter which he had received from Munosius and a manuscript copy of Copernicus' *Commentariolus*.[21] Hagecius and

20 See appendix, below, item 19.

21 Dreyer (1890), 82-3; Vetter, 9-10. Brahe, II, 428, spoke of receiving the *Commentariolus* from Hagecius at Ratisbon. The title given to the work by Hagecius is *Nicolai Copernici de hypothesibus motuum coelestium a se constitutis Commentariolus*. Hagecius was an advocate of the Copernican hypothesis and seems to have preserved a letter from Copernicus to Wapowski by presenting a copy of it to Tycho with the *Commentariolus*. Vetter says that the two works by Copernicus were presented to Tycho by Hagecius in Rome in 1575. However, he says elsewhere (page 4), that Hagecius went to Ratisbon with the imperial court in 1575 for the coronation of Rudolph, and there met Tycho. Such was the case. The Wapowski letter almost certainly was the one opposing Johannes Werner, the author of a treatise, *De Motu octavae sphaerae*. That letter was discussed by Günther (1880) and by Prowe, I, pt. 2, 221-230. It was edited in 1878 by Curtze (1878), 18-33, who collated two different manuscript copies of it; in 1884 by Prowe, II, 145-153; and in 1939, in an English translation, by Rosen. It is certain that Tycho had a copy of the letter. (See Curtze (1878), 21; Brahe, IV, 292; and Rosen, 8-9, note 15.) There was a third manuscript copy which seems to have been burned with the Strasburg library in 1870. (See Curtze (1878), 19; Dreyer, editor, IV, 507; and Prowe, I, pt. 2, 223 note, 285 note.) According to Prowe, I, pt. 2, 285 note, an 1839 copy of the Strasburg copy is preserved in the Polish library in Paris. That Paris copy has the following closing words copied from the Strasburg manuscript: "Descripta Pragae ex D. Hayetii exemplari mense Januario MDXXXI". Dreyer, editor, IV, 507, described the Strasburg manuscript as "exemplum Pragae transscriptum a. 1531 'ex D. Hagetii exemplari'". According to Prowe, I, pt. 2, 285 note, the text of the Paris copy, except for the spelling of "Coppernicus", was exactly like that of the Vienna

Tycho corresponded for many years, and when the latter wanted to move to Prague, the former interceded with the emperor for him. Another of Hagecius' correspondents was Andreas Dudith,[22] and another foreign friend was Martin Mylius.[23]

Hagecius supported the movement for calendar reform. In 1597 he was commissioned by Emperor Rudolph to examine all calendars published in Prague and to report on them, but the granting of permission for printing them remained in the hands of the archbishop, and bad calendars were issued as before.[24] Hagecius did not confine his activities to scholarly work. In 1566, as military surgeon, he took part in the expedition to Hungary against the Turks. He was present at the siege of Raab, where he wrote a note to his calendar-predictions for 1567.[25] This military expedition was but another chance for him to travel.

Hagecius was a true scientist in the sense that he was willing to change his mind when he was shown to be in error. His tract on the comet of 1577 [26] gave that body a parallax of from 5 to 6 degrees, which would place it below the moon, and Tycho dealt [27] with Hagecius first on his list of observers who believed

copy, which, according to Curtze (1878), 21, was made in 1575. Prowe believed that the Strasburg copy was made in 1531, seven years after the letter was written, from Hájek's copy. In that event, the copy must have belonged to Simon Hájek, not Thaddaeus Hagecius ab Hayck. However, it seems possible that the Prague copy was made in 1531 and that it later came into the possession of Hagecius, and that after that the Strasburg copy was made. Since the Vienna copy is like the Strasburg copy, it seems significant that the former was made in 1575, the year Hagecius supposedly brought a copy to Tycho. Tycho asked Hagecius to send him a copy of Werner's work on the eighth sphere, and Hagecius eventually did so. (See Rosen, 7-8, note 14.)

22 See appendix, below. Dudith was an opponent of astrology.

23 February 21, 1582, Hagecius was in Görlitz and was a guest for dinner at the home of Martin Mylius. See Jancke (1861a), 272.

24 Vetter, 10.

25 Vetter, 2, 4.

26 Item 48 of appendix, below.

27 Brahe, IV, 261-296.

the comet to be sublunar. Tycho showed that Hagecius had erred in his calculations and that his observations had really shown that the comet was supra-lunar. But before the publication of Tycho's book, Hagecius had published a second work on the comet of 1577 [28] in which he examined the works of Maestlin and Roeslin on the comet. Then, when dealing with the comet of 1580,[29] Hagecius rectified his errors with regard to the comet of 1577 and recognized it as supra-lunar.

Hagecius was deeply religious, feeling the power and majesty of God in celestial phenomena. He was an Utraquist,[30] that is, a member of a sect of Hussites in Bohemia, but his tolerant spirit was shown by his offer to the Moravian Brethren to write a defense of the Brethren against an attack which had been made against them. Although he wrote the defense, the Brethren, unwilling to stir up further trouble, did not use it.

Hagecius' first work on the comet of 1577 [31] begins with a poem in Latin to the pious Christian reader by Procopius

28 Item 49 of appendix, below.

29 Riccioli, II, 13, 40; Brahe, IV, 293, and Dreyer, editor, IV, 507. Delambre (1821), I, 229, said 1582, but there does not seem to be a record of any work by Hagecius on that comet. Hagecius' letter to Mylius (item 49 of appendix), printed in 1580, is in favor of astrological implications of comets and does not recognize comets as supra-lunar. The change in Hagecius' opinions seems to have been first expressed in his work on the comet of 1580 which was printed in Görlitz in 1581, namely the *Apodixis physica et mathematica de cometis*.... See Koch (1907-1910), v. 86: 54 note 1.

30 He seems not to have been Jewish, in spite of suggestions to that effect. Wolf, J. C., I, 1157, gave a name in Hebrew followed by its Latin translation and further information: "R. Taddai Chaggai Abhaget, Medicus Hebraeus, latinum de Metoposcopia tractatum scripsit an. 1561. quem MS. Bartoloccius vidit." Thus Wolf called the man he was discussing "Rabbi". Wolf was evidently referring to Thaddaeus Hagecius ab Hayck, although the latter's *Metoposcopia* first appeared in 1562. Jöcher, II, 1315, discussing Hagecius, cited Wolf. Vetter, 1, calls Thaddaeus Hagecius' father "patrician" and his mother "noble", and makes no mention of any Jewish blood in Hagecius' veins. Prague in the sixteenth century had a large Jewish community and there is the possibility that Hagecius was a converted Jew. However, in the light of Professor Vetter's article, it seems more likely that Wolf erred and confused Thaddaeus Hagecius ab Hayck with another man.

31 Item 48 of appendix.

Lupacius.[32] This is followed by the dedication to Elector Augustus, Duke of Saxony, Landgrave of Thuringia. . . ., signed by Hagecius and dated from Prague on February 24, 1578.

The narrative about the comet of 1577 begins [33] by saying that it was seen in the west at six in the evening of November 10, 1577. That night strong winds drove away the clouds which had long made observation impossible. The comet was a little above the Tropic of Capricorn in the sign of the winter solstice.[34] On the following day bad weather prevented observation, but on the 12th and 13th of the month the sky was clear and quiet. Then, especially on the 13th, many, including Hagecius,[35] viewed the comet with consternation and trepidation. It stood to the right of and higher than the moon, above the head of Sagittarius in the Milky Way, in what Hagecius called the usual and most suitable place for the generation of comets. Saturn was at that time west of the moon and below the comet's tail in the tenth degree of Capricorn,[36] where it had been in conjunction with the moon on the previous day, but having been dulled by the moon's light and being nearer the horizon, it could not be seen. That day the comet's body appeared equal to Jupiter or Venus in size. Its light was clear and pure. Its outstretched tail, alone, gave the impression of sad foreboding. In the following days, diminished in body, it appeared sadder and paler. Its motion was irregular, like Mercury's. A mane or beard, like a disordered cloud of yellowish color, was denser, thicker, and more contracted than the body of the comet, but became less dense further from the body, spreading out in little branches, like a broom, as though twigs were bunched together,

32 Lupacius was born in Prague and lived there at the time of Hagecius. He wrote, among other works, a Bohemian historical calendar which was printed in Nuremberg in 1578 and in Prague in 1584. See Zedler, XVIII, 1198, and Jöcher, II, 2605.

33 Chapter I begins on A_3 r (page 1).

34 Sagittarius.

35 Hagecius was not at home on November 12th and 13th.

36 "in X. parte Capricorni".

and ending in smoke. The tail was curved like a horn or a short Turkish or Persian sword, the back being denser, the edge thinner. Because of the tail's shape, Hagecius called the comet "Xiphias," which he said historians had described before Xerxes crossed over into Greece.[37] Others, according to Hagecius, compared the tail to a peacock's tail or to a broom of twigs. He added that God is wont to vary the forms of comets according to His own free will. Hagecius, following Aristotle's two divisions, called this comet "Barbata" or "Pogonia," because the flame projected in one direction only, pointing to the southeastern portion of the earth. The flame was so great in longitude as to seem nearly to reach the stars in the horns of Capricorn, which were higher than the end of the tail, and the tail pointed to the forehead of Capricorn. It covered more than 25° of a great circle.

On the 14th the comet appeared slightly further off; but before satisfactory observations could be made a cloud suddenly obscured it from view. On the following day the sky was overcast with dark clouds, but on the 16th there was good weather. Then Hagecius began observations with an astronomical radius, determining the distances from the comet to the neighboring stars, and finding the comet 17° 52′ [38] from Aquila [39] and 13° 13′ from the constellation Antinous.[40] Thus the comet was seen

37 Xerxes set out from Sardis in the spring of 480 B. C.

38 "17. partib. 52. scrup." A "scrupulum" has been translated throughout the summary of this book as a "minute". See chapter III, note 64, above, and chapter IV, note 116, below.

39 It is difficult to determine from what point in the constellation Aquila Hagecius took his measurements, and it seems strange that he never specified the point of origin. Since I saw his work in the summer of 1931 and no copy is now available, it is possible that he made some pertinent statement which I did not record in my notes. On November 24th, Hagecius found the comet on a straight line between Aquila and the stars in the horns of Capricorn, 17° from the ecliptic and 11° 43′ from Aquila. Hagecius' map, on B_4 r (see Hellman, pl. 8, b) shows a position for the comet, with the same relative distances if the measurement is made from the bright star in Aquila. The line through the comet and the stars in the horns of Capricorn also passes through that same bright star.

40 Or "from the second star from Aquila". The Latin reads: "A secunda stella ex informibus eius Aquilae, quę Antinoi illius pueri, ab

at 6 P. M. that day about 18° in Capricorn and 12° north of the ecliptic. From the 17th of November through the 21st the sky was cloudy; but on the 22nd it was calm and Hagecius observed that the comet was 10° 48′ from Aquila. On the 23rd the sky was again cloudy. On the 24th the comet was 11° 43′ from Aquila and 6° 28′ from the fourth star from Aquila. It appeared in a straight line with Aquila [41] and the two stars in the horns of Capricorn, being 17° from the ecliptic and about in the 26th degree of Capricorn. That day the tail stretched toward the second star from Aquila so as to pass toward the north. In the eight days from the 16th to the 24th the comet, with an eastward motion of its own, had moved about 8° in longitude and 5° in latitude, with regard to the zodiac. But Hagecius pointed out that if its motion were considered in its own circle, the comet would be seen to have traversed nearly 9°. Nor had it left its course although in the two previous days it moved only 55′. Allowing for the comet's daily motion, Hagecius calculated its position at the time of new moon, November 9th, when he thought it to have arisen. He reasoned that it was in the ninth degree of Capricorn, and nearly 9° from the ecliptic, that is, within the Milky Way. He believed that it was 43° east [42] of the sun and in the western part of the sky near the border of the eighth house. Saturn was then in the tenth degree of Capricorn and nearly 1° north of the ecliptic. Hagecius said that when Saturn was in the ninth degree, or according to Ptolemy's teaching in the tenth, it might be considered in conjunction with the comet. Hagecius admitted that some people said that they saw the comet several days before in Hungary, for in lower latitudes it could be seen better and for a longer time above the horizon.

Hadriani Imperatore in delicijs habiti, esse quidam fabulantur, 13. partib. & totidem scrupulis primis."

41 See note 39, above.

42 "In distantia verò à Sole 43. grad. secundum successionem signorum, aut 317. grad. contra ordinem signorum."

The night of the 25th was cloudy. On the 26th the comet was 13° 35′ from Aquila [43] and nearly 17° from a star in the mouth of Pegasus, being in approximately the eighth degree of Aquarius, with a latitude of 23°. Consequently, in those days it had passed, with a somewhat violent jump, through 12° in longitude and 6° in latitude. At no other time had Hagecius found such rapid motion and it aroused his wonder. On the 27th the comet was 15° 6′ from Aquila and 14° 29′ from the star in the mouth of Pegasus, or in nearly the thirteenth degree of Aquarius, with a latitude of 24°. Thus its rate of motion had decreased. For the remainder of the month the sky was overcast.

On December 1st the comet was observed 9° 28′ from the star in the mouth of Pegasus, between two dim stars in the head of Equus minor. It appeared on a straight line with the small stars in the head and in the right upper arm of Aquarius. It was a trifle below the straight line through Aquila and the star in the open mouth of Pegasus. The comet had a latitude of 25° and was nearly at the sixteenth degree of Aquarius. In four days its apparent motion with reference to the zodiac was only 3° in longitude, or 5° in its own circle. Its tail pointed at and almost touched the star in the mouth of Pegasus, somehow turning toward Ursa Minor and sending forth its vapor far beyond that constellation. On December 2nd the comet was 8° 32′ from the star in the mouth of Pegasus, in the constellation Equus minor, in the nineteenth degree of Aquarius, with a latitude of 25°. It had advanced as much in the zodiac in one day as in the four preceding ones. On December 3rd Hagecius found that the comet was 7° 36′ from the star in the mouth of Pegasus and 17° from the right upper arm of Aquarius.[44] It was in the twenty-first degree of Aquarius, but

43 See note 39 in this chapter.

44 Hagecius' map shows a star just below the right shoulder. The phrase would have no meaning without this picture in mind. For example, in Bouché-Leclercq's picture of Aquarius, the figure faces out, so that directions are reversed, and the same star is in the left arm.

its latitude was still 25°. On December 4th the sky was cloudy. On the 5th the comet was 5° 43′ away from the star in the mouth of Pegasus and 16° 5′ from the right upper arm of Aquarius, having moved 1° 53′ nearer to the star in the jaws of Pegasus, an average daily motion of 56′. The comet had a latitude of 26° and was in the twenty-third degree of Aquarius. The tail, with its peculiar curve and its back turned north, was stretched eastward toward the breast of Pegasus.

The next clear weather was on December 11th, when the comet was 4° 40′ from the star in the mouth of Pegasus, had a latitude of 27°, and was in about the twenty-eighth degree of Aquarius. The tail stretched toward the star in the right upper arm of Pegasus [45] and in the direction of the prolongation of the leg of Pegasus.[46] The size of the comet's body seemed to have diminished considerably during the five previous days. After another cloudy day the comet was visible 5° 20′ from the star in the mouth of Pegasus and 19° 22′ from the star in the right upper arm of the same constellation.[47] In the two days it had moved 1° 36′ closer to that arm. Therefore, its daily motion carried it only 48′, and it entered the sign of Pisces with its latitude unchanged. On the 14th the comet was 5° 43′ from the star in the mouth of Pegasus, just as on December 5th, and 18° 47′ from the right shoulder of that constellation,[48] its daily motion having been 35′. It was quite thin and dark in body and had a languid and very sluggish motion, as if it were already being extinguished. The comet was seen on the 22nd and 28th of December, but the first clear weather was on the 31st, when the comet was nearly 14° from the star in the mouth of Pegasus, had a latitude of 27°, and was about 10° in Pisces. Its diminished tail stretched toward the two little stars near it in

45 I do not have the original Latin for this passage, but presumably, judging from Hagecius' map, the bright star pictured in the shoulder was meant. See note 48, below.

46 Here again, without the original Latin, it is difficult to interpret Hagecius' meaning. Pegasus' right front leg was probably intended.

47 See notes 45, above, and 48, below.

48 "Ab humero autem dextro Pegasi 18. grad. 47. scrup."

the breast of Pegasus. The comet, as if wearied, had scarcely crept forward and in seventeen full days had traveled barely 8° in longitude, and in twenty days had not changed its latitude.

On January 1, 1578, bad weather did not permit careful observation of the comet, but it appeared to be in the same place as on the previous day. Observation was impossible on the following day, but on the 3rd the comet was sufficiently visible to be found 18° 34′ from the star in the left wing of Cygnus and 15° 7′ from that in the mouth of Pegasus, approximately in the twelfth degree of Pisces, and 29° from the ecliptic. It had traversed almost 2° in three days, and was a bit below the straight line between the star in the left knee of Pegasus and the bright one in the shoulder. Its small, short, shadowy and already disappearing tail was still directed toward the two stars in the chest of Pegasus. Not until January 7th did the weather again permit observation, and for several days thereafter the comet could still be distinguished. At last, on the 12th or 13th, it disappeared entirely. It left to the accompaniment of wind storms equal to if not more violent than those which marked its coming. In the last thirty days it had traveled through barely half of Pisces, but throughout the period of its duration, certainly from November 9th to January 13th, its own irregular motion had carried it over a slanting arc embracing about 66° of a great circle, the pole of which was between the first and third stars of the left hand of Bootes. Its tail was uninterruptedly turned away from the sun. The cumbersome detail supplied by Hagecius and the awkward terminology he employed, now easily supplanted by reference to fixed coordinates, enabled the average sixteenth century reader to follow the path of the comet through familiar constellations and to realize the apparent change in its rate of motion.

Next,[49] Hagecius dealt with the astrological causes of the comet. He said that the time, place, and size of the comet of 1577 could not have been predicted, but attributed the uncertainty to the fact that both pagan and Christian philosophers

49 Chapter II.

and astronomers had preferred to consider a comet divine, not natural. Its purpose is to announce God's wrath against man, and it disappears when it has discharged this function. Hagecius believed that this was the opinion of John of Damascus, and that before him Josephus, the writer of Jewish history, thought the same, when he said that a comet results from secret causes of nature. In his own age, Hagecius found learned and pious men, such as Jacobus Zieglerus [50] and Johannes Vögelin,[51] who held those same opinions. He added that whatever could be established concerning the nature of a comet had been explained by Gemma.[52] Hagecius concurred with Gemma and suggested that he also write an explanation of the comet of 1577 and that he hasten the publication of his work on meteorography. Hagecius said that he presented his own works with pleasure for the consideration and censorship of Gemma, Tycho, Johannes Praetorius, Jerome Munosius and others. Hagecius was indifferent to the causes of the comet expounded by others because, unaware of the possibilities of scientific investigation, he never ceased to believe and proclaim that the generation of all comets is hidden, and that it is not merely difficult but impossible to anticipate their appearance. He believed that the point where they appear, the question whether they are of celestial or elementary nature, and their distance from the earth can be determined by dealing with them as though they are natural phenomena and applying physical and astronomical doctrine. He believed that still further information, which had been concealed from human knowledge for centuries, could be obtained because of the grace of God, who had reserved this knowledge and finally revealed it through Regiomontanus and explained it through others.

Dealing with the distance of the comet from the center of the earth,[53] Hagecius said that he had used a celestial globe in

50 See chapter II, above.

51 See chapter II, above.

52 Cornelius Gemma. For example, see Hagecius (1574).

53 Chapter III.

determining the observed positions of the comet in the zodiac, recorded in the first chapter, for the separate days when the comet was visible. He said that he was too busy to perform the tremendous task of computing the positions by trigonometry, although students of mathematics, using distances he had noted, might practice this computation. He emphasized the importance of investigating parallax in the circle of altitude, saying that whatever can and should be known concerning comets and their passions, all those things consist in the knowledge of one parallax, and that others [who do not have this knowledge] should not claim that they have special knowledge concerning comets. To determine the parallax, applying Regiomontanus' theory, Hagecius made two observations of the comet on November 26th. At exactly 5 P.M., as determined by means of the altitude of Aquila,[54] Hagecius found that the altitude of the comet was 39° 30′ and that the azimuth [55] measured westward from the meridian was 31°. Eighteen minutes later, using a quadrant of moderate size, he found that the altitude was 38° 10′ and the azimuth 36°. Thereupon he concluded that the comet had some parallax. After deeper consideration he inferred the same from the meridian altitudes of the comet and of Aquila, and noted the slightly unlike distances of the comet. For on December 2nd, making two observations three hours apart, and on the 11th, allowing an interval of four hours between observations, he discovered that the distance of the comet from the star in the mouth of Pegasus was less than 4′ or 5′. He admitted that on the 14th of December he had not perceived such a difference but added that an error of several minutes could easily creep in. He believed that he had made definite progress in determining the parallax, which he detected as 5° and some minutes, although he discarded the minutes. From this value and the distance of the comet from the vertex, after consulting the table of parallax which he had prepared for his book about the new star,

54 It is difficult to understand how Hagecius could be sure that the time was exact unless by "Aquila" he meant a particular star in the constellation of that name. See note 39, above, in this chapter.

55 See note 130 in chapter III, above.

he concluded that the comet was a little more than eight semidiameters of the earth from its center,[56] and, what is more, lower than the moon. Evidently aiming at brevity, Hagecius suggested that those who had no copy of his *Dialexis* should consult problem ten of Regiomontanus' *De cometa.*

Hagecius' next concern was for the "true" position of the comet in the zodiac.[57] He believed it necessary here to set forth more fully the observed and true positions of the comet, and to show how they differ mutually. Therefore, he recalculated his observations for November 26th. Applying the theory of Regiomontanus' seventh and eighth problems, he located the comet's observed position in the seventh degree, thirty-seventh minute of the twelfth part of Aquarius with a latitude of 22° 49′, and its "true" place in the tenth degree, forty-fifth minute of Aquarius with a latitude of 28° 29′. The observed and actual positions for the other days he left for investigation by students of mathematics.

Finally[58], Hagecius attempted a pious and christian interpretation of the comet's signifiance. He felt certain that the nova had not yet completed its significations. In his opinion all comets indicated war, and the people and places involved could be discovered by consulting Ptolemy's doctrine. From the comet's position and the extension of its tail, he concluded that the Spaniards were about to attack the Moors, the Italians, the French, and the Belgians.[59] He urged the placing of faith in God and pointed out that the star which heralded the birth of

56 "Cum qua parallaxi & distantia Cometae à vertice ingressus tabellam parallaxeos, quam in nostra Dialexi de noua stella confecimus, statim colliges Cometen paulo plus octo semidiametris terrae à centro eius elongatum fuisse,..." The table of parallaxes is on pages 77-8 of Hagecius (1574) and gives the distance from the vertex (90° - altitude), the parallax, and the distance from the earth measured in semidiameters of the earth. Using the distance from the vertex amounts to adjusting for the horizontal parallax, which in this case must have been a little more than 7°.

57 Chapter IV.

58 Chapter V.

59 "Gallos & Belgas".

Christ brought the greatest joy to all pious men and was for them a sign of greatest good, but that for the evil and impious, such as Herod, the Jews, and like folk, it was a sign of greatest sadness and ill fortune. He held that certain phenomena could forecast good as well as evil, for he believed that pious men might witness wars and changes without being swallowed by the storms, because they commend themselves to God. Hagecius ended his book with a prayer and quoted in German a prophecy of change.

Hagecius' second book on the comet of 1577 [60] takes the form of a letter to Martin Mylius. It differs from his first book by being not so much a record of his own observations and an explanation of his theories as a discussion of the works on the same comet by Maestlin, Roeslin, and others.

Hagecius was acquainted with works on the star of 1572 by Maestlin,[61] Tycho, Thomas Digges, and John Dee, and considered Maestlin's work on the comet of 1577 no less scholarly than his earlier treatise. Hagecius admired Maestlin's ingenuity and remarked on his use of a thread held so as to connect two stars. Nevertheless, rather than follow that method of observation, he himself preferred to rely on well-constructed instruments.

Hagecius examined Maestlin's observations and calculations for parallax in detail, but felt that they were not conclusive. According to Hagecius,[62] Maestlin assumed a straight line between a star in the beak of Cygnus and another in the mouth of Pegasus and then made two observations of the comet's dis-

60 Item 49 of appendix, below.

61 See chapter II, above, especially note 266.

62 Item 49 of appendix, B_1 r: "Statuit Moestlinus stellam in rostro gallinae, & alteram in rictu Pegasi in una recta linea: & ex distantia Cometae à rictu Pegasi, in duabus obseruationibus colligit locum apparentem Cometae: quem cum propemodum eundem inueniat in utroque illo tempore: habita ratione motus proprij Cometae: hinc colligit, Cometam nullam habuisse parallaxim. In qua obseruatione Moestlini multa sunt suspecta." Presumably Maestlin measured the comet's distance from the assumed straight line.

tance from the mouth of Pegasus. Thence he reckoned the comet's apparent position, which was almost the same in both observations. He calculated the body's proper motion, and concluded that the comet had no parallax. Hagecius considered Maestlin's data uncertain and not pertinent to the investigation of parallax and his hypothesis unfounded and said that he knew no one who supported Maestlin. He himself was not then ready to defend Maestlin against Nolthius,[63] although he believed the latter's observations inaccurate. According to Hagecius, Cornelius Gemma had placed the comet of 1577 in the orbit of Mercury, however, without producing a valid or well-founded reason for this. Because Gemma wished to hear his opinion of the parallax, Hagecius had explained it to him in accurately written letters, and held no doubt that Gemma, who was honorable and truth-loving, would have changed his own opinion, had not destiny snatched him away.[64] Consequently, argued Hagecius, the opinion of Maestlin could not be established by the authority of Gemma. Hagecius recounted Maestlin's observations for November 24th and December 15th and said that for November 9th and 10th Scultetus gave a daily motion of 1° 47′ and Maestlin 5° 15′, and that they differed by nearly 12° in latitude. Hagecius also mentioned Maestlin's theory of the motion of Venus and the fact that the comet did not move like Venus. Several times Hagecius referred to "Vuitichio,"[65] whose observations, as well as those by Scultetus and Nolthius, tended to agree with his own. Hagecius summed up by saying that from a certain single observation of one comet only, whose complete revolution could not be observed, he had not been able to establish an hypothesis to cover the irregularity of its motion, let alone find one to accommodate all comets, in the manner of Ptolemy and Copernicus.

63 See the section on Nolthius, below, in this chapter.

64 Gemma died in 1579. See chapter III, note 144, above.

65 "Vuitichio" was probably Paul Wittich who studied with Tycho in 1580 and later observed for the Landgrave. See A.D.B., XLIII, 637; Dreyer (1890), 119-121.

Hagecius believed it unnecessary to say anything about Nicholas Winckler,[66] his " plagiarist ", so he proceeded to the consideration of Roeslin's book.[67] According to Hagecius, Roeslin sent this book to him and to Paul Fabricius and asked that each be willing to read it and interpret it in a friendly manner. In that book Roeslin made honorable mention of Hagecius more than once. For such a kind spirit toward him, Hagecius rendered thanks and complimented Roeslin. He said that he would give his sincere opinion of the work of Roeslin, whose zeal and skill he admired, but with whom he was not in complete agreement. Hagecius understood Roeslin's statements concerning motion but he did not see how they could be fitted to this or other comets. He applied the same criticism here that he offered against Maeslin's theory: he could not establish the mean and regular motion of phenomena whose entire revolution could not be observed. Therefore, the motion according to longitude, latitude, and declination which Roeslin attributed to the comet should be said to be invented rather than derived from nature, because doubtless it could serve his preconceived opinion. Furthermore, the positions of comets which are deduced from this motion are not exact nor do they respond to other observations, but are established only in a general way from the celestial globe, as was discovered by Cornelius Gemma himself, in whose footsteps Roeslin followed. Hagecius concluded that the comet's rate or " proportional motion," as Roeslin termed it, could not be determined from such uncertain positions.[68] Nor did Hagecius understand what Roeslin said farther on, concerning spheres, circles, poles, and axis of the world:— how through them the positions and motion of all comets can be saved. Hagecius thought that Roeslin erred in following Gemma and putting the comet in the orbit of Mercury, and re-

66 See appendix, item 109.

67 Item 93 of appendix, below.

68 Hagecius here referred to chapter II of Roeslin's treatise, and since Roeslin appears confused there, it seems reasonable that Hagecius should have been unable to follow him.

peated his opinion that Gemma would have changed his mind had he lived sufficiently long. According to Hagecius, Fabricius called Roeslin's book a huge and unformed monstrosity, but Hagecius' opinion of it was more moderate, for he thought that what Roeslin said concerning the comet's meaning was pious, intelligent, and useful. Thus Hagecius introduced the subjects of comets as portents and astrological prediction in general, which form the last part of his book. As stated on the title-page, Hagecius aimed to combat those who contended that comets signify nothing. He stressed their divine nature. To those now interested in the development of the theory of comets, this lengthy last part of Hagecius' book seems an unnecessary digression, but to Hagecius and his contemporaries it fulfilled part of an astronomer's task.

Hagecius' book was dated from Prague, and was probably not completed before 1580. Appended to it is a letter from Martin Mylius to Hagecius, dated from Görlitz on the day of the autumnal equinox in 1578. Mylius seems to have been chiefly interested in the astrological implications of comets. He specifically asked for Hagecius' opinion concerning the works of Maestlin and Roeslin, and Hagecius' book is the answer to this letter.

Among the astronomers whose work was known to Hagecius was Bartholemaeus Scultetus,[69] who was listed by Tycho[70]

69 A.D.B., XXXIII, 497-8, article by Günther.—Annalen.—Bassaeus, II, 252, 278, 345, 446.—Baumgärtel.—Brahe, IV, 296-337, VII, 61-3.—Delambre (1821), I, 229.—Dreyer, editor, VIII, 463.—Dreyer (1890), 16, 20, 131, 132-3, 171, 182, 288, 329-331, 369.—Gesner (1583), 106.—Gräve.—Hantsch. (This reproduces two maps by Scultetus and gives a sketch of his life.)—Hellmann (1883), 463, 696.—Hellmann (1891), 36-7.—Hellmann (1924), 29, 41-2.—Hoefer, XLIII, 599-600.—Jancke (1861a).—Jancke (1861b).—Jancke (1868).—Jancke (1870).—Janssen, V, 95, 345.—Jöcher, IV, 451.—Kästner, II, 382, 409.—Koch (1907-1910). (There is a picture of Scultetus opposite LXXXIII, 1.) — Koch (1916). — Kroker. — Michaud, XXXVIII, 601-2.—Poggendorff, II, 883.—Riccioli, I, xxxi, II, 13, 25, 28, 89.—Ruge.—Scheibel, 62-3, 107. — Scultetica (1837). — Scultetica (1842). — Scultetica (1915).—Vossius, 382-3 (chapter LXV, § 29).—Weidler, 377, 385, 396-7.—Witte, obiit... 1614.—Zedler, XXXVI, 765.

Schottenloher, II, 263, listed the following unobtainable work:
Epithalamia in honorem nuptiarum Bartholomaei Sculteti, mathematicarum

among the upholders of the sublunar position of the comet of 1577. He was born in Görlitz on May 14, 1540 and died there on June 21, 1614. His name was originally Schulz or Scholz, and his father's name was Martin.[71] He began his studies under his brother Zachary at Görlitz [72] but later continued them at Leipzig [73] under the mathematician Johannes Homelius.[74] He also attended the juridical lectures of Ambrosius Lobwasser,[75] but remained faithful to mathematics.

A little more than three months before Homelius' death on July 5th, 1562, Tycho Brahe arrived in Leipzig and came under the influence of both Homelius and Scultetus, and he has been called one of Scultetus' pupils. When Tycho did some astronom-

artium studiosissimi, ac Helenae Joannis Roberi civis Gorlicensis filiae. Görlitz, Ambrosius Fritsch, 1573.

Further information concerning Scultetus might be found in:

Grosser, Samuel. *Lausitzische merckwürdigkeiten, darinnen von beyden marggrafthümern in fünff unterschiedenen theilen von den wichtigsten geschichten, religions- und kirchen-begebenheiten,...* Leipzig, etc., D.Richter, 1714.

Hoffmann, Joh. Jac. *Lexicon universale historiam, chronologiam . . . explicans.* Leyden, 1698. 4 v. (This seems to contain an account by Martin Mylius. See Koch (1907-1910), LXXXIII, 57 note 1.)

Otto, Gottlieb Friedrich. *Lexicon der seit dem funfzehenden Jahrhunderte verstorbenen jetztlebenden Oberlausizischen Schriftsteller and Künstler...* Görlitz, 1800-1803, 3 v.

70 Brahe, IV, 296-337.

71 Koch (1907-1910), LXXXIII, 57; Jancke (1868), 267. It has been said that the father's name was Abraham, but Gräve, 489, disagreed with this. Kästner, I, 580, II, 382, said that Bartholemaeus was from the same family as Abraham and David Scultetus. However, Scultetus or Schulz was not an uncommon name. Koch (1907-1910), LXXXIII, 57, said that Bartholemaeus had a brother Abraham born in 1537.

72 See Weidler, Kästner, Zedler, and Gräve, 457. According to Gräve, 489-490, Zachary made a sun-dial over the apothecary's shop in Görlitz, which has been confused with Bartholemaeus' dial on the church. Koch (1907-1910), LXXXIII, 57, said that Bartholemaeus Scultetus once studied under Zachary Zimmermann, and there may be some confusion of names.

73 According to Koch (1907-1910), LXXXIII, 58, Scultetus changed from Wittenberg to Leipzig in 1559.

74 See chapter III, note 118, above.

75 See A.D.B., XIX, 56-8.

ical observing in 1564 he used a cross-staff which Scultetus subdivided for him by means of transversals, which were coming into use. Once [76] Tycho said that he learned the use of transversals (on a straight line, not an arc) from Homelius; at another time, that Scultetus had taught it to him. It seems more likely that Scultetus had learned it from Homelius and then instructed Tycho, but undoubtedly Homelius himself was not the inventor of the scheme.[77] Moreover, Gräve [78] said that Tycho attended mathematical lectures by Scultetus, and in that sense he was his pupil. The friendship between Scultetus and Tycho lasted throughout the latter's life, Scultetus being one of Tycho's principal foreign correspondents. Five of Tycho's letters to Scultetus are preserved.[79] One of these, dated October 12, 1581,[80] deals chiefly with the comet of 1577. In it Tycho, though polite and cordial, rightly disagreed with the parallax found by Scultetus for that body. In his own book on the comet, also, he discussed the parallax found by his friend.[81] According to Riccioli, Scultetus believed that some comets were above the moon and others below it.

From Leipzig Scultetus went to Wittenberg, where in 1564 [82]

76 Dreyer (1890), 20, 329-331; Kästner, II, 355.

77 See Koch (1907-1910), LXXXIII, 59 and 59 note 3.

78 Gräve, 458.

79 Dreyer (1890), 132-3, 288, 369; Kästner II, 409-411; Koch (1907-1910), LXXXVI, 53-8; Thorndike, VI, 183, n. 14.

80 Brahe, VII, 61-3. Scultetus had sent a copy of his work on the comet of 1577 to Tycho. After Tycho's criticism Scultetus, although he noted the comets of 1580 and 1582, wrote nothing further on the subject. See Koch (1907-1910), LXXXIV, 72-3, LXXXVI, 53, and Koch (1916), 29-30.

81 Brahe, IV, 306 ff.

82 This date does not agree very well with his helping Tycho in Leipzig in 1564. However, according to Gräve, 458, Scultetus went to Wittenberg and other schools and then returned to Leipzig. Gräve's statement fits well with Hoefer's, that Scultetus first went to other universities and then taught at Leipzig. Koch (1907-1910), LXXXIII, 58-9, said that Scultetus enrolled at Wittenberg in 1557, and thus may have known Melanchthon personally; that he went to Leipzig in 1559; and that he received his degree in Wittenberg on February 24, 1564.

he received the degree of master of arts. He also lectured at Wittenberg for some time. In 1567 he returned home, but made several short trips, until in 1570, at the age of thirty, he returned to Görlitz, never again to leave it, except for his map-making trips in the neighborhood. From 1570 to 1586[83] he taught arithmetic and spherics at the gymnasium. Then he entered the city administration and in 1589 to 1592 became in turn recorder,[84] judge, church-trustee, deputy burgomaster, and burgomaster,[85] which position he held at his death in 1614. His diary has proved useful to later writers on the history of Görlitz and its inhabitants.[86] He was made a noble by either Rudolph II or Ferdinand II.[87] His tombstone, erected in St. Nicolas' church in Görlitz in 1642, bears the words "Quid agam, requiris? Tabesco. Scire, quis sim, cupis? Fui, ut es; eris ut sum." He was twice married[88] and left two sons and three daughters.[89] He was a man of high character and well loved.

A scholar of considerable breadth, and the author of several theological, juridical, astronomical, and astrological works,[90]

83 Koch (1907-1910), LXXXIII, 65, gave the last date as 1584, as did also Jancke (1861a), 265.

84 Hellmann (1883); Gräve, 458.

85 *Idem*; Koch (1916), 21. According to Koch (1907-1910), LXXXIII, 174, and Koch (1916), 21, Scultetus first entered the city administration in 1578. Scultetus' positions were listed by Gräve, 459. For a discussion of them see Koch (1907-1910), LXXXVI, 46-9 and Koch (1916), 21.

86 Jancke (1861b), 280; Jancke (1861a), 265; and Jancke (1868), 267, said that this diary covered the years 1567 to 1594, or through 1593. See also Koch (1907-1910) and Koch (1916).

87 See Gräve, 462, 500-2.

88 Scultetus' first wife whom he married in 1570, was Agnes Winckler Tielen, a widow. She died in 1572. His second wife, whom he married in January 1573, was Helen Roberus. See Jancke (1868), 268-9.

89 Scultetus' wives and children were discussed by Koch (1916), 45-53.

90 Scultetus' astrology was a part of his astronomy, for, as Koch (1916), 24, said of him, he was an astronomer of his time, seeking to measure the direct influence of the stars on human life. Some of his manuscripts are extant. See Jancke (1868) and Koch (1916), 55-7. See Baumgärtel, 248-9, for a 1596 letter by Scultetus. See Hoefer and Gräve, 470 ff., for references

to manuscript notes by Scultetus, entitled *Annals of Görlitz.* These may include his diary and his *E libris rerum gestarum Gorlicensium.* See Scultetica (1837), Scultetica (1842), and Scultetica (1915). His printed works include:

Almanach und Schreib-Kalender, auffs Jahr . . . 1584 . . . Görlitz [1583] (B.M. catalogue. According to Koch (1907-1910), LXXXIV, 74 and LXXXVI, 53, these almanachs appeared regularly; see Koch (1916), 53-8. This item is probably the work which Koch (1916), 27, said was dedicated to Michael Ender and Sebastian Hoffmann, and where the Gregorian and Julian calendars appeared together for the first time.)

Brief von Tychone de Brahe (Scultetus, editor; see Jöcher).

Calendarium Ecclesiasticum & Horoscopium perpetuum. Görlitz, 1571. (Gesner (1583); Jöcher; Witte. Gräve gave a description but no date. Koch (1916), 44, called the 1571 edition the "first" (see the *Computus ecclesiasticus...*, listed below). Zedler listed a 1578 edition.)

Calendrier réformé. Görlitz, 1601. (Hoefer. The original title was probably not in French.)

Computus ecclesiasticus, in calendarium perpetuum omn. h. mundi annorum Chr. directus et extructus. Almanach u. Kirchenrechng. aller Jahr.... von anfang der menschl. Geburt ... biss ... diese Zeit. M. Anleitg.... den Computam eccl. in des A. u. N. T. Zeit ... zu verstehen. New zugericht u. in e. leichtuerstend. Ordng. gebr. Görlitz, A. Fritsch, 1574. (Jöcher; Gräve; Poggendorff; full title from Rosenthal, catalogue 168, item 631. According to Koch (1907-1910), LXXXIII, 68, this was a second edition of the *Calendarium Ecclesiasticum....*)

Curriculum humanitatis Jesu Christi in terris, continens historiam redemptoris evangelicam. Frankfort-on-the-Oder, 1690. (A.D.B. Hoefer said that this appeared in Görlitz, in 1580, in folio, and in Frankfort-on-the-Oder, in 1600, in quarto.)

De origine & curatione pestis (Jöcher; Witte; Zedler; Gräve, who cited Zedler.)

Diarium humanitatis Christi . . . (Witte; Gräve, who said that this is the same as the *Curriculum humanitatis....* However, see Koch (1916), 27, where this work was said to have been finished in the beginning of 1581.)

Gnomonice de solarijs, siue doctrina practica tertię partis Astronomiae. Von allerley Solarien / das ist / Himmlischen circuln vnnd Uhrn / wie man dieselben an die auffgerichten Planicien oder Wende / vnd in allerhand hole Instrument / von den planis sphaericis vnd vermischten superficiebus zusammen gesetzt / künstlich verzeichnen vnd representiren sol. Nach Geometrischem grundt zugericht. Görlitz, 1572. (Bassaeus, II, 345, gave the date for this work as 1573, but the earlier date was given by the Cat. Belg., 1308, the Bodleian library catalogue, Gesner (1583), B.M. catalogue, Wolf, II, 13, Gräve, 467-8, Koch (1907-1910), LXXXIII, 71, Zedler, and Sotheran, catalogue 857, item 2192. The Sotheran catalogue reported that this work is interesting for the transversal division on the circle and said that the A.D.B. statement that there was a Latin edition simultaneously with the German one is erroneous and due to the bilingual title. Weidler seems to have thought that Tycho had much to do with this work, and according to Koch

(1907-1910), LXXXIII, 71, its publication was made possible largely by the generosity of Matthias Menius.)

Immerwerender Allmanach vnd Kirchenrechnung aller Jahr / von der Geburt Christi biss zum End der Welt. Sampt einem computo Ecclesiastico, dess alten vnnd neuwen Testaments Zeit vnd Jahr rechnungen gründtlich zuverstehen. Görlitz, 1574. (Bassaeus, II, 345. This is probably the same as the *Computus ecclesiasticus* listed above.)

Inventuris non obstant inventa. Görlitz, 1572. (A.D.B.; Poggendorff; Gräve, 465, 469, where the date was given as 1574 and the printer as Fritsch; Gräve, 467, where the date was given as 1572 and it was said that there was a second edition in 1583. According to Koch (1907-1910), LXXXIII, 67, the title given above was the motto under the title of the *Calendarium Ecclesiasticum & Horoscopium perpetuum.*)

Karte von d. Lausitz. (Gräve, Jöcher, and Zedler mentioned this and other maps, i. e. of Meissen. Ortelius' *Theatrum orbis terrarum,* according to Poggendorff and the A.D.B., reproduced some of them. Gesner (1583), 106, listed *Misniae & Lusatiae Chorographia,* printed in Görlitz in 1569. Kroker mentioned and Ruge described the map of Meissen and Lausitz, and Ruge, 225, stated that the map was dated 1568, not 1569 as Ortelius had said. See Koch (1907-1910), LXXXIII, 62-3. There was also a map of Oberlausitz which appeared in 1593 and was described by Ruge and mentioned by Jancke (1861b).)

New und Alter Römischer Allmanach und Schriebkalender . . . 1601. Görlitz, Rhambaw. (Gräve, 465-6.)

Opus sciatericum (edited by Scultetus. Jöcher; Zedler).

Phaenomena novilunii ecliptici, sub meridiano Gorliciensium de coelo observata 1567 d. 8. April. Görlitz, 1567. (A.D.B.; Gräve; Poggendorff.)

Prognosticon Nouilunij Ecliptici. Das ander Theil von der Sonnen Finsternuss so im Aprill 1567. gesehen worden / darinn angezeiget werden die zukünfftigen Geschlechten auff Erden / so in den Jaren 1568. 1569. 1570. vnd den nachfolgenden sich zutragen sollen. Görlitz, Ambrosius Fritsch, 1569. (Bassaeus, II, 278; Gesner (1583), 106, where, however, a Latin title was given. Gräve gave the date as 1568 and Günther (A.D.B., XXXIII, 498), gave it as 1567.)

Prognosticon Meteorographicum perpetuum. Ewig werend Prognosticon, Von aller Witterung in der Lufft vnd der Wercken der andern Element: Soviel betrifft die ankunft, natur vnd Wirckung aller Wind, Regen, Schnee, Thaw, Reiff. Cum adnexo. Görlitz, 1572. (There were editions in 1583 and 1588. A.D.B.; Bassaeus, II, 345; Gräve, 467; Hellmann (1883); Hellmann (1891), 36-7; Hellmann (1924), 41-2; Jöcher; Poggendorff.)

Prognosticon, Uber die Mundanam Revolutionem Im Jahr . . . 1584, . . . [1583 ?] (B.M. catalogue. Gräve cited a "prognosticon" for 1595.)

Vita Christi & Apostolorum. Frankfort, 1660. (Jöcher; Zedler; Gräve, 476-7.)

Vom Vrsprung der Pestilentz / vnnd jhren zufallenden Kranckheiten / auch derselben fürkommung vnnd heilung Doctoris Paracelsi Schreiben

Scultetus was particularly interested in calendar reform.[91] He defended it, although a Protestant, and expressed his regret that some people fought against a good proposal because of hate for its initiators. Gregory XIII sought his opinion on the proposed changes,[92] and Scultetus' literary relations with his contemporaries was based largely on this interest. Kepler [93] and Peucer and the Jesuit, Anton Possevinus, sought him out in Görlitz, and Paul Fabricius was the model whom Scultetus tried in some ways to follow.[94] In 1601 he came forward with an improved calendar [95] and his *Gnomonice de solariis* of 1572 proved so important that a Dutch edition of it appeared in Amsterdam in 1670. From 1573 to 1598 he wrote at least nine, probably twenty-six, prognostications in the German language.[96] He seems to have had some notion about spiders as weather signs. His *Phaenomenon novilunii ecliptici* retained its astronomical

fleissiger mit Obersehung auss seinen Büchern zusammen getragen / auch hinzu gesetzt Summarien / auch Concordantzen. Basle, Peter Perna, 1575. (Bassaeus, II, 252; B.M. catalogue. This item may be the same as the *De origione & curatione pestis,* listed above. Gesner (1583) listed a *tractatus de peste . . . ex . . . Paracelsi libris,* printed in German in Basle in 1575, which might refer to either item.)

91 See Koch (1907-1910), LXXXIV, 70-88, LXXXVI, 67-9; Jancke (1868), 271. One *Ephemeris* is preserved only in manuscript (see Koch (1907-1910), LXXXIII, 76-7), but other works on the calendar appeared in print. See Koch (1916), 20.

92 Günther in A.D.B., XXXIII, 497-8.

93 Kepler's visit was in 1607. See Gräve, 461 note xxxx; Koch (1907-1910), LXXXVI, 59; Koch (1916), 29.

94 Jancke (1868), 272. Paul Fabricius was the imperial astrologer. See Thorndike, VI, 184 and 184 note 17, and item 39 of the appendix, below.

95 Gräve, 462 note *, and 500, thought that this appeared in 1598.

96 Hellmann (1924), 29. Hellmann (1883) listed one quarto *Practica* for 1581, printed, however, without date or place of publication; two printed in Görlitz in 1590 and two printed there in 1593 and 1594 respectively. Of the *Prognosticon Meteorographicum Perpetuum,* Hellmann (1924), 41-2, wrote: "... Scultetus ... veröffentlichte 1572 ein Buch mit dem Titel Prognosticon Meteorographicum Perpetuum, das man für eine Anleitung zur Aufstellung von Wetterprognosen halten könnte. Das ist es aber nicht, sondern nur eine Darstellung des jährlichen Verlaufes der Witterung, in der ungewöhnlich viel von schädlichem Tau, Reif und Nebel gesprochen wird."

importance for a considerable time and, like others of his works, was unspoiled by its quite conventional astrological predilection.[97] Scultetus' interest in geography, or cartography, seems to have been originally aroused by Homelius, and in that field, likewise, he attained considerable repute. He was consulted by the Russians who desired to measure the size of the earth,[98] and one of his maps was reprinted as late as 1725.[99]

Scultetus' Latin tract on the comet of 1577 [100] is not quite as learned as one might expect from an author of his scientific background, although it is couched in scholarly terms. It seems like the work of a mathematician who was not grounded in astronomy. An introductory chapter dedicates it to the senators of Görlitz, and lists twelve observations of the comet by Scultetus, which determined its position in the eighth house, the positions of the planets and constellations during the comet's appearance, and other data.

The first part of the text deals with the comet's visible diurnal revolution and its path in the "sublunar" region of the world from November 10, 1577 to January 13, 1578. Scultetus began with a lesson in spherical trigonometry, illustrated by diagrams. Then he described the comet as having a lively motion toward the west, as well as its own motion upward and northward; for, he said, it was always visible in the middle of the heaven north of the ecliptic. Although it first was seen below the equator, crossing that circle after the tenth day, it then appeared entirely north. He assumed a circular orbit for the comet,[101] his scheme thus resembling Maestlin's. He deemed

97 Günther in A.D.B., XXXIII, 498.

98 See Koch (1907-1910), LXXXIII, 80-1, and Koch (1916), 33.

99 Baumgärtel.

100 Item 96 of appendix, below. The first and second sections were fully discussed by Tycho in the tenth chapter of the *De Mvndi Aetherei... Phaenomenis* (Brahe, IV, 296-337) and will be only sketched here.

101 "... Circulus enim iam ostensus de his omnibus nos edocet, vt qui tanquam norma in nostro Horizonte expositus, de singulis illius adparitionibus rationem reddere potest...." See chapter III, note 10, above.

it necessary to observe with various instruments the comet's distance from the stars, its position in the circle of altitude or azimuth, and its latitude or almucantarath, and to apply them to the circumference of the comet's circle. Thus he could conveniently vouch for his description of the comet's observed motion, before he broached his second topic which concerned its parallax and true motion, and the third which dealt with its meaning.

Ten observations of the comet made by Scultetus from the time of its first appearance to the evening of December 12th,[102] fell along a circle which cut the ecliptic at an angle of 45° in the twenty-first degree of Sagittarius and of Gemini. This circle was the comet's path and from it he calculated that the comet's maximum northern latitude would be reached in the twenty-first degree of Pisces and its southern in the twenty-first degree of Virgo.

Scultetus first saw the comet at 5 P.M. on November 10th and recorded its position in the zodiac and also in latitude, and noted that its tail was turned away from the sun. He described the comet as resembling a huge shining spherical mass which vomited fire and ended in smoke. Scultetus' fault was not lack of information but rather the inclusion of many details which proved repetitious and many which had no bearing on the comet. These data, such as the positions of the planets or the Ptolemaic designations of the stars, which are superfluous to an astronomical discussion, were, however, useful for finding the comet's supposed astrological significance.

In the beginning, the comet was beneath Aquila and Antinous, but then it moved toward the first star in Antinous where it arrived on November 13th. By the 18th it reached the second star of that constellation. On the next day it entered the sign of Aquarius and at last it arrived at a space where there were no stars for 12°, passing south of Aquila and the Dolphin and to the north of the covering on the left hand of Aquarius.[103]

102 "ad crepusculum III. Id. Xbris".

103 "vestimentum manus sinistrae Aquarij". The picture given by Bouché-Leclercq, 145, shows this covering to be a piece of material which covers the

Meanwhile, it hastened to the equator, which it crossed at the second degree of Aquarius between the 19th and 20th of November. Thereafter it began to deviate to the north. On December 1st, when visibility was good, Scultetus found the comet in the seventeenth degree of Aquarius, an advance of an entire sign plus one degree. The comet's latitude was then 35° and its tail stretched beyond the eleventh star of Pegasus. On December 2nd the comet arrived north of the eastern star in the mouth of Equulus. Afterwards it approached the other star in the mouth which two days before had sent beams through its extended tail. Similar observations were recorded for December 4th and 8th, after which the comet decreased sensibly in luminosity and size and proper motion. Observations were also recorded for December 14th, when the comet was in Pisces. Finally, Scultetus presented a table of the apparent motion of the comet, including the position of the sun at 6 P. M., the comet's longitude, latitude and declination, and the comet's distance along its own circle from the twenty-first degree of Sagittarius.

In the second part of the tract Scultetus dealt with what he called the true motion and parallax of the comet and with the comet's position in the sublunar region. His calculation of its parallax [104] was made by trigonometry from two observations

loins of the figure of Aquarius and is held, at one end, by his left hand. In Hagecius' map of the comet's path, Aquarius is pictured from the back. Therefore, this "covering" is held in the right hand.

104 After reading Scultetus' book in the B.M. in the summer of 1931, it seemed that he had made only one determination of parallax. To confirm, or deny, this impression, it would be necessary to re-examine the treatise, a task now impossible. However, Tycho, who gave deep consideration to Scultetus' calculation of parallax (Brahe, IV, 306 ff.), spoke of only two observations for parallax, both on January 1st. Moreover, the title of Scultetus' second section is: "Partis Secvndae Descriptionis Cometae ΣΧΗΜΑΤΙΣΜΟΣ, De Hvivs Meteoricae Impressionis, [sic] Vero Motv, Magnitudine, loco, &c. in sublunari regione M. Quae ex Parallaxi, Cal. Ian. anni inuentis cIↄ, Iↄ, Lxx, VIII, observatione facta & in eandem calculo Astronomico, ad institutum doctrinae Triangulorum, directo, è sublimi concepimus: habitu exinde machinae M. duplici formato: posita nimirum Terrae mobilitate, qua Cometae non nisi vnicus motus ad ortum versus tribuitur, & eiusdem immobilitate, qua alter reuolutionis diurnae adijcitur: qua ratione praesenti diagrammate huic Meteoro dupl. faciem accessisse videmus."

of the distances of the comet from two neighboring stars on January 1st, and yielded the value 5° 22′.[105] He furnished a diagram of the earth and the sphere of the moon with the circle of the comet between them, and added eight descriptions of different observations, each accompanied by a diagram. These diagrams had been previously presented in a single composite one. Then he displayed nine triangles in which various sides or angles were given and the others sought, and which led up to a trigonometrical explanation of how the comet was proved sublunar by its parallax. A diagram and a table show what Scultetus considered the relative sizes of the earth, the moon, and the comet, the latter being about one fourth the size of the moon.

The third section is astrological and provided the occasion for many diagrams of the relative positions of the comet and the planets and constellations and for the utilization of Scultetus' knowledge of trigonometry. Scultetus gave a detailed discussion of the comet's position, form, tail and so forth. His opinions were based on book II, chapter IX of Ptolemy's *Quadripartitum* and on Cardan's commentary upon it.[106] Other writers, including Pliny and Regiomontanus, were cited. This third section, although representative of Scultetus' times, is not relevant to a study of the development of the theory of comets and received no space in Tycho's discussion of the book.[107]

Another believer in the sublunar position of the comet of 1577 who attained considerable repute was Andreas Nolthius.[108] Because he received particular notice from Tycho,

105 Brahe, IV, 325; Riccioli, II, 89. Tycho made a careful analysis of the observations. It is difficult to state the exact cause of Scultetus' high value for the parallax without a further examination of his book. Scultetus was a proficient mathematician and his trigonometrical calculations were probably accurate. The error may be one of observation and be largely due to his having made but a single determination. No allowance seems to have been made for the comet's own motion.

106 See chapter I, above.

107 Brahe, IV, 296-337.

108 Bassaeus, II, 276.—Crawford library catalogue, 324.—Dreyer, editor, VIII, 460.—Dreyer (1890), 60.—Frisch, editor, VII, 289.—Gesner (1583),

Maestlin and others, he is of importance here despite the low level of his work. However, there is little available information concerning him. He came from Einbeck, or at least lived there, and has been characterized both as a mathematician [109] and as an astronomer.[110] He wrote, in German, calendars or diaries with "praktika" or prognostications, and is known to have written practicas for the years 1579, 1580, 1581 and 1582. The first three were printed in Erfurt; for the last, no place of publication was given.

As has been seen,[111] Nolthius wrote on the nova of 1572, which he considered sublunar. Perhaps it was because of this previous work that his book on the comet of 1577 attracted the attention of intelligent contemporary astronomers and as late as the seventeenth century was again mentioned by Claramontius and Kepler during their controversy over Tycho's discovery of the supra-lunar position of comets.

The tract on the comet of 1577,[112] dedicated to Philip, Duke of Brunswick and Lüneburg, was dated from Einbeck on February 2nd, 1578. In his dedication or preface, Nolthius mentioned having predicted, in his *Prognosticon* for 1577, a fiery heavenly sign which he now identified with the comet, and which, he said, he had observed and described as diligently as his other affairs permitted. He bade the Duke consider this comet a sign of God's word, as Ludwig I, son of Charlemagne, had interpreted the comet in his time.

First [113] Nolthius explained what comets were. He thought them signs of God's wonderwork, naming as sources of this belief the "heathen" poets and philosophers. Therefore, all the

47.—Hellmann (1883), 365.—Hellmann (1924), 5, 6, 14, 28.—Kepler, VII, 248.—Ludendorff.—Riccioli, I, xxx, II, 13, 28, 89, 137, 159.—Scheibel, 75, 105-6.—Zedler, XXIII, 1102-3.

109 Riccioli, I, xxx; Hellmann (1883), 365; Hellmann (1924), 28.

110 Zedler, XXIII, 1102-3.

111 Chapter II, above.

112 Item 78 of appendix, below.

113 Chapter I.

more, said he, should Christians observe what comets are, what their causes are, what effects they produce, and what they mean. To arrive at his "definition" of a comet, he cited Aristotle, Pliny and Ptolemy as saying that a comet is formed of earthly vapors, pulled aloft by the sun, moon and stars, and that it signifies changes in the air and also among people on earth. Nolthius also cited the definition given by others, that a comet is formed by a union of planets; but this he refuted, saying that it was believed only by people who were insufficiently acquainted with the paths of those bodies, because planets never move further than 8° from the ecliptic and comets occur anywhere. Still others, according to Nolthius, thought that a comet was a star which occasionally moved out of the brilliancy of the sun, where it was otherwise hidden, and showed its long tail, while it, itself, stayed near the sun; but this he said he could disprove by the parallax. For if the tail alone floated in the air, the comet would have no sensible parallax, but only the tail would have one and would be set apart from the rest by several degrees, which has never happened. Nolthius, therefore, held to the first definition, which he ascribed to Aristotle, Pliny and Ptolemy.

Nolthius next concerned himself with the causes of comets.[114] He remarked that some learned people believed that comets did not have natural origins, but that they remained in the air as long as God wished, as notices of changes on earth, and that one could not predict their appearance from the paths of the stars. However, he, himself, agreed with Albumasar that comets governed by Mars could be predicted for the years of many planetary conjunctions. In fact, having found for 1577 many constellations in which Mars and also Saturn and Mercury held precedence, he predicted a fiery heavenly sign for that year. This natural origin of comets Nolthius believed to have been prearranged by God; and he thought that the material for comets was pulled aloft, formed and ignited chiefly by Mars, Saturn and Mercury.

114 Chapter II.

Nolthius held that the solar eclipse of 1574 in the first degree of Sagittarius extended its influence to 1577. At that eclipse Mars, in the tenth house, had the most power, and Mercury and Saturn were in the seventh house. This configuration Nolthius called the chief cause of the comet which appeared in November 1577 in the sunset glow, because the sun had again arrived at the position it held at the time of the eclipse. Furthermore, said Nolthius, in 1576 Saturn and Mars were in conjunction in the second degree of Capricorn and in 1577 they were in opposition in Capricorn and Cancer. He thought that the comet was pulled aloft by Mars, and was collected and prepared in Capricorn by Saturn. He concluded that the comet was ignited by Capricorn, where it was first seen on the 11th of November. Nolthius designated the lunar eclipse of 1576 as a third natural cause of the comet and of its path. In the fourth place he named the lunar eclipse of April 2, 1577; and in the fifth and last place, the conjunction of Mars, Jupiter, and Mercury in Virgo in September 1577. He pointed out that five kinds of configuration in which Mars, Saturn and Mercury were important had effect in 1577. He saw no basis for including in his list the lunar eclipse of September 1577, which he did not believe would achieve its effect until May 1578. So much, said he, for the astrological causes. On the other hand, he wanted to remind his readers that the comet was a sign of God's anger at their sins.

Then [115] Nolthius discussed the comet's path and the signs through which it moved. He said that it was first seen at 6 o'clock on the evening of November 11th, 1577 on the western horizon, although he himself had not seen it because of intervening buildings. On the 12th Nolthius saw it near the Milky Way in the first degree of Capricorn not far from the ecliptic in the direction of the equator, near Saturn. That evening the comet set shortly before 7, slightly later than it had set on the previous evening, which made Nolthius realize that it did not revolve uniformly with the "primum mobile" once in 24 hours. Its own slow motion from west to east, opposed to the other,

115 Chapter III.

caused it to set later each evening, as was observed despite the fact that for several days the sky was overcast and the comet was merely glimpsed through the clouds. On November 24th Nolthius discovered that the comet moved, not only following the obliquity of the ecliptic in longitude, but also in latitude from the ecliptic. It passed over the equator, through Capricorn and by the bright star of the Flying Eagle and stood in the beginning of Aquarius not far from the tail of the dolphin, almost in a straight line with the mouth and shoulder of Pegasus, so that the star of the third magnitude in Pegasus' jaws was seen half-way between the comet's head and Pegasus' shoulder. On the 25th, according to Nolthius, the comet was further from the eagle and nearer the horse's jaws, under the constellation of the Dolphin, and rose higher toward the north so that it formed an obtuse triangle with the bright star of the eagle and the horse's jaws. The comet's tail was not so long as previously, which Nolthius explained as due to the brightness of the full moon. However, during the whole time that the comet was visible, the tail stretched mostly eastward and a little southward, inclining more toward the south when it approached setting, because it and the air through which it floated were pulled and led around by the revolution of the heaven.

On the 1st and 2nd of December the comet was further within Aquarius and had risen higher, so as no longer to form a triangle with the bright star of the eagle and the horse's jaws but to be in a straight line with them. When the comet first appeared it was reddish and bright, but at the beginning of December Nolthius thought it whiter and less bright, although no smaller, because the moon had not yet risen. He observed the comet with his astronomical radius, and found that the short arm which had two hundred and four points touched the five hundred and ninety-seventh point of the radius, when the comet was 41° above the horizon. From this he calculated the number of degrees of the sky covered by the comet and the length of the comet.

On the 6th and 7th of December Nolthius found that the comet had moved further, standing not far from the jaws of Pegasus. The tail had grown thinner, darker and shorter. On the evening of the 7th Nolthius observed the comet with a quadrant and found it 41° 8′ [116] above the horizon when its azimuth west of the equinox toward the meridian was 44° 25′. After an hour, observation with a quadrant showed the comet 33° 15′ above the horizon and with an azimuth, west of the equinox toward the meridian, of 27° 30′. This was the observation from which Nolthius calculated the comet's parallax and its distance above the earth.

After the 7th of December there was much bad weather; so that the next observations Nolthius gave were for the 19th, by which time the comet stood in the beginning of Pisces. It was much diminished in size and much darkened because of the moon-shine, so that the tail could not be seen perfectly. Nolthius said that on this same evening the moon was encircled for an hour by a round rainbow of all colors, a phenomenon seen for a few days by others and followed by winds. On the 26th of December Nolthius saw the comet beneath Pegasus. It was much smaller and its tail was barely visible because of the moonlight. By the 30th and 31st it had moved further into Pisces, and, since the moon had not yet risen, a dark streamer was seen issuing from it. On the 1st and 3rd of January, the comet, much diminished, once more directed its path [117] nearer to the thigh [118] of Pegasus. Nolthius could discover, unaided by any instrument, that the rate of the comet's motion was not uniform but very rapid in the beginning, when the flame was large, and barely half as fast when the comet's size decreased. For at first the comet set later each evening but toward the end it set earlier.

116 Here, as above in the discussions of Maestlin's and Hagecius' books, a "scrupulum" has been translated as a "minute". (See chapter III, note 64, and chapter IV, note 38.) This was done by Tycho (Brahe, IV, 338). Besides, Nolthius subtracted 4° 59 scr from 5° 32 scr and obtained 33 scr.

117 "Scheibenlauff".

118 "öberschenckel".

Nolthius devoted his fourth chapter to the comet's parallax. He said that the sphere of the sun and the stars above are so far above the earth that the whole thickness and size of the earth are as small as a little dot in comparison; and that, therefore, astronomical instruments and sun clocks can be treated as though situated at the earth's center. But the sphere of the moon and what lies beneath it are so much smaller and lower that the instruments can distinguish a difference in position, which is called parallax. The higher a body is above the earth and above the horizon, the smaller is its parallax. Nolthius believed that because the comet was in the air far beneath the moon, the parallax must be determined before anything definite could be said about the comet's position. Therefore, on December 7th, 1577, he made two observations, at an hour's interval, of the comet's altitude and horizontal azimuth.[119] With the aid of spherical trigonometry and considering the two observations and the height of the pole, he found that at first the parallax was 4° 59′; that is, that the comet's position in the circle of altitude measured from the center of the earth was 46° 7′, when the quadrant showed only 41° 8′ on the earth's surface. Nolthius believed that the earth's thickness, from its center to his position on its surface, had " taken away " the difference. Calculation from his second observation yielded the value 5° 32′ for the parallax, the quadrant showing only 33° 15′ in the circle of altitude when the value was 38° 47′. This led Nolthius to conclude that during the hour between the two observations the comet had moved 7° nearer the horizon, that the earth's thickness had to be reckoned with, and that the parallax had increased 33′.[120] He believed that the values of the parallax furnished the only means of discovering the height of the comet above the earth. However, he believed that the comet had additional parallaxes, toward the zodiac and toward the equator; that when it was observed in the twentieth degree of Aquarius it was really in the twenty-fourth. Using the " doctrine of con-

119 See note 130 in chapter III, above.

120 Nolthius was obviously confused on how to handle the parallax, and consequently he was unable to express himself clearly.

vex triangles", he examined the comet's position in the circle of altitude, shown by the quadrant in the second observation on December 7th to be 33° 15′, and found that the comet was 11° 3′ north of the equator. This he termed the "observed declination"; but he gave the value of its "true declination" above the equator as 15° 36′. He supposed a similar difference between the apparent and true latitudes, measured from the zodiac.

A discussion of the distance of the comet above the earth's surface logically followed the details about the parallax.[121] Having determined the parallax[122] from the observations, that is, the difference which the semidiameter of the earth makes in the observed position of the comet, Nolthius thought it necessary once more to compare, by trigonometry, that parallax with the observation. This he believed would furnish the proportion between the comet's height and the earth's semidiameter, which he determined as "eightfold super-two-part thirds."[123] In other words, he found that the comet was eight and two thirds semidiameters of the earth from the center of the earth, which he said equaled 7726 2/3 German miles if the semidiameter were taken to be 895 German miles.[124] Accounting for all the

121 Chapter V.

122 As was stated in the preceding paragraph, Nolthius' values for the parallax were 4° 59′ and 5° 32′.

123 Nolthius called this value "octuplam superbipartientem tertias (sic)".

124 This value for the semidiameter would give the distance as 7756⅔ German miles. It is impossible to discover either what value Nolthius used for the German mile or whence he got his value for the earth's semidiameter. Tycho, see above, especially chapter III, note 28, used a value for the semidiameter, smaller by 35 German miles. The approximate ratio of German to English miles is 4½ to 1. However, assuming that Nolthius used the Brunswick mile, since his work was dedicated to the Duke of Brunswick, the ratio would be 1 Brunswick mile to 6.7140 English miles as given by Woolhouse, 66, which would make Nolthius' value for the earth's semidiameter far too large. According to Noback, 181, 529, the Brunswick mile (7419.422 meters) comes out to 4.868 times the English or London mile (1523.986 meters), or according to the same author, p. 529, the London mile is .20539 of the former German geographical mile which is the same ratio, 1 to 4.868, and which is again repeated by Noback, 530. According to Zedler, XX, 307, 11 English miles equal 3 German miles, which is equivalent to saying that 1 German mile equals 3⅔ English miles; Zedler also said, giving

fractions in his computation, he found 7762 2/3 German miles as the distance from the center of the earth at which the comet moved. But its distance from a point on the earth's surface, such as Nolthius' position in Einbeck, varied continuously because the comet's path was not about the center of observation but about the other.[125] Nolthius thought that on December 7th the comet was 7142 7/12 miles from him at his first observation and 7235 3/4 at his second, so he concluded that it had moved with the heaven and also 93 1/6 miles in the line of sight. He ended his chapter by voicing his agreement with Plato's remark that arithmetic and geometry are like two wings by which human understanding can reach the sky.

Concerned with the length of the comet,[126] Nolthius directed attention to the difficulty of determining the exact limits at its great distance, because its tail was whiter and thinner at the end. However, on December 1st, he had observed the comet with great care, using an astronomical radius and referring to Regiomontanus' work.[127] Examining trigonometrically the point which the radius showed, he discovered that the " angle of vision " was so great that day that the comet's length covered 19° 24′ of the heaven. Further calculation, with the aid of Euclid's book 6, proposition 4, and taking into account the comet's distance from the center of sight, yielded the value 2441 5/6 German miles for the comet's length.

The usual astrological discussion of the comet's effect and meaning furnished the material for the seventh and last chapter.[128] Ptolemy's works, particularly the final sentence of the

the values in feet, that a German mile is 3⅕ English miles. Brockhaus, XII, 346, said that 1 German geographical mile equaled 7420.438 meters and 1 English or London mile equalled 1523.986 meters, making 1 German mile equivalent to 4.868 English miles, which is nearly the same as the value given by Noback for the Brunswick mile.

125 the center of the earth.

126 Chapter VI.

127 Nolthius quoted Regiomontanus thus: " Non enim naturae thesauros prorsus euacuare, sed in plaerisque, scibilibus ipsi veritati propinquum degustare mortalibus conceditur."

128 Tycho (Brahe, IV, 348) called this chapter alien to his purpose.

Centiloquium, at that time attributed to Ptolemy, were heavily relied on. Nolthius recalled the information contained in his second chapter and said that, because of the influence of Mercury, Mars and Saturn and because of the constellations it traversed, the comet predicted wars, pestilences, deaths, danger to travelers, short-lived heresies, winds, heat and so forth. Not all its effects would be felt immediately, and all northern and central Europe were to feel them although its greatest influence would be in the east. Nolthius closed the chapter with a prayer that God turn aside the well deserved punishment and protect mankind.

Although Nolthius was among those who realized the importance of parallax in measuring the distance of heavenly bodies, it is obvious that he did not fully understand its meaning. He had no fixed basis, such as the horizon, for measuring the value, and thought it could be determined by a single observation. His use of parallax was, therefore, unreasoned, and his values for that quantity are far out of accord with the facts, only partly because of errors in observation.

An even less clear conception of parallax was held by George Busch,[129] an artist and amateur astronomer, or " Liebhaber der Astronomie". But he too was assured of an audience for his work on the comet of 1577 because of his previous discussions of the nova.[130] He observed both these phenomena with an astro-

129 Bassaeus, II, 293. — Brahe, IV, 365-6. — Bruun, II, 67. — Crawford library catalogue, 70-1.—Doppelmayr, * * 3, 161.—Dreyer (1890), 64-5.—Gesner (1583), 265.—Hellmann (1883), 70.—Hellmann (1924), 26.—Jöcher, Erganzung, I, 2460.—Ludendorff.—Nielsen, 450, 451.—Poggendorff, I, 350.—Riccioli, II, 28, 89, 137, 139, 159-160. Riccioli gave a detailed account of Busch's observations of the nova of 1572.—Scheibel, 64-5, 95-6.—Schottenloher, IV, 377.—Struve, I, 549.—Weller (1857-8), 322, 361.—Will, I, 156-7.—Zinner (1934), 85, citing Doppelmayr and Brahe, III, 279-88 and IV, 365-6.

130 See chapter II, note 284, above. Busch wrote at least two tracts on the nova of 1572 (Gesner (1583), 265) which have been judged to be on a very low plane (Ludendorff and Dreyer, who spoke of the "fancies" of Busch). The first of these books appeared in the following German editions:

Von dem Cometen, Welcher in diesem 1572. Jar, in dem Monat Nouembris erschienen. Erfurt [Jesaias Mechler], 1572. (Schottenloher, IV, 377; Poggendorff, I, 350.)

Von dem Cometen / welcher in diesem 1572. Jar in dem Monat Nouembris erschienen. Erfurt, 1573. (Crawford library catalogue, 70; Zinner (1934), 85).

Von dem Cometen, welcher in diesem 1572 Jar, in dem Monat Novembris erschinen. Beschriben durch Georgium Busch, Norimbergensem. Augsburg, 1572. (Struve, I, 549; Zinner (1934, 85).

Von dem Cometen, welcher in diesem 1572. Jar in dem Monat Novembris erschienen. Zu Ehren, Den—Herrn Rathismeistern und Raht der löblichen Stadt Erffurt etc. beschrieben durch Georgium Busch, Norinbergensem, der Astronomischen Künsten liebhaber, wonhafftig in Erffurdt. (at end: Anno M. D. LXXIII.) (Weller (1857-8), 322; Struve, I, 549; Scheibel, 64-5. This may be the edition which Zinner (1934), 85, said appeared in Magdeburg in 1573.)

Von dem Cometen . . . Augsburg, Michael Manger, 1573. (Schottenloher, IV, 377.)

Von dem Cometen, der in diesem vergangnen 1572 Jahre. im November vnnd December ist gesehen worden. Gezogen aus dem schreiben Georgii Busch, von Nürnberg (Scheibel, 65).

Beschreibung von dem Cometen, der 1572 und 73 erschienen. Erfurt, 1573. (Will, I, 156-7; Jöcher, Erganzung, I, 2460 citing Will; Doppelmayr, * * 3. Bassaeus, II, 293, gave the title *Georg Buschen Mahlers in Erffurdt beschreibung von dem Cometen / welcher An. 1572. im Nouemb. erschienen.* 1573. This probably referred to the edition cited by Will, Jöcher and Doppelmayr.)

The following two Danish editions, probably also of the first work, were both printed in Copenhagen in 1573:

Om den ny Stierne oc Comete / som sig haffuer ladet til siune vdi Nouembris Maanet / Aar 1572. Screffuet ved Georgium Busch Norinbergensem, Boendis til Erfurt / Oc nu vdsaet paa Danske. Prentet i Kiøbenhaffn / aff Laurentz Benedicht. 1573. (Nielsen, 450.)

Om den ny Stierne oc Comete / som sig haffuer ladet til siune vdi Nouembris maanet Aar / 1572. Screffuet ved Georgium Busch Norinbergensem, Boendis til Erfurt / Oc nu vdsaet paa Danske. Prentet i Kiøbenhaffn / aff Matz Vingaard / 1573. (Nielsen, 451; also, Stolpe, I, x, where Busch was described in the title as "Norimbergensem, not "Norinbergensem". Bruun, II, 67 also cited a 1573 Copenhagen edition, but he did not give sufficient information to show to which of the two editions cited by Nielsen he was referring.)

Busch's second work on the star has the title:

Die andere Beschreibung von dem Cometen / Welcher in dem vergangenen 1572. Jar erschienen, vnd noch jtziger zeit in diesem 73. Jar, vnter den Firmamenten sichtbarlichen vorhanden. (at end: Gedruckt zu Erffurdt / zu dem bundten Lawen / bey S. Paul.) [Jesaias Mechler, 1573]. (Schottenloher, IV, 377; Crawford library catalogue, 70; B.M. catalogue; Zinner (1934), 85. The titles cited by Weller (1857-8), 361, and Bassaeus, II, 293, probably refer to the same edition. According to the Crawford library catalogue, this seems to be an augmented edition of the earlier work on the nova, and was described by Tycho.)

labe, a quadrant, and a square. His book on the comet was the last but one summarized by Tycho in the section on observers who regarded the comet as sublunar.[131] The son and grandson, respectively, of the Nuremberg physicians, Sebald Busch, junior and senior, George Busch was born in that city, where his father practiced for more than twenty years, before moving to Erfurt in 1538. George learned some astronomy from his father and grandfather and furthered his studies by reading good books. About 1570 he gave up his painting in favor of astronomy and also moved to Erfurt, where he died about 1590.

Busch's book on the comet of 1577 [132] presented a mixture of fancy, knowledge of past cometary theory, and astronomy. In his dedication he said that, as a lover of astronomy and astrology, he wished to follow in the footsteps of the men of the past who had observed celestial phenomena, and that, also in accordance with tradition, when putting something in print, he wished to entrust it to favorable hands: therefore this dedication to the Landgrave William IV. The first section of the book, addressed " to the friendly reader ", says that God, being angry, placed a sign in the heavens to try to frighten people out of their sinful ways. Therefore, they should all unite in praying that He spare them their well earned punishment. Busch proposed to consider in four chapters what the sign was, what it was called, where it moved, how high above the earth it was, and what would happen as a consequence. Throughout the book he spoke of the "sign" rather than of the "comet."

Busch also wrote a " prognostication " for the year 1580, a quarto volume, written in German, and printed in Erfurt in 1580, (Hellmann (1883), 70; Hellmann (1924), 26), and the following tract about a lunar eclipse in 1573: *Erklerung Der grossen und gresslichen Finsternis / Welche inn dem 1573. Jahr / an dem 8. tag Decembris an dem Mond erschienen / ... Sampt derselbigen ... bedeutung / die in diesem 74. Jahr an dem 11. tag Martij jren anfang hat /* ... (at end: Gedruckt zu Erffurdt / Durch Conradum Dreher / hinder der Himmelpforten.) 1574. (Crawford library catalogue, 71; Zinner (1934), 85).

131 Brahe, IV, 365-6. According to Riccioli, II, 28, Busch believed all comets sublunar.

132 Item 21 of appendix, below.

Busch began by recording his observations. Despite much cloudy weather in November, he perceived the comet near the horizon at Erfurt at about 6 P.M. on November 11th. It set so rapidly that it was impossible to observe it with astronomical instruments. However, on the following evening, from 5 to 6 P.M., he was able to do so. At 6 o'clock it was 11° 10′ above the horizon. The altitude was sighted from the line of Busch's house, with the help of a half circle which moved from the zenith and bent in the direction of the circle of the sky through the center of the comet downwards to the horizon, and showed the azimuth to be 37° 15′. The "sign" moved 80° 50′ from the zenith and was 12° 20′ north of the ecliptic and 15° 20′ south of the equinoctial circle. Busch deduced the comet's path in longitude from these observations, although he did not consider them sufficient, but thought that it was necessary to have: 1) the line from the observer's position on earth; 2) the line from the center of the earth, both lines going threadlike through the center of the phenomenon; 3) the zenith, found with the help of the moveable circle; and 4) the pole of the earth or of the zodiac and the circle of position and also the azimuth and almucantarath degrees. Thereafter, one obtains the horizon of the district where one wants to find the true motion of the observed phenomenon. Busch explained that from these established points arise two straight stretched lines, together with several small and great circles, which all interlock through the center of the "sign," giving a true position. These measurements furnished the following values: true altitude above horizon, 11° 30′; true horizontal azimuth, 39° 36′; true distance from the vertex in a circle of altitude, 78° 20′; true distance in latitude south of ecliptic, 14° 40′; true declination from the equator in south latitude, 13° 0′. Busch elaborated these data by locating the comet in relation to the zodiacal divisions, the different constellations and fixed stars and the planet Saturn. In addition, he said that shortly after 2 P.M. the comet reached the meridian of Erfurt with the right ascension of 27° 28′,[133]

133 See chapter III, note 19.

and that its true time of setting was 7:30 P.M. He stated that another important fact to consider was parallax, but that it would not be explained in his book because the ordinary man could not understand it.[134] However, he reported that the parallax "of different aspects" reckoned from a plane triangle was 2° 20′ [135] and that from this the distance of the body from the center of the earth was found to be twenty-four semidiameters of the earth. Busch's observations were all made at 6 P.M. On the 12th of November the comet's motion from 6 to the time of its setting was too swift to allow of observing its path, but on the 16th it was 17° above the horizon and made a circle from east to west. On the 18th it was 21° 45′ above the horizon, on the 21st, 24° 10′, and on the 27th, 33° 50′, at which time it moved into Aquarius through Capricorn and went into conjunction with the body of the starry dragon and the flying swans. Busch thought it evident from these data that the comet had a rapid proper motion, for during his observations it had gained 35° in altitude and moved a little more than 2° in 24 hours.

Then [136] Busch considered what the "sign" is and how it would be called, and explained the different spheres [137] and the height of the "sign" above the earth. He divided the sky into different sections corresponding to their distance from the earth. In doing this he referred back to Alfraganus. He even said that Copernicus, although he described this differently, would have agreed that the space between the heaven of the moon and of the earth is elementary and is divided in three parts and anything appearing within them is elementary. The "sign", Busch added, is in the elementary part. The sky has another division, in ten parts, called the firmament, wherein are the seven planets, the sun, the moon, the twelve signs and all stars. Interestingly enough, Busch realized the enormous sizes of the stars. He

134 This reason was probably only a mask for Busch's own inability to elucidate the method and theory.

135 Riccioli cited Busch as recording a parallax of 2° 21′ for the comet.

136 Chapter II.

137 " vnterschiedlichen Himel ".

thought that the lowest part of the elementary region of the air reaches to a height of eleven times the semidiameter of the earth, and that in that part certain phenomena occur, such as meteors, which are dust and smoke from the earth and waters, drawn aloft by the strength of the sun, moon and stars. In the summer this region is warmed by radiation and in the winter it is cooled, for which reason meteors have different appearances. The phenomena which take place there, such as thunder and lightning, meteors, clouds, fog, rain, snow, the cloudiness of the galaxy, the color of the clouds, and the colored circle around the sun, are all heated and have earthly origins. Busch's theory held that the middle division of the elementary region extends from 11 to 22 semidiameters from the center of the earth, and is cold and frosty because it cannot be warmed by refracted solar rays, which do not reach so high, and because the sun is too far away for its rays to affect this air while passing through. Hail is formed there in the summer, and there the devil resides. The third division of the elementary region extends to 33 semidiameters of the earth from the earth's center, and is quite warm, receiving its motion from that of the tenth sphere of the upper heaven and being nearer to the sun than the two lower divisions. There bright apparitions, of three types, called comets, often appear.[138] The first of these, called "Stella Comata" or "Crinita", has a round body surrounded by rays. Often the rays are only on top, giving the appearance of a head with hair. Comets of the second type have streamers like a beard and are called "Cometae Barbati." Those of the third type stretch out streamers from one side, like a tail. They are termed "Cometae Caudati", and can have either long or short tails. In Busch's opinion, comets are also meteors, since they are formed out of hot, sulphurous, saltpeterish and terrestrial materials and are drawn up by the forces of the sun, moon and stars. Their duration depends on the amount of material in them, and sometimes they are extinguished before they are

138 Dasypodius (see chapter VI and the appendix, below) made three classifications of comets. Compare with Aristotle's two types of comets (see chapter I, above).

completely consumed, if all their material has not at first been prepared for fire, and when the remainder is entirely dried out it is ignited again and burned up. If, through the hindrance of certain constellations, the material from which comets are formed does not reach the third region, it remains in the cold region and becomes sheet-lightning and thunder-claps, and so forth. The comet of 1577 had a long tail and might be called a tailed comet. Busch felt the necessity of referring to the writings of Ptolemy, Guido Bonatti, Leopold of Austria, and others, who classified the comets according to the planets which ruled them.[139] The sun rules all the comets, as is seen in the case of

139 Ptolemy seems not to have classified comets, although a classification was associated with his name. See chapters I and II, above, especially the discussion of Grosseteste. Busch's classification, or one similar to it, was quite generally used. For example, Cecco D'Ascoli, who flourished in the first quarter of the 14th century, spoke of a comet called Milex (Thorndike, II, 961); in 1337 Giovanni Villani noted two comets, one called Ascone and one called Rosa (Thorndike, III, 287 note 16) and in 1347 he noted a comet called Negra (Thorndike, III, 316); in 1402 Jacobus Angelus of Ulm (see chapter II, above) gave nine species of comets after Bonatti and nine effects after Albertus Magnus (Collard, 85; Thorndike, IV, 83; see also chapter I, note 158, above), and Albertus probably tied up the effects of comets with the types; Simon de Phares also described a comet which appeared in 1402 and which he called "Verru" (Thorndike, IV, 78); John de Bossis, writing on the comet of 1472, described its tail as resembling the species of comet called "pavo" while its nature was a cross between "miles", "nigra" and "tentaculum" (Thorndike, IV, 424); and Angelo Cato de Supino of Benevento gave that comet the name "Pogonias" (Thorndike, IV, 427). The classification given by Busch closely resembles that of Grosseteste, who cited Ptolemy as his authority, and also that attributed to Bonatti by Jacobus Angelus, and the one used by Leopold. By Busch's classification, the first type of comet, called "Nigra", has the nature of Saturn, has a long tail, and is of a leaden color. The second, called "Rosa", belongs to Jupiter and shines like gold and silver and is shaped like a human head. The third is called "Argentum", has a silvery color, and is also ruled by Jupiter. The fourth type of comet, of horrible reddish appearance, is called "Veru", is dominated by Mars, and moves close to the sun. The fifth, "Cenaculum", has the reddish glow of Mars with an ash colored tail. The sixth, "Partica", likewise ruled by Mars, does not resemble it so closely. The seventh, "Matutina aurora", is of the nature of Mars and is very red with a long tail. The eighth type of comet, called "Miles", is ruled by Venus, is very large and moves through the twelve signs of the zodiac. The ninth type, "Dominus Ascone", is of the nature

the tailed comets, which stretch their tails in the direction of the sun's path.[140] When first seen, the body of the comet of 1577 was pale, whitish, and lead-colored, but its tail was reddish with mixed smoke colors, and the higher it rose the more it changed to dark lead colors. Busch thought that the mixture of colors showed that the comet was composed of a mixture of materials and that many planets ruled over it, the different colors belonging to the different planets. He said that God does nothing in vain and that therefore men should take warning from this comet and abstain from their sins. He added that this comet had the characteristics of the comets which he had classified as Nigra, Cenaculum and Dominus Ascone. Lead color predominated and Saturn had the most influence over the comet, as in the case of Nigra.

The third chapter attempts an explanation of the origin and path of the comet of 1577. According to Busch, comets arise through conjunctions of planets, especially Saturn and Mars, with the help of solar and lunar eclipses. On March 20, 1576 Saturn and Mars were in conjunction in Capricorn with Venus, a lunar eclipse following in Aries. This constellation received great strength through God and drew the "natural dusts" and vapors on high. All this produced the comet which was further influenced by lunar eclipses in 1576 and 1577, and the material for which increased in the intervening time and was prepared for burning by the turning of the heavens. The comet was ignited about October 10, 1577, when Mars reached the place of the conjunction. Busch explained the comet's not being seen so soon in Erfurt as due to bad weather before St. Martin's day and to the fact that the comet set too soon after sunset. He said that in addition to its diurnal motion, the comet had a path of its own, an arc in the sky moving eastwards, and that it rose in this circle 2° 20′ every day. Reckoning from this, he found that the comet was ignited in Sagittarius, and immediately

of Mercury and is smaller than the other comets. Its color is grey and leaden and it has a long tail which it throws around at random.

140 This might be interpreted as meaning that the comet's tail followed the sun.

passed through that constellation and Capricorn to Aquarius, where it was when he was writing.[141] It also traversed the constellation Antinous. Busch's value for its parallax placed it 20616 German miles from the earth's center or 19757 from its surface. Its size he described as tremendous.

Finally,[142] Busch considered the comet's "effectual" signification, which he thought was influenced by the constellations traversed by that body. He said that since the exhalations composing comets are for the most part poisonous, those bodies in general bring pestilences with skin eruptions, by which animals, birds, fish and humans die. He even listed the constellations which supposedly signified that this would happen, and added that comets also bring famine, because the crops are spoiled; and also deaths and wars and tyranny, because the happy kind of complexion in men dries up and leaves a dry, irascible temperament. The constellations indicated these effects for the comet of 1577. Furthermore, wars on land and water were to follow. Busch warned against the Turks and other disbelievers, to whom the comet's tail pointed. He advised turning to God for protection. In addition he listed the predictions which he said the Arabs drew from comets, with reference to their height above the earth, and added that the readers of his book should heed how God warned them of their sins. The book closes in prayer.

Although Busch's observations would now be boiled down to two a day, of the right ascension and declination of the comet,[143] it is greatly to his credit that he knew how to measure with reference to fixed circles. He found an amazingly large parallax for the comet, and his idea of its distance influenced his entire cometary theory. Of course, his theories are valueless. Although his book gives a clear picture of the comet of 1577 and also of those theories which observations of that comet were disproving, its influence was a hindrance to the development of the new astronomy.

141 December 1, 1577.

142 Chapter IV.

143 See chapter III, note 19.

CHAPTER V

THE COMET OF 1577: MEN WITH A SCIENTIFIC BACKGROUND WHO MADE NO ATTEMPT TO MEASURE THE COMET'S DISTANCE

STEINMETZ.—DASYPODIUS.—BAZELIUS

In the same list of writers on the comet of 1577 which held Busch's name, Tycho placed [1] Valentin Steinmetz of Gersbach,[2] with whose book he dealt at considerable length and comparatively favorably. But there is little available information concerning Steinmetz himself. In his tract on the comet he reported that he was a pupil of the late Joachimus Camerarius, and it seems that he was at one time a professor at Leipzig,[3] where his brother, Mauritius Steinmetz, was a professor of mathematics and botany.[4] Valentin has been called " Philomathesius "[5] and is the author of several prognostications or practicas.[6]

1 Brahe, IV, 363-4.

2 Bassaeus, II, 367.—Brahe, IV, 363-4.—Hellmann (1883), 479.—Hellmann (1924), 30.—Jöcher, IV, 800.—Poggendorff, II, 999.—Riccioli, II, 28, 89.—Scheibel, 97.—Zedler, XXXIX, 1719.

3 Tycho, Scheibel, and also Riccioli who classified Steinmetz with the believers in the sublunar position of all comets.

4 On A_{ii} r, A_{ii} v, A_{iii} r, and C_{ii} v of the Leipzig and Magdeburg editions of Valentin Steinmetz's tract on the comet of 1577 (items 103, 103a of appendix, below) and on A_{ii} r, A_{ii} v, A_{iii} r and C_{i} v of the Augsburg edition (item 102), he spoke of his brother Mauritius. This Mauritius or Moritz was born in Gersbach and died in Leipzig in 1584. See Poggendorff, Jöcher and other references.

5 Hellmann (1924), 30.

6 Hellmann said that Steinmetz wrote certainly five, probably seventeen, prognostications between 1581 and 1597. He listed (Hellmann (1883), 479) prognostications for 1581 (a quarto printed in Leipzig), 1582 (a quarto printed in Erfurt), 1592 (a quarto printed in Leipzig) and 1597 (a quarto printed in Erfurt). These include the following two tracts:

Prognosticum Astrologicum, oder grosse teutsche Practica auff das Jahr 1592. Leipzig, 1591. (Bassaeus, II, 367; Hellmann (1883), 479).

His book on the comet of 1577 [7] was dedicated to Valentin Meder because of the latter's interest in astronomy and other mathematical skills and with the hope that it would not only please him, but also instruct the young people who read it. All Valentin Steinmetz's observations seem to have been made in Leipzig. He and his brother first observed the comet on St. Martin's day, November 11th, shortly after 5 P.M., and he decided that it originated in Capricorn. When he saw it a second time, three days later, he found it in the Milky Way. Then he traced the origin of the comet to a conjunction of Saturn and Mars in Capricorn in the previous year. Material was added to it throughout the following year, and it was ignited because of a meeting of Jupiter and Venus in October. At 5 P.M. on the 17th of November he found the comet 24° above the horizon, almost in the thirteenth degree of Capricorn.

Steinmetz's record of observations was too vague to be of scientific value in an age when some observations were sufficiently accurate to prove the absence of parallax. November 17th the comet set at 8:15 P.M. On the 21st, shortly after 5 P.M., it was 30° above the horizon. It traversed 22° of Capricorn bearing north and was almost 4° from the ecliptic. It set at 8:45 P.M. together with the horn of Capricorn. Bad weather interrupted the observations, but the comet was perceived 5° further east and 4° higher on November 24th, and 6½° higher, or 40½° above the horizon, shortly after 5 P.M., on December 1st, when the altitude of Aquila was 33°. At that time the comet advanced eastward and reached the thirteenth degree of Aquarius. It was 54° 26′ from the sun, and had a latitude of 26° measured north from the ecliptic and a declination from the equator of 8° 20′. When it set after 8:30 P.M., Perseus was in the middle of the sky. On December 3rd Steinmetz made further observations using larger instruments. The last recorded

Schreibkalender auf das Jar . . . M.D. LXXXII. Gerechnet durch M. Valentinum Steinmetz. Erfurt, Johan Beck and J. Börners, [1581]. (B.M. catalogue and supplement).

7 This pamphlet appeared in three editions, items 102, 103, and 103a of appendix, below.

observation was made on December 7th. From November 11th to that date, the comet had traversed Capricorn and all but the last 5° of Aquarius.

Steinmetz cited Joachimus Camerarius as having written that such an unusual comet must mean evil. Like so many other writers on the comet of 1577, Steinmetz enumerated previous comets with a short account of the misfortunes which followed them. He spoke of the comet in Nero's time, and of comets in the years 340, 454, 557, 603, 676, 745, 761, 839, 876, 945, 983, 1066, 1211, 1253, 1301, 1305, 1312, 1337, 1433, 1434, 1444, 1472, 1491, 1500, 1506, 1531, 1532, 1556 (observed by his brother Mauritius), and 1558. He predicted bad times attendant on the still visible comet, wars because of the red or martial aspect of the tail, and sudden death because of the resemblance of the body to Venus or Jupiter. He cited the hundredth saying of the *Centiloquium* and ended with a short prayer to God for mercy.

A name well known in the sixteenth century was "Dasypodius", the Latin translation of a German name,[8] in all probability referring to some type of rabbit.[9] One owner of the name was Petrus or Peter Dasypodius, the Swiss humanist. He lived in Strasburg and was the author of Latin and Greek dictionaries. His son Cunradus or Conrad [10] lived in the same city [11]

8 The A.D.B., IV, 763, citing Ersch and Gruber, Wackernagel's *Litteraturgeschichte*, Grimm's *Wörterbuch*, I, xx, and Hirzel, listed the possible German names "Rauhfuss", "Rauchfuss", "Has", "Haslein", and "Hasenfratz". Blumhof, 16 note, gave the name "Rauhfuss"; Wolf (1845), 137, and Poggendorff, I, 524, suggested a choice of "Rauchfuss" or "Hasenfuss"; Hellmann (1883), 85, suggested "Hasenfuss"; and Haag listed Conrad Dasypodius under "Rauchfuss" and not under "Dasypodius".

9 The name means literally a rough foot or one covered with hair or feathers, or a hare or young hare or one with the appearance of a hare.

10 Adam (1615), I, 441-3. — A.D.B., IV, 764, article by L. Spach. — Bassaeus, I, 452, where a list of books by Dasypodius can be found. — Blumhof.—Bodleian library catalogue.—Cat. Belg., 383.—Crawford library catalogue, 151.—Frank, I, 322.—Haag, VIII, 391.—Hellmann (1883), 85.—Hellmann (1924), 14, 26.—Hoefer, XIII, 149-150.—Jöcher, II, 37.—Poggendorff, I, 524, 1554.—Riccioli, II, 28, 89.—Rosenthal, catalogue 168.—Scheibel, II, 362.—Schmidt.—Schoepflin, 292.—Thorndike, VI, 88-90.—Thou-Teissier,

and died there on April 22nd or 26th, 1600.[12]

Preparatory to teaching, Conrad Dasypodius received his mathematical education in Strasburg, Paris, and Louvain. He returned to Strasburg as professor of mathematics, succeeding, in October 1562, Christian Herlin under whom he had previously studied. There he disputed the philosophy of Ramus and taught geography and astronomy in addition to mathematics. His special interest in Greek mathematics was fostered by his knowledge of the necessary mathematics as well as of the Greek language, and he prepared and had published editions of several Greek scientific works, including Euclid's *Elements* and the *Catoptrica,* the first book of Aristotle's *Meteorologica,*[13] and part of Ptolemy's *Apotelesmatica* [*Quadripartitum*]; and, in 1572, Theodosius' and Autolycus' works. In consequence, he helped raise the general level of mathematical and scientific knowledge. In addition, he made a translation from a work by Regiomontanus[14] and wrote at least four and possibly eight prognostications in German in the years 1575 to 1582.[15] He

371-2.—Ungerer, 164-5, 489.—Vossius, 34 (chapter VIII, § 5), 6 (chapter II, § 1), 68 (chapter XVI, § 27), 111 (chapter XXVI, § 13), 192 (chapter XXXVI, § 22).—Weidler, 380-1.—Weller (1857-8), 323.—Witte, obiit... 1601.—Wolf, III, 51-62.—Wolf (1845).—Zedler, VII, 225.

11 According to Adam, Conrad was born in Strasburg. Schmidt said that he was born in 1530 although some authorities have dated the event in 1531 and 1532. Wolf spoke of him as "Konrad Dasypodius von Frauenfeld", the city in Switzerland whence his father had come, and Wolf (1845) gave 1531 as the year of Conrad's birth, adding that the senior Dasypodius had moved to Strasburg in 1530. Both Poggendorff and Hellmann designated Frauenfeld as Conrad's birthplace. However, he seems to have spent his life in Strasburg, whether he was born there or not.

12 Witte gave the date of Conrad Dasypodius' death as April 20, 1601.

13 See below (in this chapter, especially note 24), the comparison of Dasypodius' classes of comets with those defined by Aristotle in the *Meteorologica.*

14 Doppelmayr, 16.

15 Hellmann (1924) gave the full title and table of contents of the prognostication for 1582. Elsewhere (Hellmann (1883), 85), he listed three "practicas" or "prognostsications" by Dasypodius for the years 1575, 1576, and 1578, all quarto volumes printed in Strasburg.

corresponded with contemporary astronomers, including Kepler.

Teacher, author,[16] and canon of St. Thomas' Church in

16 Books by Dasypodius are listed in various bibliographies and catalogues, the most nearly complete list being Blumhof's (17-32), which, because of the rarity of Blumhof's work, is quoted below. The items marked with an asterisk were seen by Blumhof. His bibliographical references and summaries have been omitted.

"I) Euclidis Catoptrica, id est Elementa ejus scientiae, qua universa speculorumvit (*sic*) atque natura explicatur: primum Graece, antehac nunquam in lucem edita; et nunc nova translatione per *Conradum Dasypodium* in Latinam linguam translata. Argentorati. 1557...

II*) ΕΥΚΛΕΙΔΟΥ ΤΩΝ ΠΕΝΤΕ ΚΑΙ ΔΕΚΑ ΣΤΟΙΧΕΙΩΝ, ΕΚ ΤΩΝ ΤΟΥ ΘΕΩΝΟΣ etc. Euclidis quindecim Elementorum Geometriae primum ex Theonis Commentariis Graece et Latine. Cui accesserunt scholia, in quibus ad percipienda Geometriae Elementa spectant, breviter et dilucide explicantur, authore *Cunrado Dasypodio*, Scholae Argentinensis professore. Argentorati Excudebat Christianus Mylius. J564 [sic] ...

III*) ΕΥΚΛΕΙΔΟΥ etc. Euclidis quindecim elementorum Geometriae secundum: ex Theonis commentariis Graece et Latine. Item Barlaam monachi Arithmetica demonstratio eorum quae in secundo libro elementorum sunt in lineis et figuris planis demonstrata. Item Octo propositiones stereometriae, ejusdem cum praecedentibus argumenti. Per *Cunradum Dasypodium* scholae Argentinensis Professorem....

IIII*) Propositiones reliquorum Librorum Geometriae Euclidis, Graece, et Latine, in usum eorum, qui volumine Euclidis carent. Per *Cunradum Dasypodium*, scholae Argent. etc. Argentorati apud Christianum Mylium. 1564....

V) *Christiani Herlini* et *Cunradi Dasypodii* Euclideae Demonstrationes in Syllogismos resolutae. Argentorati. 1564....

VI*) Analyseis geometricae sex librorum Euclidis. Primi et quinti factae a *Christiano Herlino*: reliquae una cum Commentariis, et Scholiis perbrevibus in eosdem sex libros Geometricos. a *Cunrado Dasypodio*. Pro schola Argentinensi. Arg. 1566....

VII) *Dasypodii* Logistica. Argent. 1567....

VIII) Volumen primum mathematicum. Prima et simplicissima Mathematicarum disciplinarum principia complectens: Geometriae. Logisticae. Astronomiae. Geographiae. Per *Cunradum Dasypodium* in utilitatem Academiae Argentinensis collectum. Una cum Classium ejusdem Academiae, ordinariis Lectionibus. Arg. excudebat. *Josias Risselius*. 1567....

IX) Compendium Theoriae Planetarum.... [1567].

X) Hypotyposes orbium coelestium congruentes cum tabulis Alfonsinis et Copernici, seu etiam tabulis Prutenicis, editae à *Cunrado Dasypodio*. Argent. 1568....

XI) Euclidis Propositiones. Elementorum. 15. Catoptricorum. Opticorum. Harmonicorum. Et Apparentiarum. Per *Cunradum Dasypodium*. Arg. Apud. haeredes Christiani Mylii. 1570....

XII) De terminis geometricis per *Dasypodium* Arg. 1570....

XIII) Euclidis Elementorum Liber primus. Item, Heronis Alexandrini vocabula quaedam geometrica: ante hac nunquam edita, graece et latine. Per M. *Cunradum Dasypodium.* Cum gratia et privilegio Caesareo, atque Regis Galliae, ad sexennium. Argentinae 1571....

XIV) Euclidis varia Scripta Graece et Latine, edita a *Cunrado Dasypodio.* Argent. 1571....

XV) Euclidis omnes omnium Librorum Propositiones graece et latine: editae per M. *Cunradum Dasypodium.* Cum gratia etc.... Arg. 1571....

XVI) Euclidis Phaenomena in Sphaericis scriptoribus *Cunradi Dasypodii* —Argent. 1572....

XVII*) Sphaericae Doctrinae Propositiones Graece et Latine, nunc primum per M. *Cunradum Dasypodium,* in lucem editae, quorum authores sequens indicat pagina. Cum privilegio Caesareae Majestatis ad sexennium et Regis Galliae ad septennium. Excudebat Christianus Mylius. Argentorati. 1572....

XVIII*) ΛΕΞΙΚΟΝ, seu Dictionarium Mathematicum, in quo Definitiones et Divisiones continentur scientiarum Mathematicarum.... M. *Cunrado Dasypodio* Authore. Argent. 1573....

XIX) Euclidis Elementa Graece et Latine. Interpr. *Cunrado Dasypodio,* cum Scholiis. Argent. 1573....

XX) *Cunradi Dasypodii,* Mathematici Argent. Scholia et Resolutiones seu Tabulae in Lib. III, Apotelesmaticos *Cl. Ptolemaei*: Una cum Aphorismis eorundem Librorum. Denique brevis explicatio Astronomici Horologii Argentoratensis. ad veri et exacti temporis investigationem extracti.... [1578] ... [This was published in an edition of Cardan's commentary on the *Quadripartitum.* See B.N. catalogue, XXIII, 805, according to which the title should read "Lib. IIII" not "III".]

XXI*) Brevis doctrina de Cometis, et Cometarum effectibus. Per M. *Cunradum Dasypodium.* (Argentor. Excudebat N. Wyriot.) 1578....

XXII*) *Von Cometen, vnd jhrer würkung. durch M. Cunradum Dasypodium beschriben. Gedruckt zu Strassburg bey Niclauss Wyriot.* 1578....

XXIII) Isaaci Monachi Scholia in Euclidis Elem. VI. priores libros, per *Dasypodium.* Argent. 1579....

XXIIII*) Oratio *Cunradi Dasypodii* de disciplinis Mathematicis: Ad Fridericum II. Sereniss. Regem Daniae etc. *Ejusdem* Hieronis Alexandrini nomenclaturae Vocabulorum Geometricorum translatio. *Ejusdem* Lexicon Mathematicum, ex diversis collectum antiquis scriptis. Excudebat Nicolaus Wyriot. Argent. 1579....

XXV*) *C. Dasypodii* Heron Mechanicus: Seu de Mechanicis artibus, atque disciplinis. *Ejusdem* Horologii astronomici, Argentorati in summo Templo erecti, descriptio. Argent. Excudebat Nicolaus Wyriot. 1580....

XXVI*) *Cunradi Dasypodii* Protheoria mathematica, in qua non solum disciplinae Mathematicae omnes, ordine convenienti enumerantur: verum etiam universalia Mathematica praecepta; explicantur.. Arg. 1593....

XXVII*) *Cunradi Dasypodii* Institutionum Mathematicarum Voluminis primi Erotemata. Logisticae. Geometriae. Sphaere. Geographiae. Cum Privilegio Caesareo. Pro schola Argentinensis imprimebat *Josias Rihelius* 1593....

Strasburg, to which position he was named in 1563, and later deacon there, Dasypodius' chief claim to fame now seems to be the astronomical clock in the Strasburg cathedral, which he not only constructed,[17] but also described in writing.[18]

Although a mathematician of great repute in his own time, Dasypodius' astronomical knowledge was inferior. Tycho made a short unfavorable criticism of Dasypodius' work on comets, which he said was entirely devoted to astrological predictions and showed the influence of Aristotle and Ptolemy.[19] Riccioli listed Dasypodius not only among those who found the comet of 1577 to be below the moon, but also among those who believed that all comets are sublunar. However, in considering the importance of Dasypodius' books on the comet of 1577, his prominent position in the community and his consequent influence must be kept in mind.

Dasypodius wrote two books on the comet of 1577, one in German, the other in Latin.[20] The preface to the latter, dedicat-

XXVIII*) *Cunradi Dasypodii* Institutionum Mathematicarum Voluminis primi Erotematum. Appendix. Elementorum. Arithmeticae. Geodaesiae. Opticae. Catoptricae. Scenographiae. Theoriae planetarum. Logisticae. Astronomicae. Astrologiae. Musicae. Mechanicae. Cum privilegio Caes. et Reg. Gall. Imprimebat Jos. Rihelius. Argent. 1596. . . ."

Haag, VIII, 391, listed ten published works by Dasypodius, and several other works attributed to him, which Haag said were probably left in manuscript. Nine of the items listed by Haag were also listed by Blumhof. The other item listed by Haag was entitled: *Volumina mathematica III pro scholâ argentinensi*, and was an octavo published in Straburg in 1570.

17 Dasypodius directed Isaac and Josias Habrecht in the construction. See Schoepflin, 292, note 6. The clock was constructed in 1574. See Thorndike, VI, 88, and Ungerer, 164-5, 489.

18 *Warhafftige Ausslegung vnnd Beschreibung des Astronomischen Uhrwercks zu Strassburg im Münster / welches er anfänglichs erfunden vnd angeben hat.* Strasburg, 1580. See A.D.B., IV, 764, Bassaeus, II, 284, and Crawford library catalogue, 151. See also Dasypodius' *Heron Mechanicus* (1580) where the clock is described in Latin.

19 Brahe, IV, 361-2 (in chapter 10 of the *De Mvndi . . .*). Blumhof said that Dasypodius' two works on the comet of 1577 include nothing unusual and show that a professor of mathematics in the sixteenth century could have an exceedingly confused picture of comets.

20 Items 33a and 33 of appendix, below.

ing the tract to " CL. V. Ioanni Sambvco, Caesareae Maiestatis Historico . . .",[21] and dated from Strasburg, February 1, 1578, declared that the purpose of the book was to explain clearly the Aristotelian theories and to describe the comet of 1577 in terms of them. Dasypodius said that Aristotle distinguished two types of comets, whereas others noted three groups with different motions and times. Eight lines of verse[22] by Adamus Colbius Fagius tell that he was aware that celestial phenomena, especially bearded luminaries, signify misfortunes, but that as a good Christian he thought he could avoid them. Three pictures represent three different " kinds " of comets.[23]

Dasypodius first dealt with the material, the form, and the power of comets. He began with a quotation from Aristotle on meteors and added that comets were among those phenomena about which much had been written. He believed that although it was known that comets were engendered in the upper air, not everything could be explained by the method of sensory observation, taught by Aristotle, and that therefore it was necessary to resort to inference in explaining comets. In the opinion of Dasypodius, none of the divers explanations of comets, offered by various philosophers, approached the truth as closely as did Aristotle's, whose theory of hot, dry exhalations he presented briefly. Comets, he said, do not first come into being when they are first visible, but have accumulated gradually. He traced the notion of " stella barbata " etc. back to Aristotle, Pliny and " others." Although without attaching much significance, Aristotle had described two types of comet.[24] Dasypodius said that our daily experience teaches us that comets are of many different forms and materials. He further elaborated on Aristotle's

21 Johannes Sambucus, a Hungarian student, "physician, antiquary and poet", was born in Tornau, Hungary, in 1531 and died in Vienna in 1583 or 1584. See Green, editor, 289 ff., Hoefer, XLIII, 233-4, Thorndike, VI, 88, and Zedler, XXXIII, 1653-4.

22 a_{1v} r.

23 a_{1v} v. See appendix, below.

24 Aristotle (1923), 344^a 21 - 344^a 25. See chapter I, above, especially note 25.

theory, and then discussed the different ways in which comets can be classified: by motion, color and so forth, and what the different types can be called. He distinguished between the " stella comata ", " barbata ", and " caudata ". He believed that little by little heat and dryness exhaled by the earth collect and stick together and afterwards are set in motion by an eclipse of the sun or moon or by the configurations of the stars, and are ignited by Mars or Mercury. Discussing predictions from comets, Dasypodius, unlike many of his contemporaries, identified the " comet " which appeared at the time of the birth of Christ with that of 1572. Although he showed independence in doing this, he also showed that he did not realize that novae are stars, not comets. He then explained how the material is collected and how that which is superfluous descends again.

The second chapter, on the nature, properties and effects of comets, begins by saying that astronomers should observe a comet with astronomical instruments and note its position in longitude and latitude; whether it is in the ecliptic, which is rare, or touches the ecliptic, which is more frequent; its distance from the center of the world and from the center of the field of vision; the type of tail; and the planets which are in the region.[25] Dasypodius thought that in order to make predictions, divided by Ptolemy into two groups, general and particular, one should know the sign of the zodiac in which the comet appears, the size, form, motion and so forth, and whether or not the comet's tail is turned toward the sun. The fact that a comet's tail is always turned away from the sun was evidently not grasped by Dasypodius. He emphasized the importance of noting the part of the sky in which a comet appears, in order to predict the part of the earth which will be affected by the plague which he assumed would follow the comet. In addition, he stressed the importance, for the purpose of making predictions, of the resemblance of the color of a comet to that of a planet.

25 Nowadays, it is more usual to observe right ascension and declination than longitude and latitude. Of course, it is superfluous to see if the comet touches the ecliptic, because that information is contained in the observations of right ascension and declination. See chapter III, note 19, above.

He considered the comet's size, brilliance, and form and the time of its appearance valuable in foretelling events. Furthermore, he made the statement that the motions of comets do not follow fixed laws.

Then [26] he dealt with the effects of comets, being entirely concerned with predictions, and adding very little to what had already been said. He illustrated [27] his discussion of these effects by a recital of events which comets had supposedly presaged.

Finally,[28] Dasypodius considered the comet of 1577, reverting to such ideas as those of Leopoldus and Albumasar. The only sentences of astronomical worth say that the comet appeared about the 9th of November, 1577, and remained visible until January 1578, that it varied in color from pale to red,[29] and that the tail or " beard " reached from Capricorn to the beginning of Aquarius. According to Dasypodius, the comet was the occasion for gloomy predictions, such as the deaths of kings and important people, wars between kings, earthquakes, winds and other ills. He ended his book with a phrase, cited in both Greek and Latin, the latter being: " Impunè nunquam uisus fulgêre Cometes." [30]

Dasypodius' conception of what astronomers should observe was largely correct, but his conclusions, based on authority rather than observation, were erroneous, and his purpose was astrological. However, he furnished some data useful to astronomers.

A work of no greater astronomical value than those by Steinmetz and Dasypodius was the treatise in Latin by Nicolaus Bazelius,[31] a doctor and surgeon in the second half of the six-

26 Chapter III.

27 Chapter IV.

28 Chapter V.

29 " rutilum ", red, inclining toward golden yellow.

30 See the account of item 12 in the appendix, below.

31 Variants of the name are Baselius, Bazel, or Basel.
Bassaeus, I, 528, which listed item 10 of appendix, below.—Bib. Belg. Gand, series 1 and 2.—Biographie Nationale... de Belgique, I, 742-3, article

teenth century. Bazelius was born in Nieuwkerke, in Flanders.[32] It is not known whether he studied medicine and surgery in Louvain or in Paris. In 1578 he held a public office in Bergues. In evaluating his work it must be remembered that in the sixteenth century many men of education felt themselves qualified to discuss such astronomical phenomena as comets and meteors, even though they were untrained in astronomy, because the principal requirement for such a discussion was a knowledge of ancient literature, and this Bazelius had to a certain extent.[33] Tycho, who listed Bazelius with the believers in the sublunar

by Félix Nève.—Brahe, IV, 362-3.—Dreyer, editor, IV, 510, citing item 10. —Foppens, I, 899, which listed item 7.—Gesner (1583), 620, listing item 10. —Hellmann (1924), 31.—Hoefer, IV, 662, which listed item 7.—Jöcher, I, 876, which listed item 7.—Riccioli, II, 28, 89.—Scheibel, 101, listing item 7 and citing page 453 of Tycho's book appearing in " 1577 ".—Schotel, 106-113.

Further information might be found in the following sources, mentioned in the above listed works:

Sweertius, Franciscus. *Athenae Belgicae, sive nomenclator Infer. Germaniae scriptorum,*... Antwerp, 1628.

De Backer, or Debaecker, Louis. *Recherches sur la ville de Bergues en Flandre.* Bruges, 1849, p. 210.

Hautschilt. *Imago Flandriae.* 1604.

32 The Bib. Belg. Gand, series 2, v. I, concluded from the title of Bazelius' 1561 prognostication [see note 263, below] that Louis Debaecker (*Recherches hist. sur la ville de Bergues en Flandre,* p. 210) and Félix Nève (*Biographie Nationale,* I, p. 742) were wrong in maintaining that Bazelius was born in *Bergues-Saint-Winoc* (also called Winoxberg); that he was really a native of Nieuwkerke (Neuve-Eglise) near Bailleul or Belle; and that he was merely a doctor and surgeon in Bergues.

33 Félix Nève (Biographie Nationale... de Belgique, I, 742) said that Bazelius' studies must have been of an unusual breadth because he busied himself with scientific observations foreign to his profession. This is not a necessary conclusion. Moreover, his publications seem not to have been numerous. The B.M. and B.N. have only one book, each, by Bazelius, namely item 10 of the appendix below. However, the Bib. Belg. Gand. series 2, v. I, listed the following work by Bazelius, which Hellmann (1924), 31, called his first prognostication, the last being items 8, 9, and 10 of the appendix below:

¶ *Prognosticatie || vanden Jare ons Heeren. M.D. || LXJ. Ghecalculeert op den Meridiaen der stede || van Poperinghe in Westvlaenderen / door M. Ni= || claes Bazelius van Nieukercke / by Belle / Mede= || cijn eñ Cirurgien / Eñ der Astronomijnscher con= || sten een vast Liefhebber / woonende binnē der voor || seyder stede inden gulden helm inde Jperstrate.||*

position of the comet,[34] called Bazelius' book " careless ", and pointed out deficiencies in it, such as his failure to define either his longitude or his latitude. However, Tycho was more favorable toward the astrological explanations in the book.

According to its caption, which really applies to the whole book, the first section of Bazelius' tract on the comet of 1577 [35] deals with the time, form and position of the comet, with its observed motion, and with the general meanings of comets and the meaning of this one in particular. The comet was first seen from Bergues in the sixth degree of Capricorn, on November 14th, at 5 P.M., immediately after sunset. However, the author dated its first appearance on the tenth. He referred its longitude to the position of Saturn and its latitude to the ecliptic; and even if those are not the usual points of reference, it must be acknowledged that he defined the comet's apparent position. His next recorded observation was on November 16th. The tail, when first seen, was diffused in length and width and stretched out toward the moon, then in the third degree of Aquarius. It appeared pale and mournful, with a horrible form, and it spread southwards, for which reason, said Bazelius, citing Aristotle, the comet was called bearded. It set later each day and by the beginning of December it had moved through the zodiacal signs so that it followed the moon. It moved further and further north in latitude, toward Bazelius' zenith, its head preceding, its tail always turned toward the east. The observations made were the usual ones and the positions recorded were for November 22nd, 24th, 28th and December 28th, by which times its brilliance was considerably diminished. On the 4th and 6th of January it could scarcely be distinguished. Its path was observed throughout the period of the comet's visibility, bearing towards the place in Cassiopeia where the star of 1572 had been seen. This comment by Bazelius introduced a few further remarks on the nova visible until the beginning of February 1574.

34 Brahe, IV, 362-3. According to Riccioli, Bazelius considered all comets sublunar.

35 Item 10 of appendix, below.

This was stationary, as Bazelius pointed out, and thus unlike the comet.

Next, Bazelius concerned himself with ancient and modern opinions on the causes or generation of comets. Then he discussed the nature and significance of the hairy or bearded star of 1577. He continued this discussion at considerable length, basing his opinion on book I, chapter III of Ptolemy's *Quadripartitum*. The future of the year 1578, said Bazelius, would be full of misfortunes and these were predicted not only by the comet of 1577 but also by other astronomical events. For example, the lunar eclipse of the night of September 26th to 27th, 1577, would have its effects from the beginning of May 1578 to the end of August. After enumerating different phenomena, Bazelius made predictions for 1578, such as religious and political changes. The comet, he said, was a forerunner of all these evils, and he enumerated its effects.[36] In this connection Gemma's *De naturae divinis characterismis*[37] was cited, bits of poetry dealing with the effects of comets were quoted, and a poem by "Jovianus Pontanus" on the general meanings of comets was presented.[38] According to Bazelius, Aratus confirmed these predictions; and Manilius wrote extensively on the subject in his first book. Bazelius further concerned himself with the signification of the comet of 1577, relying on Ptolemy's theories. He gave no astronomical information. Since the comet was west of the sun, said he, its effects were to have a late beginning and were to continue into subsequent years. He spoke of the localities where the comet's influence would be felt, for example, India, Macedonia, Italy, Albania, Greece, Hesse, Thuringia, Styria, and the Orkney Islands, in all of which

36 They were pictured on the recto of A_{ii}. See the description of item 10 of the appendix, below.

37 This work appeared in 1575.

38 It consists of the first fourteen and the last two lines of the second of the two sections on comets in Pontano's *Meteororum Liber* (Pontano (1902), I, 215-7).

places there was to be a maximum of dryness and sterility. He also listed the towns to be affected, and recounted the prophecy found in Eeckhoutte,[39] which was illustrated by the figure of a woman nursing wolves.[40] Bazelius brought his book to a close by elaborating his ideas on predictions from comets without any reference to the particular comet which prompted him to write. Any comet would have suggested the same remarks.

39 See the description of item 10 in the appendix, below.

40 See the description of item 10 in the appendix.

CHAPTER VI

THE COMET OF 1577: PREACHERS AND POETS WHO USED THE COMET MERELY AS A THEME TO FURTHER THEIR OWN PURPOSE

CHYTRAEUS. — PAULI. — SELNECCER. — HEERBRAND. — KREIDWEISS. — ROCCA

MANY of those who described the comet of 1577 were churchmen whose main interest in the comet was in its use as a theme for their sermons. Prominent among these was David Chytraeus (Kochhafe),[1] who was the last of the "Fathers of

1 Chytraeus, the son of a Lutheran clergyman, was born in Ingelsingen on January 26, 1530. (Krabbe, 551 note. Thorndike, V, 397-8, citing Krabbe, gave the same date. Dreyer, editor, VIII, 454, placed Chytraeus' birth in Swabia in 1530. Paulsen, 8, gave the year 1530 and it was implied by Chytraeus in his tract on the comet of 1577 (see below, note 14). The A.D.B., IV, 254, gave the date February 26, 1531 for Chytraeus' birth.) He died in Rostock on January 25, 1600. At a very early age he attended Tübingen University. (According to the A.D.B., IV, 254, this may have been at the age of nine. Döllinger, II, 521, Adam (1653), 682, and Thou-Teissier, 402, called Chytraeus a pupil of Camerarius. Joachim Camerarius left Tübingen in 1541. Klatt, 5, gave 1539 as the year Chytraeus went to Tübingen.) There he studied law, philosophy and philology, and later proceeded to theology. Having received the degree of bachelor and, at the age of fifteen, that of master of philosophy, he went to Wittenberg, where he came into close association with Melanchthon. (According to Paulsen, 8, Chytraeus went to Wittenberg in 1544.) He went to Heidelberg in 1546, to Tübingen in 1547, and returned to Wittenberg in 1548, where, following Melanchthon's counsel, he taught rhetoric, elementary astronomy, and Melanchthon's *Loci communes,* learning as he taught. After traveling in Switzerland and Italy (Zedler, V, 2311, said that Chytraeus went to Italy in 1550), he was called to the university of Rostock by the Dukes Henry and John Albert of Mecklenberg. (Paulsen, 11, gave the date of Chytraeus' call to Rostock as 1550. Freher, I, 314, gave it as 1551.) There, on April 21, 1551, he began his lectures on the Christian catechism and the books of Herodotus. He gave theological lectures from 1553 on, but first became a full professor of theology in 1561, after becoming a doctor of philosophy on April 29th.

Further information concerning Chytraeus can be found in the following works: Adam (1653), 681-696.—A.D.B., IV, 254-6, article by Fromm.—

Bassaeus, I, 27-8, 455-6, 631, II, 47, 285-6, where lists of books by Chytraeus can be found.—Brahe, IV, 366-7.—Calinich, 239.—Döllinger, II, 521-531. This gives much information concerning Chytraeus' religious activities.—Dreyer, editor, VIII, 454.—Dreyer (1890), 24, 242.—Frank, I, 166, 221-2, 300.—Freher, I, 314-5. There is a picture of Chytraeus opposite page 311.—Gesner (1583), 186-7.—Janssen, IV, V, VI, passim.—Klatt. This is excellent and has bibliographies of Chytraeus' writings and a list of books about him. —Krabbe, 550-7.—Lindesiana... Luther no. 1412, listing a book by Chytraeus. —Paulsen. This has bibliographical notes but no mention of any astronomical writings.—Pressel. This biography emphasises Chytraeus' relations with the Church.—Realencyklopädie, IV, 112-6, article by George Loesche. This article contains excellent bibliographies of works by and about Chytraeus, as well as an interesting account of his life and work.—Riccioli, II, 136-7.—Scheibel, 96.—Schottenloher, I, 118.—Thorndike, V, 397-8.—Thou-Teissier, 402-9.—Witte, obiit... 1600. This has a short bibliography of Chytraeus' writings.—Zedler, V, 2311-2.

The following references are cited in the above works:

Bacmeister, Lucas. *Leichpredigt bey dem Begrebniss des Ehrwürdigen, Achtbaren und Hochgelarten Herrn Davidis Chytreai*... Rostock, Stephan Müllmann, 1600 (Klatt. For information concerning Bacmeister see Döllinger, II, 532 note 15.)

Becher, Dr. Otto, *Das Kraichgau und seine Bewohner zur Zeit der Reformation. Oratio von David Chytraeus*... Karlsruhe, 1908. (Klatt; Schottenloher, I, 118.)

Chytraeus, Ulrich. *Vita Davidis Chytraei, Theologi summi. Historici eximij, Philosophi insignis, Viri optimi & integerrimi. Memoriae posteritatis, Orationibus et Carminibus Amicorum, justisque Economiis consecrata,* Rostock, Christophor Reusner, 1601. (Klatt said that this does not contain any biography by Ulrich of his father, but is a compilation of various works.)

Geiger, Ludwig. In: *Göttinger gelehrte Anzeigen*. 1870. (Klatt.)

Goldstein, Johannes. *Oratio de vita, studiis, moribus & morte reverendissimi & clarissimi Domini D. Davidis Chytraei*... Rostock, Myliandrini, 1600. (Klatt.)

Hausmann, Richard. *Studien zur Geschichte des Königs Stephan von Polen*. I. Teil. Diss. Dorpat. 1880. (Klatt.)

Kohfeldt, G. *Der akademische Geschichtsunterricht im Reformationszeitalter mit besonderer Rücksicht auf Dav. Chytraeus in Rostock*. In: *Mitteilungen d. Gesellschaft für deutsche Erziehungs- und Schulgeschichte.* Jahrg. XII, Heft 3 (1902), 201-228 (Klatt; Schottenloher, I, 118).

Krabbe, Otto. *David Chytraeus*. Rostock, 1870. 2 v. (Klatt.)

Lisch, G. C. F. *Des Professors Dr. David Chytraeus zu Rostock Bericht von der Kirchenordnung an den Herzog Ulrich von Mecklenburg. 1599.* In: *Jahrbücher des Vereins f. meklenb. Geschichte u. Alterthumsk.* v. 18. (1853): 187-8 (Schottenloher, I, 118).

Lisch, G. C. F. *Beiträge zu der Geschichte der evangelischen Kirchen-Reformation in Oesterreich durch die Herzoge von Meklenburg und die Universität Rostock, namentlich durch Dr. David Chyträus.* In: *Jahrbücher*

the Lutheran Church ". He was a follower of Melanchthon, but also a worker for the middle way between the sects. He was no fire-brand, but a true scholar. His influence in theological circles was evident from 1555 on, when he edited his *Regulae Vitae.*[2] In 1561 he was in Naumburg at the time when the *Corpus doctrinae Saxonicum* was under discussion there. At Rostock he was interested in the improvement of the university. He took an important part in the theological disputes and congresses so frequent in Germany, and must have had an enormous influence on the Lutherans of his time. He fought against the dogmatism of Peucer.[3] Some of his theological books were condemned,[4] together with other writings of reformers, but such

des Verein f. meklenb. Geschichte u. Alterthumsk. v. 24. (1859): 70-139. (Schottenloher, I, 118; Klatt).

Schnell, H. *Ein Zeugnis des Rostocker Theologen David Chyträus über den Abendmahlsstreit.* In: *Neue Kirchliche Zeitschrift.* 11. (1900). S. 175-180 (Schottenloher, I, 118).

Schütz, Otto Friedr. *De vita Davidis Chytraei.* Hamburg, 1720. 4 v. (Klatt).

Strobel, Georg Theodor. *Neue Beitr. z. Litt.*, I. Bd. 1. St. S. 150 ff. (A. D. B., IV, 255-6; Schottenloher, I, 118).

Struvius, Burkhard Gotthelf. *Davidis Chytraei ad principes Anhaltinos excusatoria (6. Juni 1587).* In: *B. G. Struvius, Acta litteraria.* II. (1717). S. 360-2 (Schottenloher, I, 118).

Sturz, Christoph. *Oratio memoriae Reverendi, Clarissimi & Excellentissimi Dn. Davidis Chytraei*... Rostock, Myliandrini, 1600. (Klatt.)

Veesenmeyer. *Anmerkung über des Chyträus Geschichte der Augsburgischen Confession.* In: *Literarische Blätter.* 6. (1805) Sp. 305-9 (Schottenloher, I, 118).

Wachler, L. *Gesch. der histor. Forschung,* I, 193, 214, 232, 238, 256 (A. D. B., IV, 256).

Wegele. *Geschichte der deutschen historiographie.* Munich and Leipzig, 1885 (Klatt).

Westphalen, Ernst Joachim de. *Monumenta Rerum Germanicarum, Cimbricarum et Megapolensium Tomus III.* (Klatt; A. D. B., IV, 256).

See also: *Epicedium in Obitum D. Davidis Chytraei. scriptum a M. Johanne Neovino Superint Sver.* (Klatt. This can be found in Ulrich Chytraeus' *Vita Davidis Chytraei*...)

2 A.D.B., IV, 255, and Krabbe, 555, gave the date for this work as 1555, with a Wittenberg imprint. According to the B.M. and B.N. catalogues, there were editions printed in Wittenberg in 1556 and 1570 and in Leipzig in 1558 and 1561.

3 It is possible that Peucer also wrote on the comet of 1577.

4 Calinich, 239.

censorship is illustrative of the times and tells nothing of Chytraeus himself. In 1576 with others, including Selneccer, he took part in the composition of the Torgau Book. He was the author of many theological tracts and was an historian of considerable ability. He is especially noted for his history of the Augsburg Confession, of which there were many editions, and for his Saxon chronicles and shorter yearly chronicles.

As was natural in the sixteenth century, Chytraeus' religious convictions colored his astronomical conceptions. This is obvious in his interpretation of the new star [5] and the comet. Tycho thought of him primarily as a theologian, and discussed him last on his list of observers of the comet of 1577.[6] The ideas of Chytraeus must have had wide circulation because of the high position of their author in both theological and academic circles. The German translation of the comet tract was made for the purpose of bringing the book within the reach of those who did not know Latin. That fact is indicative of the growing importance of the vernacular tongues and of the participation of the people in the literature of the time.

In the first part of the Latin edition [7] of his book on the comet of 1577, Chytraeus dealt with the nova which he believed was first seen on November 8, 1572. He said that it remained stationary near Cassiopeia until the beginning of 1574 when it vanished. He referred to Deuteronomy 32, the chapter giving Moses' song which describes God's mercy and vengeance. Although he used the term " star ", Chytraeus did not classify the nova as a fixed star, but seemed to treat it as a comet and quoted Aristotle on comets. He described the comet of 1556,

5 He wrote about the new star of 1572 and its significance in matters of church and state in his commentary on Deuteronomy (Thorndike, V, 397-8).

6 Brahe, IV, 366-7. Tycho and Chytraeus were friends. Tycho had studied in Rostock in 1566, and when he revisited that city in 1597, Chytraeus wrote to him, expressing regret that ill health kept him from paying his respects to Tycho. (See Dreyer (1890), 24, 242.)

7 Item 29 of appendix, below. Unfortunately, because of the war, the Latin and German editions cannot be compared at present. The summary of the Latin text was made from notes taken in 1931 and cannot be as full as the summary of the readily available German edition.

giving its position and what he considered were the disasters resulting from it, and listed, without discussion, the comets of 1500, 1506, 1531, 1532, and 1533, and mentioned the comets of 454, 557, and 594 and their accompanying disasters. He also spoke of "the terrible comet" in 676 during the rule of Constantinus Pogonatus,[8] and told of two comets seen simultaneously in 729. He associated certain comets with certain historical events, such as the "tragedy under Henry IV".[9] He referred to St. Matthew 24, the chapter in which Christ foretells the destruction of the temple and signifies the approaching judgment; and he advised that since that time and that hour are unknown, people should watch like good servants, expecting the master's coming at every moment. Chytraeus also referred to St. Luke 21.2, which tells of the poor widow casting two mites into the treasury; Peter 2.2, which deals with false teachers; Timothy 3;[10] and Deuteronomy 28, which deals with the blessings for obedience and the curses for disobedience. He talked of God dealing out punishment, and said that there was a possibility of allaying this punishment if the king and people would reform. Making further reference to Deuteronomy 32, he said that men return to God and are converted through punishment. Then he cited Pliny and Hipparchus on the apparent motion of the stars.

The whole first section was dated from Rostock, December 19, 1572. It is immediately followed by a section dealing with the comet of 1577,[11] which, according to Chytraeus, was first seen after sunset on November 11th. In the next days, after

8 Constantinus IV, called "Pogonatus", was emperor from 668 to 685.

9 Chytraeus did not say which Henry IV he meant; but the German Emperor was probably the man in question, and the "tragedy" to which reference was made must have been Henry IV's surrender to the Pope at Canossa in 1076. This would have been tragic from the point of view of Chytraeus.

10 Probably 2.3, which describes the enemies of the truth and commends the holy scriptures.

11 This section is entitled "De Cometa, qui proximo mense Novembri, anno 1577. conspici coepit". It was dated from Rostock, December 2, 1577, i. e. "Datum Rostochii, IIII. Nonas Decembris, Anno 1577".

the moon emerged from conjunction, quickly following Aquarius, the comet turned its path toward the little dipper and passed by the constellation [12] of Antinous on the 25th of November, at which time it was equidistant from Aquila and the Dolphin. The tail, already rare and clear, was seen to develop toward Pegasus. At the end of November, standing between the Dolphin and the constellation Equuleus, the comet extended its burning streamers through the mouth of Pegasus as far as the two stars of the head and nearly to the mouth of Pisces. Chytraeus said that the comet was of great size and extent, comparable to the one which preceded the earthquake when the towns of Helice and Bura were swallowed up, and the Lacedemonians were defeated at Leuctra.[13] Chytraeus then enumerated six comets seen in his lifetime, which he said was nearly forty-eight years.[14] The first was reddish, appeared in August, 1531, and wandered with great speed through the constellations Leo, Virgo and Libra. The next he described as darker, broader and more lasting than the first. It appeared in October 1532 and was visible during the two following months. The third, in 1533, radiant and large, was far from the zodiac, next to Perseus, directing its long tail westward over four constellations. In March 1556 a livid black and red comet stretched its tail northward through Virgo, Bootes, Cepheus and other constellations. The fifth comet was that of August 1558. The sixth "comet" was the nova of 1572. Chytraeus then discussed comets and stars in general, quoting from Seneca and Apollonius. He said, and rightly, that the lack of observed parallax for the star of 1572 left no doubt that it was in the highest region. He discussed whether comets are hot, dry exhalations collected together and burning, and then made the usual statement that all experience testifies to the fact that comets bring calamities, supporting this contention with a quotation in Greek

12 The Latin is "Aram Antinoi" and might also be translated as the "altar of Antinous".

13 371 B. C.

14 This points to Chytraeus' birth in 1530 instead of 1531.

from "Ptolemy, book two", undoubtedly book two of the *Tetrabiblos,* concerning the effects of comets. He said that comets from Mars and Mercury signify war, and also that comets are signs from the Son of God of punishments which can be averted or made easier by reforming. Then he called his readers to renewed piety. The book closes with a Latin poem or song concerning the comet by Johannes Frederus, relating some of the calamities mentioned in the text.

Chytraeus' German book [15] on the comet of 1577 is largely a translation, by Jacobus Praetorius, of the Latin tract. In the preface, dated December 16, 1577, Praetorius said that comets are signs put in the sky by God to warn men to be better, and that he was translating the book so that those unacquainted with Latin might read it and know the warning of God and become good Christians.

The first section, dated, as in the Latin edition, December 19, 1572, deals with the nova of 1572. It opens with a quotation, in German verse, of Seneca's description of the comet visible near Cassiopeia in Nero's time. The words spoken by the Empress Octavia to Seneca to describe that comet, which was followed by great evils, were quoted by our author because, he said, for the past forty days there had been an unusual star set afire soon after the conjunction of Saturn and the sun in Scorpio. This new star was first noticed November 8th, stationary near Cassiopeia, forming a diamond shaped figure with the stars in the breast of that constellation. It looked like Jupiter. Although, according to Chytraeus, laymen saw no difference between this unusual star which remained stationary for a while and then disappeared and true natural stars, both those that are fixed and those that wander and are called planets, it is certain that planets at no time deviate from the zodiac whereas the new star was stationary 50° north of the zodiac almost in the zenith. Fixed stars, said Chytraeus, are always visible at definite times and have their constant rising and setting and

15 Item 30 of appendix, below.

never become contracted or extinguished. There follows a summary of Aristotle's views on comets: earthly exhalations kindled in the atmosphere, of two types, those in which the exhalations are pressed together and which spread their flames out evenly and often remain stationary like the star in question, and those whose material is packed closer in one place and more thinly elsewhere where the matter is drawn out lengthwise so that they extend their fiery streamers on one side. Chytraeus described the comet of 1556 and the tragedies supposed to have followed after it. He also mentioned the comet that appeared two years later and listed important personages who died shortly thereafter. He said that comets in all times preceded ills and gave accounts of the comet of 557, during the time of Justinian while all Thrace and Greece were torn by sword and fire; of the comet of 594, during the time of Gregory the Great; of a comet that appeared during the time of Constantinus Pogonatus; of one that came in 729 during the time of Leo the Isaurian; and of three comets that pointed to war between the grandsons of Charlemagne. During the reign of Henry IV, said Chytraeus, there were many comets and there was one the year before Henry V succeeded his father. Eclipses and stars and no doubt comets also, he added, are signs of misfortunes. Citations were given, as in the Latin edition, from Matthew 24, Luke 21.2, Peter 2.2 and Timothy 3, with a warning to the people to reform. Chytraeus said that Pliny told how Hipparchus after seeing a new star was moved to study all other stars; and it was hoped that the readers of this tract would observe the heavens with understanding eyes.

The second section deals with the comet of 1577, which, according to Chytraeus, was first seen on November 11th, St. Martin's Day, shortly after sunset, not far from the horizon. It was shining, bright and clear, in Capricorn, not far from Saturn and it had a long tail, just like thick fiery smoke. On the following day it appeared right after the new moon close to Aquarius and stretched northward. The text continues, as in the Latin edition, describing the comet's appearance and position on November 25th and at the end of November and comparing it in size and extent with the comet of 371 B.C. He said that Anaxagoras considered that unusual torch in the sky to be rather a burning meteor than a comet. As in the Latin edition, Chytraeus discoursed upon the six comets which had appeared during his lifetime of forty-eight years. He thought that, although the cause and nature of these phenomena cannot be completely investigated, it is nevertheless useful to bear in mind these wonderworks of God, and signs of coming misfortune. According to Chytraeus, Seneca counted comets among the everlasting natural stars, and supported this with the testimony of Apollonius, who learned his astronomy in Babylonia, that not only the seven planets but also comets have their definite and usual irregular paths in the sky. Comets were supposed to be special stars, of unequal color, form and size and to wander through the top of the heavens and first to appear when they come to the lowest point in their circles and therefore were clearer and larger when seen from near. However, when they raised themselves to the upper part of the circles and withdrew from the observers, they supposedly seemed smaller and darker and finally disappeared completely, as had the star of five years before. But, according to Chytraeus, the present comet was not an everlasting incorruptible star but a meteor or high-soaring matter set afire in the upper portion of the air which, when the matter shortly thereafter was contracted, would be entirely extinguished and disappear. He added that the natural scientists taught that comets first arise out of dry vapors pressed tightly together in the air, sticking together and increasing in bulk, and finally ignited by the sun and other planets. He repeated that comets have always been followed by wars, pestilence, famine and other changes and, as in the Latin edition, referred to Ptolemy's *Tetrabiblos,* where comets with the nature of Mars and Mercury are discussed. Chytraeus urged men to heed the warnings in the sky and lead Christian lives. Germany, he said, was being wasted in wars, and the common man was godless and giving himself to Epicurean and beastly lusts.

Hence it was not surprising that God was sending punishments of which men were warned by heavenly signs and comets. Because there was a comet at the time of the battle of Salamis, 478 B.C.,[16] Chytraeus was moved to write this memoir of the comet of 1577. He cited Seneca as saying that no one could find out anything greater or more glorious about God or learn anything more useful than what concerned the nature and powers of the stars, and added that everyone should study this.

Included in the German edition of Chytraeus' tract on the comet of 1577 is a tract by Simon Pauli,[17] who was born in the town of Schwerin in the Duchy of Mecklenburg, October 28, 1534. He studied in Rostock and Wittenberg and was professor of theology at the much frequented university of Rostock. He was rector of that university four times and was superintendent from 1565 to 1573, and then again after 1574. He was a staunch follower of Luther and Melanchthon and was devoted to Chytraeus.[18] He died in Rostock on July 17, 1591.[19]

The portion written by Pauli and published with Chytraeus' tract [20] is theological and has little astronomical importance.

16 This battle is usually considered to have taken place in 480 B. C.

17 A.D.B., XXV, 273-4, article by Krause.—Bassaeus, I, 82-3, II, 161-2, where books by Pauli were listed.—Calinich, 238.—Döllinger, II, 531-5.—Freher, I, 284-5. Pauli's picture is between pages 272 and 273 (the first time those pages, which are repeated, occur).—Krabbe, 635-6.—Zedler, XXVI, 1456-7, where there is a list of ten works by Pauli.

Further information concerning Pauli may be found in the above works and in the references cited in them, including the following two which were cited by Freher:

Bucholzer, Abraham: *Index chronologicus*... Görlitz, 1599.

Quenstedt, Joannes Andreas: *Dialogus de patriis illustrium doctrina et scriptis virorum,*... Wittenberg, 1654. 2nd ed. Wittenberg, 1691.

18 Pauli's works include:

Auslegung derer Episteln und Evangelien. Magdeburg. 1572.

Auslegung derer Deutschen geistlichen Lieder D. Martin Luthers und anderer. Magdeburg, 1588.

Calinich, 238, seemed surprised that the first of these escaped the censors.

19 Krabbe, 677; A.D.B., XXV, 273. Freher gave the date of Pauli's death as August, 1591.

20 This section has the title "Vom selben Cometen Erinnerung / warnung vnd vermanūg D. Simonis Pavli, so er in einer Intimation vnd etlichen Predigten gethan hat."

It was written in 1577. In it Pauli said that the comet had long streamers and consequently was called " stella crinita ". There was no doubt that it heralded war and bloodshed, robbery and burning, and heresies. Pauli made continued reference to Matthew 24 [21] and also referred to Apocalypse 6.[22] He thought that the comet should serve as a warning to false teachers. First seen in and around Rostock on November 11th, St. Martin's Day, but supposedly seen elsewhere on the 10th, on which day Luther was born ninety-four years before, this comet announced the downfall of the new mistaken teachers. Pauli said that he would leave to the astronomers the task of writing of the position of the comet in the sky and of its rising and setting, how it first appeared near the stars in the " Eagle " and always stretched its streamers toward " morning " and " midday ", and what further it predicted. He spoke of the wars in the Netherlands, and added that God, through his prophet Luther, had already given warning for sixty years, but, since men did not take heed, He had then placed a fiery prophet in the sky. Pauli urged men to take warning from Jonah and the fall of Nineveh and added that it was more than time to return to God in order not to meet the punishment of the Jews.

Another churchman who described the comet of 1577 was Nicolaus Selneccer.[23] As an astronomer he was of no impor-

21 See the discussion above, under Chytraeus.

22 The revelation of St. John the Divine, 6. Pauli gave, in German, the quotation: "And the stars of heaven fell unto the earth, even as a fig tree casteth her untimely figs, when she is shaken of a mighty wind ".

23 Adam (1653), 663-4.—A.D.B., XXXIII, 687-692, article by v. Egloffstein. — Bodleian library catalogue. — Calinich, 7, 237-8, 239. — Dibelius. — Döllinger, I, 503-4, II, 344-364.—Frank, I, 143, 220-1, 242 ff., 256, 296.—Freher, I, 286-7, citing Adam as a source.—Heppe, II and IV passim.—Houzeau, I, 601.—Janssen, IV-VIII (See indices at the close of each volume.)—Jöcher, IV, 494-5.—Kluckhohn (This is an interesting account although Selneccer plays but a very minor part in it.)—Le Long (1709), II, 677-8.—Mützell (See index in volume III.)—Planck V-VI (See index in volume VI.)—Preger (This reference was given in the A.D.B. and, although an interesting account of the period, does not bear directly on Selneccer and is not important for this study.)—Ranke (This volume was

consulted because of a reference in the A.D.B., but no mention of Selneccer was found. If there is one, it cannot be important. However, there is probably a discussion of Selneccer somewhere in Ranke's writings.)—Realencyklopädie, XVIII, 184-191, article by Wagenmann and re-edited by Dibelius.—Scheibel, 107-8.—Wackernagel, IV.—Will, III, 670-686. (This gives a good sketch of Selneccer's life with a bibliography of his writings.) — Zedler, XXXVI, 1715-7, citing Adam and Freher as sources.—Zevmeri (This work is given as a reference in Jöcher, but the Nicolaus Schellerus discussed there is not Nicolaus Selneccer or Selneccerus.)

Further material may be found in the articles listed in Schottenloher, II, 267.

Born on December 6, 1530 (A.D.B., Jöcher, and the Realencyklopädie; however, Zedler, XXXVI, 1715, Adam (1653), 663-4, and Freher, I, 286, gave the date as December 6, 1532) in Hersbruck near Nuremberg, the son of a notary, by his second marriage, Selneccer early diplayed a talent for music and at the age of twelve played the organ in the town chapel of Nuremberg. In 1549 he went to Wittenberg, where he zealously studied theology and was soon able to lecture on the subject. Through his father he came into personal contact with Melanchthon, under whom he studied. He received his master's degree in Wittenberg in 1554. In 1558 (Zedler and Freher gave the date as 1557.) Elector August I called him to Dresden as chaplain and, perhaps, as tutor to the electoral prince Alexander, who was then, however, but a baby. There Selneccer remained until March, 1562 (Zedler and Freher gave the date as 1561.) when he became a professor in the newly founded university of Jena. This move was brought about by the fact that he had offended the Elector by alluding in his sermons to the latter's passion for hunting, and had also made enemies of Melanchthon's powerful adherents at the Dresden court. He had been, for a long time, a follower of Melanchthon and he himself said that he had long lain ill in the Calvinist hospital from the poison of the sacrament revolt. But after Melanchthon's death (1560), Selneccer drew closer to the Lutherans, making his position in Dresden distasteful to him. He left Jena in 1568 because the professors there were replaced by followers of Matthias Flacius Illyricus, who were Lutherans more extreme than Selneccer. In 1567 Duke Johann Friedrich, protector of Melanchthonians, had been succeeded by his brother Johann Wilhelm who was, in contrast, a zealous Lutheran and had replaced the Jena professors by men of his party, who had been forced out of Jena in 1561. Selneccer returned to the lands of Elector August and took over the position of general Superintendent and pastor of St. Thomas' in Leipzig.

In 1570 Selneccer took his doctor's degree at Wittenberg, and in the same year he was summoned by Duke Julius v. Braunschweig-Wolfenbüttel as chaplain. He established the Protestant church ritual in the Guelph lands and in Oldenburg. In 1571 he had a share in founding the university of Helmstadt. But his activity in Lower Saxony lasted just four years. As in Jena, he felt the influence of the Lutherans, and in 1574 (Zedler, XXXVI, 1715, and Freher said 1577.), thanks to the help of his father-in-law, the Lutheran Superintendent Daniel Greser in Dresden, and to the intercession of the

tance, but he must be considered in a study of this kind because of his influence on laymen and clergy in the second half of the sixteenth century. His name appears in most treatments of six-

Elector August's wife, Anna, Selneccer returned to his position in Leipzig. At the same time the Elector withdrew his favor from the Melanchthonians, whom Selneccer helped him to uproot, and bestowed it upon the Lutherans. One reason why the Melanchthonians fell into disfavor was the unceasing accusations of their opponents, none of whom had more zeal than Selneccer. The more hostility that was shown him because of his former associations with the Melanchthonians, the more animated did he become in his hostility toward them and the more did he want to see broken the power of their party, which hated him as a deserter and had tried to thwart his return to the Electorate of Saxony.

In June 1576 Selneccer wrote on the Lord's Supper, in connection with the Torgau Article which dealt with it, and in the autumn of that year he tried to win over Caspar Peucer, Melanchthon's son-in-law, to strong Lutheranism. He belonged with Chemnitz and Andreae to the clerical commission, which, from June to the autumn of 1577, exhorted the teachers of August's lands to insist on the new type of teaching. Later, Chytraeus, among others, was added to this commission. On June 25, 1580, the fiftieth anniversary of the presentation of the Augsburg Confession, the Concord formula was initiated in Dresden. At this time Selneccer had great prestige and at the end of that year Jacob Andreae was dismissed, probably at the instigation of Selneccer, who succeeded him. However, in January 1586 the Elector August died and was succeeded by Christian I who favored the oppressed Melanchthonians. As might be expected, Selneccer became the subject of their wrath. In 1588 Christian forbade mentioning the name Calvinist, and in the following year Selneccer was dismissed from his position and went into exile with other opponents of the Melanchthonians. He lived in Magdeburg upon donations of his party followers until in 1590 he obtained the position of Superintendent in Hildesheim, whence he was called to Augsburg to regulate Protestant church management. On his return to Hildesheim in 1592, he learned that Christian I had died on October 5, 1591, and that thereafter the Lutherans had regained their power in the Electorate of Saxony. He had felt ill on his return trip to Hildesheim, but nevertheless set out for Leipzig, where he died four days after his arrival, May 22 or 24, 1592.

Selneccer seems not to have had a particularly pleasing or agreeable personality. He was of small stature, with short legs, and was subject to ridicule on that score. He was continually changing sides in his controversies, wabbling from one extreme to another. But he did try to steer a middle course. Justice must be done to him by pointing out his desire for peace and concord between the sects. The Concord Formula played a real part in his life. Future generations were more in sympathy with his desire for harmony than was his own and could have a better comprehension of his standpoint. Some of his songs indicate his feeling for unity within the Church.

teenth century Protestantism. His book on the comet of 1577 was written at the request of August of Saxony,[24] and makes no pretense of being other than a prayer. At his death in 1592, he left behind him a very large number of treatises, prayers and church songs. He had real ability as a writer of hymns.[25]

In Selneccer's book on the comet of 1577 [26] the first of two prayers makes no mention of the great comet except as a sign of God's wrath. The comet's appearance is taken for granted to be a sign of His displeasure and, beyond that, it has no significance. The prayer is well written with a fine choice of language. Its motives were both political and religious. It says that men have seen God's sign of anger and are, indeed, sinful and worldly. It begs for protection and that men be led into righteous ways. It asks that they be made faithful followers of their earthly rulers and that they be made to abstain from sin.

The second prayer does not even use the comet of 1577 as an excuse for praying. It merely says that men see God's anger. It asks for mercy and begs that God's wrath be turned on the unbelievers instead of on the members of his flock.[27]

24 Janssen, VI, 441, citing Weber's *Anna von Sachsen*, said that because of the "terrible sign of God's wrath" the Elector August of Saxony had Selneccer and Jacob Andreae write a church prayer and distribute it in all parishes. See the title-page of Selneccer's book. No record of a separate prayer by Andreae has been found. (See appendix, below.)

25 Two tracts by him, listed in Ioachim Laymann's book catalogue, were confiscated by the censors, as well as a Bible translation by Selneccer (Calinich, 237, 239). Lists of his books can be found in many places (for example, Bassaeus, I, 530, II, 141-4; B.M. catalogue; Bodleian library catalogue; A.D.B.; Jöcher; Zedler). In addition to the book or books on the comet of 1577, he dealt with natural phenomena in *Libellus Sphaericus*, which seems (Scheibel, 94) to have been published as an addition to his *Propositiones ... in VIII libros ... Aristotelis* ... in Leipzig in 1577.

26 Item 98a of appendix, below. It is possible although not probable that this little pamphlet is the same as those numbered 98 and 99. In any event, it is hardly likely that the latter are of more astronomical value than the one discussed here.

27 One phrase is very reminiscent of a famous hymn or prayer by Selneccer, written before 1572. The phrase is "Lass vns nur dein sein vnd bleiben" and the hymn begins: "Las mich dein sein vnd bleiben, du trewer Gott vnd Herr".

Another Protestant theologian whose sermon on the comet of 1577 must have had a wide audience was Jacob Heerbrand,[28]

28 Adam (1653), 668-681.—A.D.B., XI, 242-4, article by Schott.—Bassaeus, I, 43-5, II, 81-2. This gives lists of books by Heerbrand.—B.M.. catalogue.—B.N. catalogue.—Bök, 77.—Calinich, 76-7.—Döllinger, II, 385 and note 23.—Dreyer, editor, VII, 83, 407, VIII, 457.—Frank, I, 243-4.—Freher, I, 311-2. There is a portrait of Heerbrand opposite page 288.—Janssen, V, 321, 348, 377-8, 382, 406, 448, 450, 457, 460, 461, VI, 440, where items 53 and 54 of the present thesis are discussed.—Jöcher, II, 1433-4.—Lalande, 103.—Realencyklopädie, VII, 519-524, article by Wagenmann, reedited by Bossert. This is undoubtedly the best account of Heerbrand's life and work and gives excellent lists of books by and about him. One statement is, however, not clear. Reference is made to *Erinnerung aus einer Predigt in Tübingen nach Luthers Tod* which supposedly can be found in *Ein Predig v. d. erschrockenlichen Wunderzeichen*, Tübingen 1577 Bl. A. 4. It is not there.—Schaff-Herzog, V, 198-9, article by J. Bossert.—Scheibel, 90, 104.—Schottenloher, I, 21.—Witte, obiit . . . 1600.—Zedler, XII, 1079-1082.

Other works which might be consulted are:

Cell, Erhard: *Oratio funebris de vita, studiis, laboribus, officiis et morte reverendi . . . Dn. Jacobi Heerbrandi Giengensis, . . . habita ab Erhardo Cellio* . . . Tübingen, Cellianus, 1600. (B.N. catalogue).

Fischlin, Ludwig Melchior: *Memoria theologorum wirtembergensium ressuscitata, h. e. Biographia praecipuorum virorum qui . . . in ducatu wirtembergico verbum Domini docuerunt* . . . Ulm, G. W. Kühnen, 1709-1710, 2 v. There is also a supplement to this. (B.N. catalogue.)

Hafenreffer, Mathias: *Leuchtpredigt über dem Absterben . . . des . . . H. Jacob Heerenbrands,* . . . Tübingen, Cellianus, 1600, (B.N. catalogue.)

Schottenloher, I, 331, II, 348, where works by Heerbrand and sources concerning him are listed.

Catalogues of the big libraries, such as the B.N., the Bodleian, and the B.M., contain lists of books by Heerbrand.

Heerbrand was born in Giengen in Swabia on August 12, 1521. His father, Andreas Heerbrand, was a weaver with education, a musician and arithmetician, who was familiar with Luther's writings. Jacob Heerbrand studied in Ulm in 1536, and, from 1538 on, in Wittenberg under Luther and Melanchthon. He obtained his master's degree in 1543, and then became deacon in Tübingen. He was removed from that office in 1548, because he did not accept the Interim, but continued studying in Tübingen, learning Hebrew from Oswald Schreckenfuchs. In 1550 he received the doctorate in theology. At the end of that year he was named pastor in Herrenberg. In 1551 he subscribed to the "Confessio Wirtembergica" and in March 1552, together with other Wittenberg theologians, he attended the Council of Trent. In 1556 he went to Pforzheim on the invitation of the margrave of Baden-Durlach to become pastor and director of the State Church, reformed on the basis of the Würtemberg agenda. In September 1557 he returned to Tübingen as a professor. He was rector of the university eight times and in 1590

one of the Wittenberg reformers. As a youth he was a diligent student and was called the Swabian night owl. However, he was easily prejudiced. For example, he opposed calendar reform on the grounds that behind it lay Satan, the Roman Antichrist.[29] Like many preachers, he enjoyed discussing natural phenomena from the pulpit, in an effort to better his congregation. His 1577 tract[30] on the comet of that year calls for renewed prayers and asks the people to reform. The first part is based on the ninth chapter of Matthew, and tells of two acts of God: of the death of a young woman for whom death meant liberation from pain after twelve years of suffering; and of the saving of the young daughter of the head of the school, who was snatched from the clutches of death. Heerbrand said that the comet or peacock-tail, visible at the time of the sermon, the twenty-fourth Sunday after Trinity Sunday, was another sermon to which he and his congregation ought to listen. He said that he would

became its chancellor and provost, succeeding Jacob Andrea for whom he wrote a funeral oration. (Andrea or Andreae, see appendix below, may have written on the comet of 1577.) Heerbrand went through several dogmatic phases of theological thought. (Calinich, 76-7, gives an interesting example of his preaching.) He was a beloved preacher. In 1597 his wife, to whom he had been married fifty years, died. The following year, because of infirmities, he gave up his professional duties. At his death in 1600, he left behind him a *Compendium theologiae.* (The B.N. catalogue listed this with a Tübingen 1573 imprint. Zedler, XII, 1080, listed many other editions.) According to Jöcher, II, 1433-4, from whom the accusative case was copied, his writings included "*librum de ecclesia, patribus & conciliis; libellum contra Petr. a Soto; Comment. in Pentateuchum; Refutationem errorum Ge. Gotthardi; Refutationem defensionis assertionum jesuiticarum de ecclesia Christi; Hyperaspisten disputationis de precatione; apologiam contra hyperaspisten friburgensem; Spongiam contra aspergines apologeticos Greg. de Valentia de adoratione ejusque speciebus; apologiam explicationis causarum, cur cum Greg. de Valentia de hominum mortuorum statuarum adoratione non amplius sit disputandum; Disp. contra purgatorium; de festo corporis Christi; de autoritate cathedrae romanae; de fine bonorum operum; de ciborum delectu; de erroribus Pontificiorum de Ecclesia;*" and others. His theological text book was translated into Greek by Martin Crusius.

29 See Heerbrand's *Disputatio de Adiaphoris et Calendario Gregoriano . . .*, Tübingen, 1584. (B.M., catalogue.) This work was mentioned to Tycho by Johannes Major, although the title was not given. (See Dreyer, editor, VII, 83, 407.)

30 Item 54 of appendix, below.

tell, first how the people were to look at it and what it signified, and second how they were to guard against it. First one should look at the comet as a horrible sign of the terrible wrath of God against the world because of the people's sins. Heerbrand compared God's act in sending the comet to the acts of judges in leaving a sword on the table in court before the criminal. He said that the comet was bringing misfortune. From it was coming poisoning of the air which would be followed by the death of cattle and the spoiling of fruits, unusual drought and heat and an unfruitful time, from which scarcity would arise. Wars and pestilence also would follow. The comet, he said, signified the sword of God, the Judge of the world, which He stretched out and showed to the world. Heerbrand said that scholars commonly held that comets come from vapors of the earth which are drawn aloft by the stars and are ignited there by the heat and the motion. He added that the sins of men are also vapor and smoke that climb up. He spoke of the comet which preceded the downfall of Jerusalem, and of the comet in the seventh year of the reign of Vespasian about which, he said, Pliny spoke,[31] and of three comets following close upon each other about a hundred years before the date of his sermon. However, the information which Heerbrand gave concerning those comets was only a list of the misfortunes and miseries which were supposed to have come as a result of them. Heerbrand also said that he saw a comet fifty years before, which was followed by much strife. In addition, in 1556 and 1558 there were two comets. According to Heerbrand, the larger a comet and the longer its duration, the greater are the ills which follow. The comet of 1577 was such as had not been seen in a hundred years.

Then Heerbrand, making many Biblical citations, considered at length the question of how people should react to the comet. He said that men should not do as children who, seeing their angry father with a rod, begin to laugh. He said that the comet showed that there was a great fire, that God's anger was burn-

31 The comet of 76 A. D.

ing. He urged men to bring water to quench it, the water which comes from the eyes; and he said that the people should all recognize their sins, regret and bemoan them. Much prayer and repentance would save them.

In this instance, the comet of 1577 served merely as an excuse for writing a sermon imploring the people to become devout. Similarly, Heerbrand in his *Erndt vnd Herbst Predig* [32] showed that his interest in the comet was entirely that of the preacher, attempting to enforce morality and religion by it.

All that is known about Vitalis Kreidweiss,[33] who based a poem on a sermon by Heerbrand, probably the one just discussed, is that he was the schoolmaster in Leonberg, as is told on the title-page of the poem. Like the sermon, this poem [34] pleads for reform to allay the evils of which the comet warned. It says that that body was ignited in the sky as a warning from God because of the sins and misdeeds in the world. Evils such as war and pestilence were predicted and it was stated that Jerusalem fell because people in that city did not seriously interpret the comet. Other examples of misfortunes following comets were listed. According to Kreidweiss (or Heerbrand), God alone knew what was signified by the comet of 1577, which He had sent to fulfill the purpose of a sermon, because people were deaf to sermons. The author believed that by abstaining from their sins men could cast off the burden placed upon them. He described the comet as a vapor full of poisons and compared it to " Mahmet's " teachings, because it was like a star although

32 Item [51] of appendix, below. There Heerbrand said that one should thank God for the rich harvest of 1578, a year for which such a bad beginning had been prophesied, and even predicted by God with the horrible comet of the previous year. Heerbrand's work on the comet of 1577 was not discussed by Tycho.

33 The works consulted in search of information were: Adam (1615).—Adam (1653).—Adam (1705).—A.D.B.—Biographisches lexicon der hervorragenden ärzte. — Freher. — Hoefer. — Jöcher. — Poggendorff. — Realencyklopädie. — Zedler.

Janssen, VI, 440, who spoke of Kreidweiss in connection with the poem, gave no further information, although he quoted from the poem.

34 Item 53 of appendix, below.

not a star, and Mohammed wanted to be considered a God although he belonged to the devil. The comet was also compared to a Turk. The sins of the time were enumerated, but seem to have been merely jollity and revelry. Attention was also focused on the pestilence then raging in Germany. To be blessed, continues the sermon, one must tread the narrow path and call upon God to show the right way. It closes with a "daily" prayer to God.

Angelo Rocca [35] was more interested in the subject matter of astronomy than were the preachers Heerbrand and Selneccer, and was not a Protestant like them and Chytraeus. However, like them and like many educated men, he was one of those men of erudition who, although not trained in astronomy, felt themselves qualified to contribute something on the subject of comets. His attitude, too, was that of the preacher. The point of view of sixteenth century scholars was unlike that of our modern savants. The lines of distinction between subjects of discourse could be easily overstepped. And, indeed, comets in the sixteenth century had an utterly different meaning for scholars and unschooled alike, from what similar phenomena would have today. To most men a comet was a thing of wonder, a fearsome sign. The writing of a dissertation on comets required a knowledge of past literature and lore, and it often took the form of a sermon rather than an astronomical treatise. For such a task Rocca was truly qualified.[36] For forty years he collected

35 Bodleian library catalogue. — B.M. catalogue. — Catholic encyclopaedia, XIII, 100. This cites Chalmers and adds no information.—Chalmers, XXVI, 309. This cites Nicéron and adds no information. — Enciclopedia Italiana, XXIX, 520, article by Luigi Giambene. This contains nothing not in the other references.—Gesner (1583), 52.—Haym, 227. This gives the title of one work by Rocca. — Hoefer, XLII, 450-1. — Jöcher, III, 2150-1, citing Nicéron and giving a short account of Rocca's life and writings, Fortsetzung VII, 165-9, giving a sketch of Rocca's life and a bibliography like that given by Nicéron.—LeLong (1709), II, 674.—Michaud, XXXVI, 197.—Nicéron, XXI, 91-106. This gives an excellent account.—Riccardi, II, 198, 384, Correzioni II[a], 146. — Scheibel, 97. — Tiraboschi, VIII, 59 ff. — Witte, obiit . . . 1620.—Zedler, II, 258.

36 Rocca was born in Rocca-Contrata, now Arcevia, near Ancona, in 1545. At an early age he was sent to the Augustinian monastery at Camerino

a remarkable library,[37] and his own writings include works on asceticism, theology, morals, philosophy, and liturgy.[38] However, although he had read much, he often was found lacking in judgment and powers of criticism.

Rocca began his book on the comet of 1577 [39] by telling that its first section would deal with the natural effects of the comet; the second with the questions of how and in what part of the sky the comet was formed; the third with the relevant philosophical and astronomical prognostications.

Rocca said that there are those who believe comets to be acts of God and not of Nature, but also those who inquire into the causes of comets. He described the old theory of condensation of vapors in the air and maintained that the events of the year were due to natural causes and were not supernatural. However, he did not entirely surrender the idea of miracles, and

where he took orders in 1552. It is because of this circumstance that the surname "Camers" is often applied to him. He studied philosophy and theology at Perugia, Rome, and Venice, and, according to some sources, received the doctorate in theology at Padua on September 9th, 1577. (Tiraboschi, VIII, 59, doubted that Rocca received his doctorate at Padua or that he later taught there.) After becoming a doctor, he taught the humanities in Venice to young Augustinians. It was during this period of his life that the comet of 1577 was seen.

In 1579 Rocca became secretary to Augustin Fivizani, superior-general of the Augustinians at Rome. In 1585 he took charge of the Vatican printing office, which was preparing editions of the Bible and the writings of the Fathers. In 1595 he was appointed by Clement VIII to succeed Fivizani who had died in January of that year, and in 1605 he was made titular Bishop of Tagaste in Numidia. He died in Rome, April 8, 1620.

37 In 1595 he received permission from Clement VIII to leave this library to that monastery of his order which he deemed fitting. The permission was renewed by Paul V in 1609. Subsequently, Rocca gave the library to the monastery of St. Augustine in Rome on the condition that it be available to the public. The act of donation was dated October 23, 1614, and the library perpetuates Rocca's name.

38 An incomplete collection of them was published in Rome in 1719 and 1745, and lists of his writings were given by Nicéron and others. In addition to his *Discorso Filosofico*, written apropos of the comet of 1577, he seems to have written a book called *Commentarius philosophicus, ac theologicus de Cometis*..., which appeared in Venice in 1577.

39 Item 91 of appendix, below.

took the opportunity to quote Pliny. He gave an explanation of miracles, or marvels and portents as he preferred to call them and linked them directly with natural causes, thus taking a step in advance of preachers like Selneccer, Heerbrand, or Chytraeus. He constantly referred to historical persons, such as Pliny, Livy, and Aristotle, exhibiting his wide reading. Of course, he cited the writings of St. Augustine, particularly the *City of God.* He discussed philosophically man's interpretation of sensory perceptions and the influence of devils. Such discussion requires a strange mixture of reasoning. Rocca brought in much extraneous information and even mentioned the star which appeared at the birth of Christ, which he called a comet. He gave five reasons why it was formed in the air close to the earth. However, he considered the comet of 1577 different from the star of the Magi and said that it belonged in the skies.

In the next section of the tract he dealt with the cause of the comet and the means and place of its generation. His first step was to say that the sky influences these in three ways. He proceeded to discuss light, with a little diversion as to cats' eyes and to planting, and with references to the influence of lunar and solar light, and to the saltiness of the sea. In a paragraph dealing with the ebb and flow of the sea, Rocca quoted Averroës to the effect that this is caused by the moon, not the sun. Rocca thought that by his discussion of these diverse phenomena he had proved that celestial bodies influence earthly things; and he ended the paragraph by mentioning hot, cold, dry and damp vapors, which have been kindled on high and called comets.

The third section deals with the location of comets. Following Aristotle, Rocca divided the air in three parts. He put comets in the third division and said that they are not vapors but hot and dry exhalations, thick and viscous, which come from the earth, and are drawn up by the heavenly bodies and become inflamed, and acquire their motion from that of the heavens.

The fourth section discusses the motion of comets. Rocca asserted that a comet moved with rapid circular motion relative to the proper motion of the primum mobile, from east to west, in a day, and he introduced the discussion of other celestial motions. He believed that the motion of comets varies with the locality of their generation and with the stars which attract it. Just as a star has one motion from east to west and another from west to east, so a comet, he thought, moves in two ways.

According to its heading, the next section should be concerned with the different shapes, names and colors of comets and with the time when they are born and with their duration; but separate sections are, in fact, made for the last two subjects. Rocca thought that comets differ in quantity and shape, and in color, which depends on the material of the comet. This may be much or little. He quoted Paul of Venice on the size of comets. He believed that their color depends on their rarity, on the fire of whose nature the comet partakes and on the velocity. Rarity would cause whiteness, density, redness. Rocca upheld his reasoning by reference to Algazel and Avicenna. He said that the material of the comet is dense in the middle and rare at the outside, whence the name " crinita ", and that it has the properties of dilatation by heat and contraction by cold. After explaining " tailed " and " bearded " comets, he said that the rarity and the density are the cause of the diversity of names and colors. However, he added that astrologers thought that the different colors of the comets were due to the different natures of the planets and that different effects arose in accordance with these different natures. It is now known that comets can be identified only by their orbits but Rocca, at least, attempted to find a natural cause for their differences in appearance, and to distinguish them that way.

Rocca gave Pliny's *Natural History* and Sessa's [40] *Meteors* as the sources of his information concerning the times when comets appear. They arise both in the north and in the south. A large part of their matter is changed into wind and into fall-

40 He doubtless referred to a commentary by Agostino Nifo of Sessa (1473-1546) on Aristotle's *Meteorologica*, which was published in Venice in 1530. (B.N. has 1540 edition.) See chapter II, above, and Thorndike, especially V, 71-5, 162-3.

ing stars and thunder, and they appear in the spring and the autumn, not in times of excessive heat and cold when dissolution of the exhalations takes place. They occur, Rocca stated on the authority of Ptolemy, at the times of eclipses or in eclipse years. He cited Pliny to the effect that comets last from seven to eighty days. He also referred to the comet in Nero's reign, described by Seneca, which lasted six months and to the comet at the time of the destruction of Jerusalem, described by Josephus as lasting one year. Like his predecessors, Rocca thought that a comet lasts until all its material is consumed, that this may be increased, and that the burning depends on the heat of the sun. Among the holders of various opinions concerning the end (or purpose) of comets, Rocca cited Aristotle to the effect that they forecast ills, and Pliny to the effect that they may presage joyous events. The next paragraph purports to show the meanings and natural causes of comets. It is the usual astrological discussion of the subject, bringing in the current or ancient opinion of a comet's generation, and listing the ills which follow in a comet's wake. The comet of 1556, the comet at the fall of Jerusalem and the comet of 448 are among those used as illustrations. The founding of Venice is mentioned and applauded.

Rocca seems to have been delighted to arrive at the next section, which deals with the comet of 1577, first seen November 8, 1577. He likened it to a hot and dry exhalation, and said that like everything sublunar it would be corrupted. He called it " barbata ", describing it as white and pale, because of its rarity. He said that its head was rosy, or golden, or silvery and had the shape of the star under which it appeared and whose motion it followed. Rocca cited Sessa as following Aristotle and saying that the comet which is generated in the lowest part of the highest region of the air will appear alone, but that the one generated at the top will be pulled by some star and will seem like that star bearded. Besides all the now unimportant details which Rocca retold, he did say that the tail was long and large and that on December 20th it was scarcely visible. Although he stated that this comet was not a star, he did not make the proper distinctions between comets and stars. However, he added that it was not an example of the splendor of wandering stars, or planets, in conjunction, as Democritus and Anaxagoras would have held, for it was far from the zodiac, near which the planets are confined. Before the appearance of the comet two lunar eclipses had occurred, which, Rocca agreed, were signs of comets. He thought that the comet was the sign of the dryness which began in the autumn and ended "this" day, January 14th. He also ascribed to the comet the misfortunes in two cities in the march of Ancona, namely Ripatransone and Ascoli, and the inundations of the sea.[41]

Rocca's last paragraph deals with the nature of the comet of 1577, which, he said, must be ascertained from the color. A comet of the nature of Saturn is of an azure or lead color; of the nature of Jupiter, white, almost silver; of the nature of Mars, of blood color; of the nature of the sun, gold and silver; of the nature of Venus, silver; of the nature of Mercury, different colors; of the nature of the moon, the color of lead. The comet, visible from November 8, 1577 to January 6, 1578, or sixty days, when first seen was a black color, then white, according to some, or leaden colored, or red according to others, and was generated by several planets. Rocca thought that the effects of the comet were to be taken from Mars and Saturn. They were to be in the east, because that is where the beard, or tail, was directed, and they were to be important because the comet lasted so long. Rocca ended by marveling at man's ability to make predictions from observations of the stars.

41 He quoted the "Poet" [Pachymerēs, 149], as follows: "Cometa non già mai fù uista in Cielo, Che non portasse al mondo qualche danno..." See item 12 of appendix, below.

CHAPTER VII

THE COMET OF 1577: TRACTS BY PERSONS OF GENERAL CULTURE, ILLUSTRATING THE WIDESPREAD INTEREST IN NATURE

RASCH. — MARZARI. — FIORNOVELLI. — TWYNE. — DE BILLY

MANY persons of general culture, who were well read but had no special aptitude for a study of comets, were, nevertheless, sufficiently interested in such general phenomena to write about the comet of 1577. One of these authors was Johann Rasch, " citizen of Vienna ", who flourished in the last three decades of the sixteenth century.[1] His talents were varied. He was a composer,[2] and in addition was the author of prognostications, a book on earthquakes, a book on wine, and the *Cometen Buech*.[3] He was also an organist in Vienna and cantor in the Benedictine convent there.[4] Thus a diversity of works [5] are attributed

1 A.D.B., XXVII, 316, article by Robert Eitner.—Günther (1890), 238-244, 248-255.—Haselbach.—Hellmann (1883), 398.—Hellmann (1924), 15, 29. —Zedler, XXX, 894.

The available information concerning Rasch is confused.

2 He composed church music. His musical works were published in Munich in 1572 by Adam Berg. See Günther (1890), 238. Johann Rasch, the composer, was described in the A.D.B., XXVII, 316, and by Zedler, XXX, 894.

3 Item 87 of appendix, below.

4 According to Hellmann. At first it seemed that Hellmann had here confused two men, as he had done in the case of Roeslin (see above, chapter III, note 78).

5 Rasch's writings, other than his *Cometen Buech* and his musical work published in 1572, are:

Bluetfluss. In unser lieben Frawen Kirchel zu Walperspach am Stainfeld in Osterreich, in disen jarn 1585. 86. 87. so noch auff heuntigen tag geht und gesehen wird... Munich, Adam Berg, 1588. (Günther (1890), 251.)

Calendarium romanum aethaicae vetustatis. St. Gallen, 1584. (Günther (1890), 249.)

De Cometis. Munich, 1573. (Scheibel, 76.)

De terrae motibus et terrae hiatibus: opus variorum auctorum et tractatuum, quorum Catalogum versa pagella exhibet:... Viennae Austriae collectum a Joh. Rasso. Strasburg, Bernard Jobin [1581 ?]. (Günther, 1890), 254.)

Ein Neu: AllJäriger Calender. Munich, Adam Berg, 1584. (Hellmann (1924), 5. Günther (1890), 239, said that this was for the year 1583 and thus must have appeared a year earlier, although he said, 250, that it was printed in 1583. He called it, 239, "ein unbedeutendes Machwerk".)

Erdbidem Chronic Nach art eines Calenders, sambt einem kurtzen Bericht vnd Catalogo Autorum. Munich, Adam Berg, [1591]. (Hans P. Kraus, Vienna book dealer, in a postal dated February 11, 1935; Günther (1890), 243-4, 254.)

Fasten - Reim. Munich, 1584. (Günther, (1890), 249.)

Fasten Lob. Guete nütze verständliche Catholische erinderungen, ainfeltiger bericht... von der viertzktagfasten, auch von allen andern allgemainen... Fastägen des gantzen jars,... [In verse.] Munich, A. Berg, 1588. (B.M. catalogue; Günther (1890), 249.)

Folge der österreichischen Fürsten. 1615. (Haselbach, 175.)

Gegenpractic. Urthail und allgemainer khurtzer bericht wider etlich aussgangene weissag, prognostic, practic und troeschrifften, auss den zuefällen des 84. unnd 88. wunderjaren, sunderlich des Misocacs, von undergang hoches Geschlechts und der Röm-Clerisey, von änderung der Reich und Religion, Von Antichrist, von lester zeit und end der weld. Munich, Adam Berg, 1588. (B.M. catalogue, supplement.)

Gegenpractic, wider etliche aussgangen Weissag, Prognostic und Schrifften, sonderlich des Misocaci, uber das 84. und 88. Jare,... Munich, A. Berg, 1584. (B.M. catalogue; Günther (1890), 250. Günther (1890), 240, described the tract.)

Genesis Austriaca. Genealogia Serenissimorum Austriae Archiducum... Carmina item plurima, in Caesarum, Regum et Archiducum . . . nativitates, coronationes,... [Vienna ? 1580 ?] (B.M. catalogue; B.N. catalogue.)

J.G.N. Osterreichischen Wesens und Landsachen unterschiedliche Bücher. Rorschach, no date. (Günther (1890), 249.)

Neu Kalendar. Das erste büch. Von computistischen Kirch Calenders besserung und wunder, von neues Gregorischen Ostercycli änderung,... Rorschach am Bodensee, 1590. (B.M. catalogue. According to Günther (1890), 251, there was a Munich 1586 edition of this work with a slightly different title.)

Practica auff das jar Christi 1579. Mit viel guten vnd nötigen Erinnerungen, vmb lustigers lesens vnd mehrer übungswillen, reimweise gestellet. [1578], quarto (Weller (1862-4), I, 335. Günther (1890), 249, gave the date of publication as 1579.)

Practica Auff das grosswunder Schaltjar. 1588. Munich, Adam Berg, 1588. (Hellmann (1891), 33 ff., where several quotations from the book were given, and Hellmann (1924), 5 and 15, where the full title was given with a description of the book. Günther (1890), 249, gave the place of publication as Gratz.)

to one man, who, because of his modesty or because of a mature attitude, called himself simply a " citizen of Vienna ".[6]

Schotten closter 1158. Stifftung und Prelater unser lieben Frauen Gottshaus, Benedicter ordens, gennant zu den Schotten, zu Wienn in Osterreich. Anno Domini 1158. [Vienna,] 1586. (B.M. catalogue.)

Vaticiniorum liber primus... Vienna, 1584. (B.N. catalogue; Günther (1890), 249).

Vier Stuck. Nicht wehrt. 270 Nützliche...viertailige lehrpuncten der alten Weisen, von betrachtung der tugenden und mancherlay weltsachen,... Munich, A. Berg, 1589. (B.M. catalogue.)

Von Erdbiden, etliche Tractät alte und newe hocherleuchter und bewärter Scribenten...Durch Iohan Rasch an tag geben. Munich, Adam Berg [1582.] (B.M. catalogue, supplement. According to Günther (1890), 240, 251, this was Rasch's first book on earthquakes. Günther (1890), 240-2, 252-3, described it.)

Weinbuch. Das ist : vom Baw und Pflege des Weins / wie derselbig nützlich sol gebawet,... Munich [1600 ?] (B.M. catalogue.)

Weinbuch / das ist / vom Bauw und Pfleg dess Weins / wie derselbig nützlich sol gebauwt werden / darneben wie man allerley Kråuter vnnd Brandtwein / Essig / Meeth vnnd Bier machen / erhalten / vnd welche abgestanden / wie denselbigen zu helffen sey. Munich, 1581. (Bassaeus, II, 355. This is probably an early edition of the item listed above. It is also possible that the books are the same and that either the doubtful date 1600 from the B.M. catalogue should be 1581 or that Bassaeus made an error. The date given by Haselbach, 175, is 1582. Günther (1890), 249, said that there were several editions of the book.)

Weissag der Zeit. Allgemeine Himels und Weldpractic, . . . 1596. (Günther (1890), 251. Günther (1890), 240, called this work anti-astrologic, because Rasch tried to separate the true from the false knowledge of events foretold by the stars.)

The following book was translated by Rasch:

Drey greuliche weissagung Daniels des Propheten, nem̃lich vom fall des Geistlichen lebens : von abnemung der kirchischen würdigkeit : von undergang des Catholischen Glaubens : Auch von zukunfft des Antichrists und vom End des Welt...verteutschet durch J. Rasch.—Ein ander Christliche Predig des H. Hippolyti vom Antichrist,... Munich, Adam Berg [1580 ?] (B.M. catalogue, see Vincent [Ferrer], Saint. Günther (1890), 249, listed a Munich, 1597, edition of this, and, 250, a 1582 edition.)

The following book was edited by Rasch:

Hauss Osterreich. Von Ankunfft, Ursprung, Stam̃en und Nam̃en der alten Grafen von Altenburg und Habsburg, darauss die heutigen Fürsten von Osterreich seind entsprossen. Auss J. Stumpfens Schweizerchronic und andern historicis gezogen durch J. Rasch. Rorschach am Bodensee, [1600?] (B.M. catalogue; Günther (1890), 249).

6 See Günther (1890), 239, 248-9.

Johann Rasch was probably born in Pechlarn about 1540.[7] He studied at the universities of Wittenberg and Vienna and later traveled in Germany and Switzerland,[8] spending some time in Thuringia, where he observed an earthquake in 1556.[9] In 1570 he became organist at the Benedictine college in Vienna.[10] At one time he was in the book business in Vienna,[11] and because of his interest in research in old books he wrote several historical works, including a genealogy of the house of Hapsburg.[12] He died later than 1615, in which year his *Folge der österreichischen Fürsten* appeared.[13] His *Weinbuch,*[14] part of which is in rhyme, gives a good account of the manner in which wine was made in the sixteenth century, and marks the high point of the production of wine in Austria. His prognostications covered the years from 1579 to 1588, and numbered at least six, probably ten. They were written in German, that for 1579 being in rhyme.[15] He seems to have taken his task rather lightly and to have known little or nothing about weather forecasting.[16] His earliest earthquake book seems to have been carelessly and speedily assembled,[17] but his second one was carefully worked over,[18] and his astrology became more and more reserved with the years.[19]

7 Haselbach, 174.

8 *Idem.*

9 Günther (1890), 244, 249, 255.

10 Haselbach, 174.

11 *Idem.* According to Calinich, 238, not one of the seventy-one numbers in the catalogue of the Vienna book dealer, Rasch, was confiscated by the censors.

12 Haselbach, 174-5.

13 *Ibid.*, 175.

14 A summary of the book was given by Haselbach, 175-8.

15 Hellmann (1924), 29; Hellmann (1883), 389, according to whom the prognostication for 1586 is a quarto volume and was printed in Munich in 1584.

16 Hellmann (1924), 5 ("Berechnung der Witterung").

17 Günther (1890), 241.

18 *Ibid.*, 243.

19 *Ibid.*, 239.

Rasch's *Cometen Buech* [20] has a slightly different form than the majority of the treatises on the comet of 1577. It is stated on the verso of the title-page that the catalogue was gathered together because of the current interest in that comet, and that the material was taken from "all" [21] the authors who had written on the subject. The dedication [22] is followed by a preface to the reader, where it is stated that books on comets are not all of equal value, and that it would require an entire work to discuss some of them, such as *Corn. Gemma in Epistola ad D. Hagecium,* [23] because of their great difficulty. After his preface, Rasch explained that this book about comets was to be divided into questions, articles, and points, for better understanding and comprehension, and that it was not for the learned but for the common man.

The first part of the book deals with philosophy, physics and meteorology, or the nature of comets, new stars and other phenomena. Philosophy, said Rasch, is a praiseworthy study which includes theology and is taught in the schools. Physics or " naturalness ", he added, teaches the why and wherefore of natural phenomena and substance, and is a study for melancholy people. This first part is divided thus:

1. If it is natural to judge from heavenly signs.
2. Concerning sky, air and earth.
3. Concerning the location of portents and comets in the sky.
4. How various miracles happen to us.
5. Concerning the name comet.
6. Of the nature of comets and if a comet is a star.

20 Item 87 of appendix, below.

21 Rasch mentioned, among others, John of Damascus, Vögelin, Schöner, Apian and Hagecius.

22 The dedication, described in the appendix, below, says, A_{ii} v – A_{iii} r, that the book was originally completed at the end of December, 1577, but had remained unprinted.

23 This particular work has not been found listed elsewhere and has not been included in the appendix, below, although the two men concerned figure prominently there. The title given by Rasch may refer to a letter printed in Hagecius (1574), 169-174.

7. How various comets appear.
8. Which comets are truly natural comets.
9. Of the figure of comets. (In this section Apian's discovery of the direction of a comet's tail is noted.)
10. Why, whence and how comets arise.
11. Concerning the prophetic knowledge of comets.
12. In which part of the heavens comets are mostly kindled.
13. At what time of year comets prefer to burn.
14. How long comets usually shine in the sky.

The second part of the tract concerns the mathematics and astronomy of comets. It contains the following list of what an astronomer should observe:

1. Concerning the kindling of comets, according to the time.
2. What should be daily observed about comets.
3. How the observations take place, with what instruments and so forth.
4. How the observations are to be presented.
5. Concerning that for which astronomical observations are useful and necessary.

Rasch also gave the following list for the mathematician:

1. Whether the comet moves before or after the sun, is seen late or early, before sunrise or after sunset.
2. Concerning the place in the sky and the constellation in which the comet is situated.
3. The configuration of a comet with the planets and other star pictures.
4. The distance and height of a comet from the earth or from the heaven.
5. If it floats in the heaven or in the air.
6. The size and form of its body and streamers.
7. Daily notice of its path.

These, said Rasch, were not observed by our ancestors.

The third part of the book considers the astrology and history of comets. According to Rasch, astronomy only regulates and measures stars, but astrology teaches prediction from the stars. He gave nine points tending to justify the astrologer and to differentiate him from the mathematician, the astronomer, the philosopher, and others. He also listed seven ways of prophesying,[24] which he described under the titles: *Prophetia, Oraculum, Vaticinium;* and the following for which he said that study is necessary: *Astrologia, Augurium, Haruspicium,* and *Sortilegium.* Prognostication, he added, must depend on the particular comet involved. He then stated and discussed the following twenty points:

1. If prediction requires knowledge and skill.
2. If a prognostic can be considered true and be acted upon.
3. If everything which occurs or is seen in the sky has a meaning or if everything on earth happens without the influence of the firmament.
4. If an astrologer can predict a comet.[25]
5. Whether or not man on earth can know beforehand and have an opinion of the strength and will[26] of a comet in the sky.
6. Whether comets should be observed or laughed at.
7. If comets only prophesy or themselves have influence.
8. Comets also give information of past and not always of future events.
9. How comets should be judged.
10. What is to be judged concerning comets.
11. If the astrologer should prophesy and warn about spiritual affairs, such as the personalities of certain preachers.
12. What and about whom or how astrologers ought to speak, such as
13. Namely, concerning physics or naturalness.
14. Concerning written history and world happenings.
15. Astrological concerns in accordance with signs and paths and so forth.

24 Rasch seemed to feel the divine power in the happenings which he wished to foretell.

25 In a fifteenth, unlisted, section of the first part of the book, Rasch said that comets might be predicted from knowledge of eclipses and planet configurations.

26 "krafft vnd wôllen". If this does not ascribe "free will" to the comet, it at least implies a certain directive force within the comet.

16. What and whom or in what direction the comet threatens.
17. When the effects will take place.
18. How long the effects will last.
19. Whether the comet also brings something good with it.
20. If the prescribed evil cannot be avoided or averted and so forth.

Rasch said that comets do not all announce future evils but also remind men of past events. He closed the subject of "comet practice" by saying that comets, being set forth by God, have both secret and open meanings. In an epilogue to the kind-hearted reader, he said that this comet gave him occasion to examine many little books on comets, and that he wished to offer this book until he could write a better and more useful one.

This treatise, which handles the material rather differently than do most of the contemporary popular treatments, is more valuable,[27] in that it stresses the observations which the mathematician and astronomer should make. However, the author did not record nor list in detail any such observations. The first section of the book merely repeats previous conceptions. The third part is purely astrological. Even the question of whether it is possible to predict a comet, although raised, is meant as a purely astrological inquiry and does not even border on the observations of the periodicity of comets which were later to revolutionize cometary theory. Besides, in this volume purporting to discuss the nova of "73" and the comets of 1577 and 1581, no reference is made to those phenomena, nor any mention made of anything especially characteristic of their appearance.

Giacomo (or Jacopo) Marzari, an historian, was another man of general learning who wrote on the comet of 1577. He was a member of a fairly illustrious family, but there is little

27 Günther (1890), 239-240, said of it that it "wenn auch natürlich nicht frei von abergläubischen und übertreibenden Behauptungen, doch mancherlei ganz beachtenswerte Aufschlüsse bietet", and he especially mentioned Rasch's notice of Apian's discovery of the direction of a comet's tail.

available information concerning him.[28] Outside of his books on that comet, he published nothing on scientific matters.[29] He be-

28 Angiolgabriello, V, CCXV-CCXXIII.—Baudrier, II, 196-7.—B.M. catalogue.—B.N. catalogue.—Bodleian library catalogue.—Haym, 54.—Lozzi, II, 487-8.—Riccardi, pt. I, v. II, 131. Correzioni ed Aggiunte, series V, 102-3.—Rumor, I, 18-9, 360.—Tiraboschi, VII, pt. I, 433.

Tiraboschi mentioned Marzari among those who wrote on the comet of 1577. Baudrier listed the French edition of his work on that comet. Angiolgabriello mentioned other members of his family and listed his books. Riccardi said of him that he came from Vicenza and lived in the second half of the sixteenth century; but even this is not valuable information, since the title-page of the *Discours . . .* (item 71 of the appendix, below) calls him " Vicentino " and the dates of his books indicate his floruit. Rumor referred to the following eight page work by Guiseppe Pieriboni, no copy of which is available:

Ritratti di cinque uomini illustri della famiglia Marzari... Vicenza, Tremeschin, 1838.

29 His books, other than those on the comet of 1577, are:

La Historia di Vicenza, del sig. Giacomo Marzari,... divisa in due libri, nel primo, si tratta della vera origine, fondatione e denominatione della città ... nel secondo, de' cittadini suoi chiari e illustri ... nuovamente posta in luce, con due tavole... Venice, G. Angelieri, 1591. (Bodleian library, B.N. and B.M. catalogues. The latter said that the date of the colophon is 1590. The 1591 edition seems to have been the first. See Lozzi, II, 487.)

La Historia di Vicenza,... nuovamente posta in luce... agiontovi la Città, con alcune antichità che in essa si ritrovano. Vicenza, G. Greco, 1604. (B.M., B.N. and Bodleian library catalogues; Haym, 54. The B.M. catalogue said that this is a duplicate of the preceding edition with a different title-page and with variations in fol. a4, and with three plates. This edition seems to give the date 1590 in the colophon (Lozzi, II, 488), although Angiolgabriello said it was dated from Venice in 1591.

La Prattica e Theorica del Cancellierie,... Vicenza, 1593. (B.M. and Bodleian library catalogues).

La Prattica e theorica del cancelliere,... Vicenza, G. Greco, 1602. (B.N. catalogue. Angiolgrabriello listed three editions by Domenico Amadio, 1593, 1602, and 1616.)

Scelti Documenti in dialogo a scholari bombardieri cc. Vicenza, Perin, 1579. (Riccardi, pt. I, v. II, 131).

Scelti Documenti in dialogo a scholari bombardieri cc. 1594. (Riccardi, pt. I, v. II, 131.)

Scelti documenti in dialogo a' scholari Bombardieri,... Vicenza, 1595. (B.M. catalogue; Riccardi, pt. I, v. II, 131; Angiolgabriello).

Scelti Documenti in dialogo a scholari bombardieri cc. (*In Vicenza, / Appresso gli Heredi di Perin Libraro. M.D. IVC*) (Bodleian library cata-

gan his tract on that comet [30] by stating that, in order to aid the public, he had decided to discuss this matter which has to do only with mathematics, the principal part of which, in his opinion, is astrology. He said that the good philosopher, in solving his problems, establishes the facts, and that therefore he would give a short history of the comet. This, according to him, was seen in the western part of the sky, after sunset on November 8th, and set two hours after the sun. It resembled a rather pale, rare fire. It was rarified in the upper part toward the east, but was thicker and tied like a knot hanging down in several strings toward the west. It was visible for sixteen days, but, he added, it must have changed its motion because it set progressively later.

According to Marzari, astrologers always err in attempting to treat the causes and natural qualities as though they can be mathematically measured. He said that our lower world is dependent on the world above, whose operations on the lower are diverse, including heating by the sun. He said that there are two types of exhalations, the humid, and that from dry matter; and that from the former comes rain. He added that a comet near the horizon gave birth to the tale of the Argonauts and their golden fleece. Hot, dry exhalations, he explained, mount and retain their fire, and comets come from these. They are composed of matter with several degrees of fire and in addition something of another element. Pointing out that Aristotle's divisions of comets are only such as " hairy ", " bearded ", and so forth, Marzari, however, made four classifications, those which have their greatest density in the lower parts, toward the top, toward the right and toward the left. This division, he added, follows that of the motions.

The book proceeds by saying that the comet of 1577 was a true exhalation which contained something coarse. This was indicated by the great quantity of autumn fruits in that year,

logue; Riccardi, pt. I, v. II, 131, and Correzioni ed Aggiunte, series V, 102-3).

30 Item 71 of appendix.

especially cold and watery apples, which betoken an over-abundance of humidity at the end of the summer. Consequently, part of this exhalation, joined with a large quantity of dry exhalation and kindled, was able to keep the fire burning for a long time. Marzari supposed that the comet would operate according to the virtues it had received from the above-described causes, and that its effects, dependent on dryness and a certain amount of density, would include corrosion and hemorrhoids. He outlined rules for taking care of oneself, at the time of this comet, in regard to meats, medicines and exercise.[31] He believed that there were other comets similar to this one, especially the one mentioned in the verses of Statius and that described by Herodotus. In order to learn more about such phenomena, he suggested reading Athenaeus on the comet in Antioch, Cassiodorus of Miletus, and Arrian on the comet in the time of Alexander.

Marzari's treatise has no scientific value. His description of the comet of 1577 is most meagre.[32] But it furnishes an additional example of the fanciful writings on comets which flooded the market.

A somewhat obscure writer on the comet of 1577 was Giovanni Maria Fiornovelli,[33] about whose life there is little avail-

31 For meats and medicines he referred the reader to Galen, but, he said, use purgatives with discretion. Use the same moderation in exercise, and be careful not to talk too much because talking attracts the blood to the veins and the spirits to the arteries. Besides, be careful not to heat your head either by the sun or by fire.

32 What Riccardi said of the second Italian edition can well be repeated here: "Opus. fisico-astrologico di ben poca importanza scientifica."

33 The volumes in which bibliographical references to Fiornovelli have been found are: Baudrier, II, 196.—Houzeau, 5594, 5598.—Lalande, 107.—Riccardi, pt. I, v. I, 462.—Riccioli, I, xxxviii, II, 10.—Scheibel, 102.—Struve, II, 550.—Tiraboschi, VII, pt. I, 433, calling him Giammaria Fornovelli.

No record has been found of books by Fiornovelli other than items 40, 41, and 42 of the appendix. Scheibel, who mentioned the *Opusculum de Cometis*, gave as his source Weidler's *Bibliographia Astronomia* which, he said, took the title from Riccioli. Riccardi, who listed the *Discorso . . .*, gave the further information that it is mentioned in Costabili's and Libri's

able information, although his books had sufficient circulation to secure their mention in various bibliographies and catalogues. He began his tract on that comet [34] by saying that, according to Leopold, a comet is only a vapor from the earth, composed of large particles well united which mount to the sky, and that it signifies changes in kingdoms and other events. He added that, according to others, a comet is formed by the conjunction of two planets; or is formed by a vapor "joining" with the luminosity of a planet or star. He related another view that comets are merely certain celestial bodies, which appear at different times, and are commonly called haired comets because of their appearance. He also said that according to Albertus Magnus comets are vapors and exhalations raised from the earth, which mount to the sky and lift the hearts of kind men. Fiornovelli added that Saturn, Mars and Mercury signify battles, and that predictions are made from those planets. Citing Ptolemy as his authority, he listed the same nine kinds of comets which Busch listed, a classification quite generally used in medieval times.[35] Fiornovelli gave some of them their Latin, others their French names,[36] but he gave them in a different order and was more specific than Busch concerning the characteristics and effects of each particular type of comet. He said, for example, that "Miles" is of the constitution of Venus; is large and re-

catalogues. According to Riccardi, information concerning Fiornovelli can be found in the following, at present unavailable, work:

Ughi, Luigi. *Dizionario storico degli uomini illustri Ferraresi* . . . Ferrara, 1804. 2 v., I, 224.

Scheibel, Riccardi, and Riccioli said that Fiornovelli came from Ferrara, and, indeed, his tract on the comet of 1577, the first edition of which was printed in that city in Italian by Baldini, speaks of that comet as the one which appeared over the lands of Ferrara. Riccioli, II, 10, mentioned an account of the comet of 1558 by Fiornovelli, but may have taken his information from the latter's tract on the comet of 1577. Fiornovelli may not have observed the earlier comet himself, although he spoke of it in his tract on the later one (item 41 of appendix, B_2 v).

34 Item 41 of appendix.

35 See chapter IV, above, especially note 139.

36 Probably in the Italian edition the Italian names were used.

sembles the moon; casts sparse rays behind; and passes through the twelve signs; that its tail is a sign of the worst ills; that it gives kings and powerful nobles cause for fear; and that it causes to be removed from the world those men who wish to introduce new laws in place of the old.

Fiornovelli next talked of great comets of the past. Among these were " Pertiqua ", which was seen about November 15th of the previous year [1577]; one before the death of Caesar; one before the death of Octavius Augustus in 14 A.D.; one at the time of Attila in 454; one in 538 accompanied by a famine, when Belisarius was sent to Rome by Justinian and delivered it from a siege of the Goths; another in 570 when Lombards went into Italy; one lasting three months in 676 when Sicily was destroyed by the Saracens; two in 738; and several others in 829 which fell from the skies killing many men and beasts and which were accompanied by other notable events. In 1006, he added, a comet appeared in the south and was followed the next year by famine and pestilence. Another appeared in 1067, when the Normans conquered Pouille and a part of Campagne. The comet of 1347 was followed by a great famine; Philip [37] of France was conquered by Edward [38] of England; more than 20,000 men were killed in battle; and in the following year there was a great pestilence. A comet in 1402 lasted a month. In 1533, the narrative continues, a comet called " Noire ", of the nature of Saturn, came from the north, with a short tail turned toward Africa and was seen for thirty days. The deaths of Pope Clement and Duke Alphonse then occurred, and in 1535 Charles V went to Africa. In 1558 another comet, " Veru ", of the nature of Mars and Mercury, appeared in the north with its tail turned toward Rome.

The last part of the book gives a prognostication based upon the comet of 1577. That comet, called " Pertica ",[39] was de-

37 Philip VI of Valois.

38 Edward III gained a great victory over King Philip of Valois at Crecy in 1346. The English took Calais in 1347.

39 The change of spelling is that of Fiornovelli's tract.

scribed as having " vn gros rayon " and a head like a star, " au vent dit Afrique ou Garbin ", in the sign of Capricorn, in conjunction with Saturn. This, said Fiornovelli on the authority of Ptolemy, signifies sickness and misfortunes in the west, upheaval of Persia, and ill " to the king of all these peoples ". Fiornovelli added that it also signifies dryness and barrenness. He quoted the second part of the work on prognostics of the pest and corruption of the air by Nicolaus of Conti, Count Chevalier of Pavia, to the effect that the comet " Pertica ", of the nature of Mars and Mercury, signifies illness and plague.[40] Albumasar was quoted on the subject of comets in Capricorn. Fiornovelli advocated prayer to avert the effects of the comet. He ended the tract by citing John of Damascus in support of his own argument that this comet appeared to announce the deaths of kings and princes, all of which comes by the will of God, whose anger is appeased when men are converted to him and ask for indulgence and mercy.

This tract gives none of the data for which astronomers look. However, the information given is evidence both of the author's knowledge of past literature and of his reliance on authority rather than on observation.

The comet of 1577 did not cause as great a stir in England as had the nova of 1572. The books which were written there about the later phenomenon were not of as high a caliber as those about the new star. There were at least three dealing with the comet: one by Hooker,[41] no copy of which has been located; one by Laurence Johnson,[42] written in Latin, which followed the Aristotelian tradition and contained few records of observations; and, finally, one by " T. T.",[43] who was probably Thomas Twyne, which was based entirely upon the astrological aspects of the phenomenon.

40 See chapter I, above, which cites Thorndike, IV, 250-2.

41 Item 56 of appendix, below.

42 Item 59a of appendix.

43 Item 105a of appendix.

Twyne[44] wrote many books,[45] some of them signed by his initials only. He was particularly interested in astrology and was a friend of John Dee. He was also a poet and translator. In addition he published "Almanacks" and "Prognostications" which are no longer extant. His *Discourse*[46] on the earthquake which took place on the evening of April 6, 1580, has been considered the most important of the pamphlets on that subject and has been said to be remarkably free from exaggeration.[47] He tied up the earthquake with other natural phenomena, and in his pamphlet on the former,[48] wrote that " Our strange and hot and

44 D.N.B., LVII, 403-4, article by Norman Moore.—Bodleian library catalogue. — Emmanuel college catalogue. — Hazlitt (1867), 620. — Hazlitt (1876-1903), 2nd series, 611, 3rd series, 251.—Hazlitt (1893), 173, 766.—Johnson (1937), 186, 310. — Ockenden, editor. — Twyne. — Witte, obiit... 1613.—Wood, I, 318, 354, 355, 383, 464, II, 130-2, III, 108.—Zedler, XLV, 2144.

Thomas Twyne, or Twine, a brother of Laurence, was the third son of John Twyne, master of the Canterbury free school. He was born at Canterbury in 1543 and died at Lewes in 1613. He received his education at Oxford, becoming a scholar of Corpus Christi College in 1560 and obtaining a fellowship in 1564, in which year he received his B.A. He received his M.A. in 1568, and a few years later studied medicine. He settled in Lewes, Sussex [Zedler said that Twyne practised in Dorchester] and acquired a large medical practise. In 1571 he married Joanna Pumfrett. In 1593 he received his M.B. degree from Oxford, and later his M.D. from Cambridge, becoming a licentiate of the College of Physicians in 1596. He had one son, Brian, who became well known as an antiquary.

45 Twyne completed a metrical translation of the *Aeneid* which Thomas Phaer had begun, and which was published in 1573 and 1584. He made translations of works by Petrarch and Bullinger and of Lambert Daneau's *Physice Christiana,* a work which tried to reconcile science and religion (see Thorndike, VI, 346-9). He also published a compilation called *The garlande of godly flowers,* which appeared in 1574, 1580, and 1602. Other works by Twyne are listed in the sources cited in note 44, especially in Ockenden, editor, 5.

46 *A shorte and pithie discourse. concerning the engendring, tokens, and effects of all earthquakes in generall: particularly applyed and conferred with that most strange and terrible worke of the Lord in shaking the earth, not only within the citie of London, but also in most partes of all Englande.* (1580).

47 Ockenden, editor, 6.

48 Twyne, 38.

drie tokens seene of late time, as the wonderfull blazinge Starre,[49] and the rare exhalations, shew that hee [God] wil come shortly to consume all with fire ". In the same pamphlet [50] Twyne used the expression " Exhalatiue impressions " when he was apparently referring to comets.

In the short dedicatory preface to his tract [51] on the comet of 1577, Thomas Twyne, if indeed he was the author of that tract, said that the purpose of the tract was " to confer some euentes with the purporte of the Blasinge Starre ". He did exactly that, looking back over the year which had passed since the first appearance of the comet. He said that the phenomenon appeared in the southwest part of the heaven on November 10, 1577, that it was soon thereafter extinguished bit by bit, and that its observers imagined both good and evil of it. He added that, although whole books had been written about the comet, some dealing with the origin of all " meteores " and particularly that of 1577, and some with its effects, his book would deal wholly with the effects. The authors he listed were Gemma, Dasypodius, Bariona, Roeslin, and Maestlin, all of whom wrote in Latin,[52] Dauid de Maudin, who wrote in French,[53] and " Frederike Nause ",[54] whose work dealt with the " signification " of all comets and was translated from Latin into English. Twyne also listed an anonymous pamphlet in English,[55] and a prognostication for 1579 by " Maister Securis ",[56] who spoke of an author of a work on the comet of 1577.

49 Ockenden, editor, 38 note 41, thought that Twyne meant the comet of 1577 by the above expression. However, there is a possibility that Twyne was referring to the nova of 1572.

50 Twyne, 28.

51 Item 105a of appendix, below.

52 See appendix, below.

53 Item 72b of appendix.

54 See the discussion of Nausea in chapter II, above.

55 Item 2a of appendix.

56 According to Broadbent, item 815, John Securis wrote *An Almanacke and Prognostication for the yeere of our Lorde God M.D.LXXVIII . . .* which was printed in London by Richarde Watkins and James Robertes.

In order to compare what did happen with what had been foreseen, Twyne discussed at considerable length the anonymous English pamphlet. The different points of prediction there set forth formed the basis for his own study. The first point concerned " the nature of all firie impressions " and " bearded " or " tayled " " Starres ", caused by a substance set on fire in the "Elementare region of fyre", and burning until consumed, such as the comet of 1577, which lasted until the end of January, 1578. Twyne believed that the end of the comet was dependent, not upon the consumption of its material, but solely upon the pleasure of God, and that it was the token of a judgment which God intended to execute on earth. The second point set forth was that the earth had got rid of poisonous exhalations and pestilential vapors. The consumption of these vapors, Twyne thought, was the cause of the drought in the summer of 1578, which prevented too much moisture and made possible such of God's blessings as abundance of corn, fruits and other foods. Also, there was moderate moisture in the autumn. The third point was that the corrupt vapors not " carried vp " were the cause of the pestilence of the summer of 1577, and of that of 1578. Twyne acknowledged that the mortality was God's " rod of correction ", but he also said that it was due to the fact that the fine vapors were removed and the heavier ones remained. In addition to the " naturall effects " portended by the comet, Twyne thought that one should also contemplate God's power and the care he exercises over his chosen people, namely the Church. This thought formed Twyne's fourth and fifth points. He was of the opinion that the Church would gain members because of the awe of the Lord inspired by the comet, and he believed that many who had been lax in their religious observance had been " brought home to the trueth ". He remarked that in Flanders God was preserving, defending and enlarging his Church. He noted that the comet's tail stretched in the direction of the moon. In the sixth section Twyne dealt with the effects of the comet resulting from its having appeared in the seventh heavenly house, that of marriage, wars " and so forth ", and in approximately the twentieth degree of Sagittarius. Twyne saw in this a symbol of the union of Queen Elizabeth with the Church, and also the prediction of a long life for the Queen. He thought that Venus with the participation of Mars, and not Saturn, dominated the comet. So, in his seventh section he compared the prosperity and happiness in England with the scarcity and war in other countries. The joys of the English were due, he pointed out, to the administration of God's chosen Queen. The advancement of women since the comet's appearance, due to the influence of Venus, was Twyne's eighth point, and he left it to the consideration of each individual man, although he believed that in this respect "also" the comet would have fulfilled its promise. However, he pointed out shameful happenings to women in countries east of England, such as the birth of an illegitimate child by John of Austria. In his ninth section Twyne considered the events which were generally supposed to follow the appearance of comets. For example, he noted that the threatened places were those to which the comet's tail pointed, and that the deaths of princes were supposed to follow comets. Therefore, he pointed to "Barbary" and the Low Countries, noting the deaths of three kings in the former and of John of Austria in the second. Twyne also pointed eastward to the Turks, the enemies of Christendom. In the tenth section Twyne remarked on the rape of a virgin in Picardy, followed by a massacre, and on the general pestilence which particularly affected the Spaniards in the Low Countries. In all of this Twyne saw the hand of God, and he expected worse to follow. The next section is in the same vein. But the twelfth section speaks of both good and evil following comets and of attempts to predict the appearance of comets, especially the prediction, by one of the men, identified solely as having been named in the beginning of the tract, of a comet in 1583. No basis for the prediction was mentioned by Twyne. In the thirteenth section Twyne enumerated evil deeds perpetrated by wickedly disposed people who felt that the occasion had arisen to do mischief when the minds of men were troubled and disquieted at the time of a comet's appearance. In the fourteenth section he discussed hygienic measures to prevent the plague. In the fifteenth he stated that the position of Mars in relation to the comet portended ill health for many, but he was not certain how far this prediction had been fulfilled. In the next section he asserted that the Lord was on the side of England and would maintain her honor, peace and wealth. Furthermore, he added that comets were sent by God in mercy to remind mankind to repent and improve. In the last section he explained that he had reviewed the points in the anonymous English pamphlet, although he felt that events had not completely carried out the " forewarnings ".

No one seems to have written about Himbert or Himbertus de Billy,[57] and almost all that is known about him is what he himself has written on the title-pages of his books.[58] According

57 Aa, pt. II, 191-2. — Baudrier, III, 184-7. — B.M. catalogue. — B.N. catalogue.—Delambre (1821), II, 530.—Du Verdier, 237-8.—Guides-Joanne: Lyonnais, 208.—Montucla, v. I, pt. III, section IX.—Scheibel, I, 305.

His name appears in alphabetical order in the B.N. catalogue under Himbertus. However, the B.M. catalogue placed the name under Billy. Delambre (1821), II, 530, merely mentioned the fact that a man named de Billy wrote a description of the path of the comet of 1577. Scheibel, I, 305, referring to Montucla, said that Himbertus de Billy was on the Church committee for the calendar. Montucla, v. I, pt. III, section IX, dealing with calendar reform in the sixteenth century, did not mention Himbert de Billy, but only the very important men connected with calendar reform. It is hard to say where Scheibel got his information.

58 Some of the books attributed to de Billy, other than the two on the comet of 1577, are:

Almanach pour l'an mil cinq cents quatre vingt et deux, avec la prévoyance et ample prédiction selon le cours et influence des astres. très diligemment supputée et recueillie par M. Himbert de Billy,... Dédié et consacré à messieurs les eschevins, conseilliers et bourgeois de la ville de Lons-le-Saulnier, au comté de Bourgongne. Lyons, Rigaud [1581]. (There is a copy in the B.N., Rés. p. V, 385, which, however, has not been consulted for the present study.)

Almanach || Povr L'An || M.D.LXXXVII. || Auec ses amples predictions du changement || & mutation de l'air, || selon le cours & influē- || ces des Astres sur les Lunaisons des douze || mois de l'An tres exactement calculees, sui- || uant la reformation Gregorienne, || sur le midy de vray coeur du Cōté de Bour- || goigne par M. Himbert de Billy, || natif de Charlieu en Lionnois, excellent & || renommé supputateur en Ephemerides ce- || lestes, bourgeois &

habitant de S. Amovr || *audit Comte de Bourgongne.* || *Dedie & consacré a Monseigneur le Reuerendissi-* || *me Archeuesque de Besançon Prince de* || *l'Empire Romain, &c.*

De tous les Almanachs voicy le plus certain,
Fait part de Billy, qui te fera voir l'annee,
Selon qu'elle promet bien ou mal fortunee,
Et selon que l'ait doit estre obscure ou serain.

A Lyon, || *Par Benoist Rigaud.* || *Et* || *A Paris,* || *chez Iean Cauelat, à la Salmandre, du* || *consentement dudit Rigaud.* || *Auec Priuilege du Roy.* (This title is taken from Baudrier, III, 184-5. There is a copy of the tract in the B.M., 531. a.24.)

Almanach || *Povr l'An Bissextil* || *M.D.LXXXVIII.* || *Auec ses amples & merueilleuses Predictions du* || *changemēt & mutation de l'air, selon le cours* || *& influence des Astres sur les Lunaisons des* || *douze mois de l'An : tres exactement calculêes* || *suyuant la reformation Gregorienne,* || *par M. Himbert de Billy, natif de Charlieu* || *en Lyonnois, excellent & renommé supputa-* || *teur des Ephemerides celestes.* || *Dediée & consacree a Illustre, haut & puissant seigneur,* || *Monseigneur de la Fin, Cheualier de l'ordre du* || *Roy, Conseiller en son conseil priué & d'estat,* || *Capitaine de cinquante hommes d'armes des* || *ordonnances de Sa Maieste, &c.* || *A Lyon,* || *Par Benoist Rigaud.* || *Et* || *A Paris,* || *Chez Iean Cauelat à la Salmandre, du* || *consentement dudit Rigaud.* || *Auec Priuilege du Roy.* (This title is taken from Baudrier, III, 185.)

Almanach et procnostication pour l'an . . . mil cinq cens quatre vingts seize. Composé par M. Imbert de Billy . . . Rouen (?) [1595]. (This title is taken from both the old and the new editions of the B.M. catalogue under Billy and under Ephemerides. It had the shelf mark PP 2389, r, which was changed to PP 2400 q. However, it is possible that there were two editions of the work, one printed in Rouen and one for which the place of printing was not given, and with the two shelf marks given above.)

Prédictions pour cinq années des choses . . . lesquelles nous sont dénoncées advenir par les révolutions des années, grandes conjonctions des plus hautes planettes . . . commençant ceste présente année mil six cens deux, composé par le sieur de Billy, . . .—Paris, pour N. Rousser [sic pour *Rousset*], *jouxte la copie imprimée à Rouen,* 1602. (There is a copy in the B.N., Rés. p. V. 217, which, however, has not been consulted for the present study.)

Presage General, Et Sommaire Discovrs Prognostic, Svr l'annee 1578. Tresdiligemmēt calculé, supputé, & recueilly par M. Himbert de Billy, natif de Charlieu en Lyōnois, disciple de M. de Montfort, dict de Blocklād, Docteur en Medecine, excellēt Astrologue, & Mathematicien Stichtois, Dedié & consacré à Monseigneur le Reuerend Prieur de Mouthe, Coligny &c. Conte, & Chanoine de S. Iean de Lyon. A. Lyon. Par Benoist Rigavd. M.D. LXXVIII. Auec permission. (There is a copy in the B. N., V 21084. The book has nothing whatsoever to do with the comet of 1577. Although comets are not even alluded to, and no mention is made of the events of the preceding year, 1577, a great many passages are copied verbatim from the author's *Description, et ample discovrs . . .*, item 12 of the appendix, below.)

to these, he was a native of Charlieu [59] in Lyonnais, and a disciple of M. Corneille de Montfort, "dict de Blockland", a doctor of medicine, an excellent astrologer, and a mathematician " Stichtois ".[60] However, the available histories of Lyonnais and Lyons do not mention him, and it is possible that de Billy and Corneille de Montfort were one.[61]

Sommaire description de l'effroyable meteore, et vision merueilleuse, n'a gueres veuë en l'air au dessus du Chasteau de l'Aubepin, proche de la ville de S. Amour, en la Franche Conté de Bourgongne [sic.] *Par M. Himbert de Billy, natif de Charlieu en Lyonnois, disciple de Noble Corneille de Montfort, dict de Blockland, &c. A. Lyon, par Benoist Rigavd. 1577. Auec permission.* (There is a copy in the B.N., LK^{7} 3596. The book has nothing to do with the comet of 1577. There are no passages in it like those in the *Description, et ample discovrs* . . . and the *Presage General, Et Sommaire Discovrs* . . .)

The books attributed to Corneille de Blockland include:

*Instrvction || De Mvsique || Par || C. De Blockland, Natif || De Montfort En || Hollande || * || A Lyon, || Par Iean de Tovrnes, || Imprimevr Dv || Roy. || M.D.LXXIII.* (Cartier, 555. The tract was dated from St. Amour, May 30, 1571. The B.N. has a copy of this edition of the tract.)

Le second Jardinet de Musique, contenant plusieurs belles chansons Françoises, à quatre parties, dediées en general à Madame de Creyssia Gabrielle de Dintreville, et chacune particulierement à quelque Damoiselle de sa connoissance par Corneille de Blockland. Lyon, Jean de Tournes, 1579. (Cartier, 594.)

*Instrvction || Methodiqve || & fort facile pour appren- || dre la Musique Pra- || ctique, || * || sans aucune Gamme, ou main, || parauant & iusques auiour- || d'huy tant accoustumee de plu- || sieurs Musiciens. || Reueue et corrigee en diuers endroits, par || Corneille de Montfort, dit de Block- || land, Gentilhomme Stichtois, excellent || Musicien. M.D.LXXXVII. || Par Iean De Tovrnes, || Imprim. Dv Roy, || A. Lyon. ||* (Cartier, 668. The tract was dated from "Lons le Saulnier", August 24, 1586, but is really a reprint of the 1573 edition. The B.N. and B.M. have copies of this edition.)

59 According to Guides-Joanne: Lyonnais, 208, Charlieu is a small town not far from Lyons, now famous for the remains of an abbey founded in the ninth century.

60 Sticht is an early name for a bishopric or diocese and, in the Netherlands, has been more particularly applied to the Diocese of Utrecht. (See Prins, XV, 370.) Montfort is in the Diocese of Utrecht. (See Prins, XII, 297.) Thus de Billy described himself as the disciple or pupil of someone from the neighborhood of Utrecht. " Sticht " and " Stifft " have sometimes been used interchangeably. (See Zedler, XL, 9.)

61 There is another angle to the question of the identity of Himbert de Billy. Du Verdier, 237-8, discussed Corneille de Blockland, saying that he

Himbert de Billy, in his book on the comet of 1577,[62] said that eclipses, meteors, visions and comets are but harbingers of plagues, wars, famines, changes of kingdoms, rebellions of people, deaths, droughts and deluges, which overtake the pitiful human race which remains obstinate in its evil ways. God was imploring people to return to the fold, he added, just as when, before the destruction of Jerusalem, a comet like a sword was seen for a whole year. The day after St. Martin's day, he continued, November 12, 1577 at 5 o'clock in the evening, God placed within view of all a new comet of great brightness, in the west in the sign of Capricorn not far from Saturn, in the eighth heavenly house close to the ninth, and traveling towards Aquila, a fixed star of second magnitude, of the nature of Mars and Jupiter.

According to de Billy, others saw the comet on the fifth day of the same month. He quoted Iunctinus (or Giuntini)[63] as saying that, at Gray in Burgundy, on the 9th of November

was a native of Montfort in Holland, a doctor living in St. Amour in Burgundy. (This has been confirmed by several other authorities including Aa, pt. II, 191-2, and Zedler, IV, 1438.) Du Verdier listed two books on music by Corneille de Blockland, printed by Jean de Tours, or Tournes, at Lyons. (Aa, pt. II, 191-2, listed these same two books. Cartier, 555, 594, 668, listed three books by Blockland, printed by Jean de Tours. Two of these were but different editions of the same book.) Du Verdier went on to say that Corneille de Blockland also wrote several diaries and almanacs under his own name and later under the name of "Imbert de Billy tailleur d'habits du Sieur de Perez Compte de S. Amour, Baron de Corgenou, &c.", which were printed at Lyons by Benoist Rigaud. Baudrier, III, 185, cited this and listed (III, 187), in the section dealing with undated works printed by Rigaud, "Diaires et almanachs [dates diverses] par Corneille de Blochland", and again citing Du Verdier, said that C. de Blockland was the author of the almanacs published under the name of Imbert de Billy, which he, Baudrier, had described on pages 184-5.

Du Verdier's theory of the identity of the two men is supported by the fact that at the time Corneille de Blockland's books were dated from St. Amour, those by de Billy were dated from the same place, and when Corneille de Blockland's books were dated from Lons-le-Saulnier, those by de Billy were likewise dated from there. However, this is not conclusive proof that there were not two men.

62 Item 12 of appendix, below.

63 See items 62, 63, 63a and [64] of appendix, below.

after sunset, he saw a fire in the air, traveling east, from which he believed the comet took its origin. At Turin, on the 11th of November, said de Billy, Bon, clerk of the postmaster-general of Piedmont and Savoy, together with many mathematicians saw the comet move with the motion of the primum mobile and set where the sun set. However, because at St. Amour it rained and was windy and cloudy on the 11th, de Billy did not see the comet until the 12th. Then it was losing matter bit by bit and getting smaller and smaller until on December 30th it disappeared. He said that at first the tail stretched eastward, then little by little southward, sometimes sloping toward the north but always returning between the east and the south. He described the color of the end of the tail as pale and leaden, but said that toward the head the star was very bright. From this he concluded that the comet was of the nature of Saturn and Mercury or of the moon. He predicted that the effects of the comet would begin about Epiphany in 1578, because then the sun would have the place in the sky where he thought that the comet was engendered and first appeared. DeBilly referred to Ptolemy's *Quadripartitum,* book II, chapter 54, and said that the evil omens would last for a long time, more especially as the comet was seen in the evening in the east. This comet, continued de Billy, showed itself out of season, in a humid, cold air ill fitted for the generation of a comet made of hot and dry exhalation. At this same time, he pointed out, Belgium was torn by wars. Who then, he asked, could deny that the comet was a herald of increased ills in the above mentioned district and that it was a work of God to show his irritation at human obstinacy and vices? He thought that if, as astrologers said, the significance of comets depends on the part of the zodiac where they are formed and the direction in which their heads and tails stretch, then this comet would produce its effect in the west where it first appeared and the east and south where its tail turned, causing wars and plunder in the eastern and southern Mohammedan countries. For the people of these latter he predicted greater ills than for Christians. There would be a great pestilence and many deaths in the east. Nevertheless, they would make preparations to ravish Christendom. De Billy advised the Christian princes to unite against the machinations of the Turks in order to be victorious. He thought that this would be a good opportunity for them to increase their kingdoms in Africa and Asia. Nine years, seven months, and fourteen days at the latest, after January 1578, there would be vast changes in government. In case of earthquake, Thrace would be in danger because of the increased violence of the winds and the rains and the inundations. The desolation would be great and there would be strange savage beasts abroad. Grasshoppers and locusts would consume whatever was growing, which would be little, because of the comet. Furthermore, he predicted that not only the Mohammedan countries but also Europe would feel the effects of the comet; for example, Savoy, Switzerland, and Piedmont. De Billy enumerated cities to be affected, citing as corroboration of his prediction the position of Mars in relation to the comet, the lunar eclipse of September 20, 1577 and the one to take place September 15, 1578, both in the ninth celestial house. The disorder, he said, would be terrific. He added that the comet's appearance in the sign of Capricorn, which is the domain of Saturn, indicated ills for a very large number of the cities of Europe. Therefore, he urged everyone to pray to God to appease his wrath. De Billy said that in many cities the air would be corrupted. The book ends in a prayer.

The astronomical worth of the book lies in the description of the appearance of the comet, the direction of its motion, the direction of its tail and the dates of its visibility; but the observations are not sufficiently accurate to be valuable. The predictions are more daring than most because they are so specific, but, needless to say, they have no greater importance.

CHAPTER VIII

THE COMET OF 1577: AUTHORS WHOSE TRACTS WERE PRIMARILY ASTROLOGICAL AND PREDICATORY

LIBERATI. — A. PRAETORIUS. — MÜLLER. — BRUNFELS. — CREAT. — P.S.T.A.F. — ANONYMOUS

NATURALLY, astrological explanations of the comet of 1577 were also made by professional astrologers. One of these was François or Francesco Liberati, a Roman who, in about 1584, lived in Paris. He worked on preparing French calendars and wrote several mathematical tracts.[1] His work on the comet of 1577 [2] is purely astrological and was written the day after the first appearance of the comet. Although it has little value astronomically, it shows that the author was versed in astrology.

1 This much information concerning Liberati can be found in Jöcher, II, 2418, and Zedler, XVII, 783. Zedler gave three references which have not been available in preparing the present sketch. The information seems to apply to the author of the tract on the comet of 1577, who is described, on the title-pages of the four editions of the tract, as an astrologer from Rome. Furthermore, two editions of the tract were printed in Paris. (See items 67, 67a, 67b and 67c of the appendix.) Zedler ascribed the following two works to the astrologer he was describing:

La perfettione del cavallo. Rome, 1639.
Perfetto Maestro di Casa.

Both the B.M. and the B.N. in cataloguing these works ascribed them to a second man of the same name. This seems reasonable in consideration of the disparity of the subjects treated and the great difference between the dates of the books. (The earliest date given for the second work was 1658.) Thus it appears that Zedler fused information concerning two different men into a biography of one man. However, the following two works seem to be by the author of the tract on the comet of 1577:

Almanach et amples prédictions pour l'an mil cinq cens quatre vingts et cinq, avec le jugement de l'éclipse du soleil, lequel sera du tout obscurcy vers l'angle d'occident, le soir du 29 jour d'avril, composé par M. François Liberati... Paris, C. de Montre-oeil, (no date). (B.N. catalogue.)

Prediction et discovrs astronomiqve povr l'an de bissexte mil cinq cent quatre-vingt,... Paris, Iean de l'Astre, (no date). (Crawford library catalogue.)

2 Items 67, 67b, 67c of appendix, below.

The comet was seen in Paris about 6 P. M. on November 11th. It was of the nature of Mars and Jupiter and was in the west near the star called flying eagle, in the eighth house of the sky, in the sign of Capricorn, while Saturn was in the same sign. Ptolemy, especially book 2, section 64 of the *Quadripartitum,* Aristotle, and John of Damascus were cited, and the line by Pachymerēs,[3] quoted by other authors on the comet of 1577, was also quoted by Liberati. Pontanus, Regiomontanus, Haly, and Cardan were listed as writers about comets.[4] Liberati thought that according to Ptolemy's precepts the comet was of the nature of Mercury. He also thought that it signified divisions and heresies in Catholicism and Mohammedanism, riots and plagues, tempests and winds, abundance of water in the west and misfortunes for kings and princes both in the east and in the south. Finally, he said, the comet signified lengthy war in the west. The heresies were to be due to a young northern prince, dominating in the west, born under Mars in the sign of Capricorn in the place where the comet began. The significations were to be particularly dangerous for those who were born when Saturn was in the twenty-first to twenty-seventh degrees of Libra, or when Mars was in opposition to Saturn. Liberati also took into account the supposed effects of the lunar eclipse of September 1577.

Many trivial and inconsequential books on the comet of 1577 had considerable circulation on the continent and although they may not have hindered the progress of astronomical thought, certainly did nothing to enlighten the populace. To secure a balanced view of the level of opinion about comets, it might be well to review some of these tracts.

3 "Impunè nunquam visus fulgere Cometes." See the summary of Dasypodius' work on the comet in chapter V, above, and the description of items 12 and 80 in the appendix, below.

4 In place of the closing sentences where Pontanus, Regiomontanus, Haly and Cardan are listed, the text of item 67 ends with a few sentences on comets in general, briefly outlining the Aristotelian theory of their earthly origin and speaking of them as signs from God. At the close of that edition there are a Latin song to "D. H. P. A." and a sonnet to the same in Italian with a French translation.

All that is known concerning Adelarius Praetorius, the author of one of these,[5] is what is told on the verso of the title-page of his book on the comet of 1577, namely that he had a position in the church in Erfurt.[6] That book opens with a prayer in poor, sing-song, German verse, calling attention to the "star" which was bringing in its train pestilence, war, and so forth. The prose, also in the form of a prayer, carries out the same ideas, saying that the sign had not been placed in the heavens in vain but as a warning. It says that there had been other similar signs in the past. These the author enumerated: for example, the comet which stood over Jerusalem for a year before the downfall of that city, a comet which lasted four months in 1337 and another in 1339, and the comets of 1400, 1401, 1402 and 1403. The recitation of these dates of comets is accompanied by a full record of the significant events which followed, supposedly as a result of the comets. The events were mainly tied up with wars and religion, such as the tyranny of Tammerlane and the burning of John Huss. The other comets mentioned were those of 1500, 1526, 1531, 1533, 1556, 1558 and the nova of " 1574 ".[7] The conclusion which the author reached was that because in this year [1577] God not only gave solar and lunar eclipses, but also the comet, men should acknowledge their sins and ask the forgiveness of Christ, and the people should peer into their hearts and see that their bad living had aroused God's anger, and should pray that the well-deserved punishment be turned aside. As emphasis, the author, giving the sources for his information, further enumerated comets and their "attendant" results, including the comet of 51 A.D., the

5 Item 80 of appendix, below.

6 Adelarius Praetorius is undoubtedly the A. Praetorius to whom Houzeau attributed both item 80 of the appendix, below (Houzeau, 5604), and a work on the comet of 1580 (Houzeau, 5615), a copy of which can be found in the N. Y. P. L. When Houzeau said that "A. Praetorius" was equivalent to "A. Richter", he was merely stating the Latin and German forms of the same name.

7 From photostats of the Vienna copy of this book, it can be seen that in that copy the " 4 " has been crossed out and " 2 " written above it.

comet in the fifth year of Justinian's reign, and two comets in 729 A.D.

Samuel Müller is another writer about whom there is little available information [8] other than that given on the title-page and in the dedication of his book on the comet of 1577.[9] He was a natural philosopher and doctor, a native of Kempten in Bavaria. According to his tract, the comet was first seen on the ninth of November. It was visible after sunset, its tail stretched in a south-easterly direction. Müller stated the relative positions of the sun, the moon, Mars, Jupiter and Venus and said that the comet passed close to the bright star in Lyra [Vega], and through the last 29° in Capricorn, through all of Aquarius and up to the third degree of Pisces. It disappeared on January 14, 1578. Its daily motion was 1°. He said that in November, before the comet's appearance, there had been storms and thunder and lightning, and that on November 13th there was a " chasm " in the sky, and that, while the comet lasted, the sky was fiery and red. He recorded a meteor for December 20th. History, he said, shows comets to be evil omens; for example the comet in Capricorn in 1527, when the Turks went to Vienna, was such. Furthermore, he added, at the time of the comet of 1531 the Turks went to Hungary, and the lunar eclipse of 1530 in Aries had similar effects. This comet also, he predicted, was an omen of great misfortunes and a warning to keep from doing wrong. The slight astronomical value of the tract lies in the description, meagre as it is, of the comet's path and of its tail, and the dates of its visibility.

The identity of Otho Brunfels, author of another pamphlet [10] on the comet of 1577, is a mystery.[11] This pamphlet says that

8 Hellmann (1924), 28, listed a prognostication or weather forecast by Müller for 1583, written in German. Hellmann (1924), 28, referred to Hellman (1883), but Samuel Müller was not mentioned in that work, although two " Practicas " by Tobias Müller were cited for the year 1583, besides many other works by that author. Squarcialupus, item 37(3), of appendix, below, C_3 v (page 46), referred to Samuel Müller.

9 Item 75 of appendix, below.

10 Items 20b and 20c of appendix, below.

11 He must not be confused with the well known botanist and theologian, Otto Brunfels, who died in 1534 and whose deeds are recorded in many

signs in heaven and in the air are not in vain, but predict punishment by God in the form of pestilence, war, and so forth. Brunfels said that because the Jews paid no attention to the warnings of Christ and his apostles, God preached to them with miraculous signs, earthquakes, wind, and an unusual solar eclipse, and because they continued in their ways, forty years later, he sent a comet, like a sword, which stood for a year over Jerusalem. And afterwards, the city was besieged and the temple burned, and many people perished by sword, hunger, and pestilence, and the rest were scattered in all lands and the whole Jewish disciplinary collapsed. There was a comet for four months in 1337 in the reign of Emperor Ludwig of Bavaria; and, before it disappeared, another was seen. As it vanished a third appeared in 1339, followed in the next year by a terrible plague throughout the world, bringing changes in the Roman Empire, so that the Pope put Emperor Ludwig under ban and the Electors chose another Emperor, whereupon followed great disunion, war, and bloodshed. In 1400, 1401, 1402, and 1403, the narrative continues, there were four comets, and thereafter a tyrant in Tartary with 1,000,000 men on horse and on foot ravaged the Orient. In 1409 the Emperor Sigismund suffered defeat by the Turks in Hungary and the following year there was the Hussite war. There was another comet in 1500 followed by the descent of the Tartars on Poland, and a plague in Germany, and other calamities. After the comet of 1526 the Turks

biographical dictionaries. Our author is not mentioned in any of the following: A.D.B.—B.M. catalogue.—B.N. catalogue.—Biographisches lexikon der hervorragenden ärzte.—Doppelmayr.—Ersch and Gruber.—Hoefer.—Jöcher. —Michaud.—Poggendorff.—Will.—Witte.—Zedler.

The Brunfels of the second half of the sixteenth century is mentioned by Hellmann (1924), 26, as having written a weather forecast or prognostication in German in 1582, a copy of which was in Munich. The fact that there may have been a Nuremberg edition of the pamphlet on the comet of 1577 has not served as a clew to the identity of the author, since he was mentioned by neither Doppelmayr nor Will. It is possible that he used the name of Otho Brunfels to give weight to his pamphlet, or that he was an obscure person who could rightly claim the name, or that Brunfels merely referred to the town whence he came.

went to Vienna. One knows, said Brunfels, that after the comets of 1531 and 1533 there was much misery in Westphalia, Hungary, Denmark, England, France, and Italy. There were misfortunes in the Netherlands, France, and Poland following the comets of "56" and "58". They were also followed by famine in Germany and elsewhere. This, according to Brunfels, was God's punishment. Because God, that year [1577], showed signs on the sun and moon, namely horrible eclipses, there was no doubt but that He was moving the stars and the heavens, in order that man should take heed. Brunfels said that, because the comet which appeared on November 11th of this 1577th year was in the heavenly sign of Capricorn in the evenings between 5 and 6 o'clock, and was burning larger and brighter with the new month, great punishment and misfortunes would follow, particularly for the lands lying under Capricorn. Men should take warning, he added, acknowledge their sins, fear God's wrath, and by improving their lives stave off the punishment. He said that, unfortunately, the masses are not bettered by such signs, but scorn them in Epicurean fashion and become more abandoned and meaner. So, he added, when we observe this comet we should remember that God is angered by our sins, and we should look into our hearts and think. We should better our lives and pray that Christ avert the punishment, that we be ruled by His spirit, and that we appear before the Judgment Seat of Jesus Christ on Doomsday, and that we become blessed. Amen.

This little tract cannot be called astronomical. It was written by a supposedly learned man, not necessarily a theologian, and gives various past examples of the effects of comets and some astrological arguments. It is an example of the non-scientific approach to the subject, and it makes very vivid the horror, awe, and fear which the comet of 1577 inspired in the masses, although, according to Brunfels, they were not bettered by it.

There seems to be no record of Johan Creat.[12] His tract, in German, on the comet of 1577 [13] states that the comet was first seen on November 11th; that it had a large tail like a broom; and that it followed after the sun, with which it caught up. He thought that the comet boded evil, and he said that one should look at Poland and Turkey for examples of the effects of the star of 1572. Furthermore, he said that the significance of the star of 1572 was not yet at an end. He listed astronomical phenomena of the previous hundred years and their supposedly attendant disasters; in 1489 there was a comet, and there were many wars in that period; the comet of 1491 and the eclipse of the sun under Taurus on May 8th were followed by the attack of the Turks on Dalmatia and a long winter of severe cold; in 1497 there was a lunar eclipse on January 18th and a solar eclipse on July 29th, and the disasters which followed, such as the wars, lasted until 1499. The author enumerated the deaths of kings and princes. The year 1516, he said, saw lunar eclipses on January 19th under the sign of Leo and on July 13th under Capricorn and a solar eclipse on December 23rd also under Capricorn. The results, he added, were wars and other disasters, such as the advent of Luther and the occurrence of violent storms. On January 20, 1519, he said, there was a conjunction of Mars and Saturn under the twenty-sixth degree of Capricorn. There followed the Pope's bull against Luther, wars, deaths of important people and other unpleasant events. On February 17, 1527, in Kauffbeuren in Wurtemberg, three suns were seen enclosed in two rainbows, and disasters, such as famine and deaths from pestilence, followed. He told that the comet-like

12 The fact that he existed is attested by the volume about to be discussed, but no reference has been found to any other work by him. Nor does the story of his life appear in any of the well known biographical dictionaries, and since no publisher nor place of publication was given for his book on the comet of 1577, not even a guess can be made as to the probable locality from which he came. The name is probably a pseudonym, since, according to Zedler, VI, 1555, "Creat" means a young person admitted to an academy to learn horsemanship.

13 Item 31 of appendix, below.

midnight star of '72-'73 was followed by the " Spanish fury " in Antwerp, and by rape, murder and fire. In 1575 in Thuringia and Saxony, he said, four suns and two rainbows were seen in the sky from 6 to 8 o'clock in the morning, bringing good times, for which thanks should be offered to God. In 1576, he continued, there was peace and plenty. However, about 5 or 6 o'clock in the evening of St. Martin's day, 1577, a horrible comet came as a warning from God. Its purpose, he added, was to make the people cease their sinning. If the people failed to improve thereafter, punishment would follow, for as sins increase punishment becomes more severe. In the year 1577, Creat reported, there were two eclipses. He lamented that brotherly love no longer existed and that eternal goodness was no longer held above material things. The author begged the people to do penance so that the comet should not have appeared in vain; and prayed that God avert the well earned punishment and that He remain with the people, and that the sun illumine the hearts of men with holy light.

The most important astronomical fact given in the tract is that the comet was seen between 5 and 6 P.M. on St. Martin's day, November 11, 1577, following after the sun. Furthermore, for those interested in recording eclipses of the fifteenth and sixteenth centuries, it may give interesting though not unique information.

A French writer on the comet remains just as obscure as Creat. The use of the initials P.S.T.A.F., in place of his name, has not drawn aside the curtain of anonymity in the case of the author of the tract about to be discussed.[14] Although he seems to have been familiar with ancient literature, his book is of the same low caliber, in regard to astronomy, as those items listed as anonymous. Although probably written because of the comet of 1577 and then reprinted because of the comet of 1580, it deals with generalities and does not describe any particular comet, despite the fact that it speaks of " ceste comete qui apparoist maintenant".

14 Items 38a and 38b of appendix, below.

P.S.T.A.F. considered that a comet was a sign of God's power, but he did not believe that it necessarily presaged evil. The name comet, he said, was derived from the Greek word for hair. The Latin name, he added, is crinita. Citing Aristotle, he said that the truth cannot be known unless one knows the cause. The efficient cause (of comets), he added, is the light of the sun and the stars, which make the matter move. He further stated that the immediate cause is an eclipse or failure of light; that the material is a viscous exhalation accumulated by the movement of the celestial bodies and is different from the usual exhalations from the earth; and that the same star which generated the material, sustained it. In addition to the efficient and material causes, the author explained the formal and final causes. He did not want his readers to think that a comet was formed by chance. Comets are formed, according to P.S.T.A.F., in the uppermost part of the elementary region, or the region which, he pointed out, was sometimes, but not correctly, referred to as that of fire. The author was aware that this was not the opinion of all men, but he said that it was the true opinion. In this connection, having previously mentioned the names of Aristotle and Dinocrates, he further exhibited his learning by sketching the theories of Anaxagoras, Democritus, the Stoics, the Pythagoreans, Pliny, and Epicurus. Refuting the argument that if the comet were in the air it would move with the movement of the air, our author said that the part of the air nearest the sphere of the moon moved with the movement of the celestial spheres. Similarly, he refuted other arguments which did not agree with his theory. To say that stars became inflamed in the sky because of dryness, he thought was like speaking of the corruptibility of the eternal. Comets, he added, cannot be planets because they are seen outside the limits of the zodiac. Furthermore, he was sure that they were not fixed stars because they moved and because they were not always of the same size but increased and decreased. He told the interesting myth, which he said he took from Hyginus' book of fables, that Merope, one of the seven Pleiades, daughters of Atlas, did not dare show herself because her sisters had married gods and she had married a mortal. Having been chased from the company of her sisters, she wore her hair down and flowing out as a sign of sadness and shame, and was called comet. Following this tale, the author returned to the subject of the forms of comets and called the different types by names, mostly those used by Pliny. The latter, said P.S.T.A.F., stated that comets lasted from seven to twenty-four days. However, he cited Geoffrey of Tours as telling of a comet lasting a year. Their motion, he continued, is like that of the planets, except that some move from east or north, so that one might say with Seneca that they have no path but go where the view of their pasture leads them.[15] The tract gives more bits of the theories of Aristotle and Pliny and closes with a long and unusual passage tending to disprove the then current beliefs in the effects of comets, for which beliefs the author used the term "badineries".

Another anonymous tract,[16] dated on St. Catherine's day 1577, was addressed to N. Labieno, a good friend, who had asked for a report on the comet. The author modestly said that such a report was not within his power. Nevertheless, he collected " Significationes ". He related that two total eclipses in 1577 were bringing great misery for the lands lying under Aries, Cancer, Libra, and Capricorn. He predicted that God would remove, by death, several lights of the Christian Church because of the ungratefulness of Christian men for His Word and for His faithful servants. Revolution also was predicted. The author added that a comet had appeared in the fifth degree of Capricorn at the end of 1577 and had moved, at the time the pamphlet was written, to the twenty-sixth degree plus several minutes. He said that this comet would affect the freedom and security of men's lives. He quoted from " Ptolemy's " *Centiloquium,* cited Pontano, and deduced therefrom that the comet of 1577 would bring a war and also death, if not to the nobility,

15 The French reads: "... tellement qu'on peut dire auec Seneque, qu'elles n'ont aucun chemin, mais qu'elles se trainent la part où la veuë de leur pasture les conduit." This means that they follow their supply of fuel.

16 Items 3 and 3a of appendix, below.

at least to the people on the land. If the comet started in the ninth sign, he said, death and pestilence would come as predicted by the eclipses of 1577, with effects lasting into 1578. In addition, the comet would excite wars and would embitter the hearts of people. He added that, since the motion of the comet was from west to east, and according to astronomers nearly two whole degrees, there would be trouble with a foreign enemy and much bloodshed. This comet, bigger than any in the memory of man, said the anonymous author, was not bringing anything ordinary. The air had already lost its strength for man and beast, he continued, due to the two recent eclipses so close together; and the following year, on March 14th, there was to be a conjunction of Saturn and Mars in Capricorn, where the comet first appeared. Thus Mars was to rule in that year and, undoubtedly, there would be death, war and hate. The duration of the effect, like that of the star in Cassiopeia which was still daily exercising its influence, was to be ten years, and it was a simple matter to determine where it would be felt. The countries and cities affected by the comet were enumerated. However, the author added, what lay in store for individuals would have to be found from their nativities. Nevertheless, even those whose horoscopes did not menace them ought to live in fear of God. The author concluded by promising that, if he was shown more when the comet was extinguished, which he hoped would be soon, he would write again to Labieno.

It becomes apparent that the quality of the writings on the comet was uneven. The heights of reasoning were reached by such men as Tycho and Maestlin, but the lesser authors still struggled along in ignorance. It was the important work which, when digested and expanded, raised the level of astronomical theory and eventually brought the information to that class of people who in the late sixteenth century were reading the valueless tracts we have just described. The advances in theory were largely made by those who gave a major portion of their time to astronomy.

CHAPTER IX

CONCLUSION

THERE can be no doubt that the study of the comet of 1577 put observation of such phenomena on a new and sounder footing. However, no claim can be made that immediately after the appearance of that comet all who wrote on astronomy completely dropped Aristotelian and similar theories and immediately recognized comets as celestial bodies moving according to the general laws governing the motion of such bodies. These laws, indeed, as applied to the solar system, were first stated in their present generalized form by Kepler in the beginning of the seventeenth century. Furthermore, as Dreyer said, "It was not until Hevelius had again shown from accurate observations that comets are much farther off than the moon that the opponents to their character of heavenly bodies were finally silenced some sixty years after Tycho's death, and not till 1681 that the parabolic form of their orbits with the sun at the focus was discovered by Dörfel." [1] Although shortly thereafter Halley made his famed prediction of the return of the comet of 1682, necessitating a closed orbit, astronomers had to wait until the winter of 1758-9 to test this periodicity. The mere fact that several earlier books about comets were reprinted in 1577 and the years immediately thereafter shows that the new information and ideas were not yet assimilated. However, the observations of the comet of 1577 provided an impetus to the formation of the pres-

1 Dreyer (1906), 415-6. The discovery that comets move in a parabola with the sun at the focus first made it possible to formulate a theory in which these bodies follow the laws governing the planets. Hevelius, using improved and very large instruments, including a telescope, although he used his quadrants and sextants for measuring angular distances, made such accurate observations of comets and studied the observations of past comets so carefully, that it became apparent that many comets followed parabolic orbits and that their obliquity to the ecliptic was large. See Hevelius (1665), Hevelius (1666) and Hevelius (1668). It was his friend, Halley, who realised the possibility that some comets might move in elliptic orbits. But, because of the simpler mathematics involved, their paths are still first computed as parabolas.

ent theories about comets and accelerated their gradual acceptance.

The size of the bibliography[2] of treatises on the comet of 1577 has shown that a far greater body of data was accumulated concerning that comet than any previous one. The large number of observations of the comet made at widely separated places was in itself a help to the establishment of a theory of comets.

At the close of the sixteenth century the most important developments in the theory of comets were accomplished by Tycho, Maestlin, and Roeslin. Their calculations of the orbit of the comet of 1577 were the first attempts to compute the orbit of a comet.[3] Other observers, including Gemma and the Landgrave, also did valuable work. The assertions concerning the absence of parallax for the nova of 1572 and for the comet of 1577 had a dramatic quality. Many thinking people felt it necessary to discard the belief that a comet was an atmospheric phenomenon. And, because it was conclusively shown to be further from the earth than the moon, the comet was among those bodies which should and could be carefully observed.

Careful observations of the comet of 1577 were made. The accuracy of Tycho's work depended in a large measure on his use of improved instruments. Although Maestlin rejected the use of many of these, believing his use of a thread less apt to introduce errors,[4] it is apparent that a new attitude toward instruments had arisen. Accuracy of observation required not only bigger and better instruments but also the ability to correct for their errors. Similarly, and partly as a consequence, the need for accurate star catalogues was answered by sixteenth century astronomers. Their redeterminations of the positions of the stars proved important in raising the caliber of observations.

Observers in 1577 were able to take advantage of the new knowledge accumulated during the preceding hundred years.

2 Appendix.

3 See chapter III, note 10, above.

4 See chapter III, above, and page 21 of item 70 of appendix, below.

Had not Regiomontanus [5] set forth his theory of parallax, Tycho and the other observers of the comet of 1577 would not have had at hand the method or tool necessary to prove that the comet moved among the heavenly bodies. Had it not already been established that a comet's tail always points away from the sun, much effort would have been wasted in observing the direction of the tail of the comet of 1577. In the scientific writings on that comet, the early sixteenth century discovery of the direction of a comet's tail was accepted. In fact, the attention of observers was no longer primarily focused upon the tail. The shift of emphasis to observations of the positions of the comet's head was important in gathering the data for a formulation of a theory of its motion. The value to cometary theory of the knowledge of the direction of a comet's tail could not be fully realized until after physical astronomy had advanced to the stage where the formation of the tail could be explained. However, sixteenth century observers, such as Tycho and Maestlin, recognized the importance of the sun in that formation. They believed that the tail was caused by sunlight passing through the body of the comet.

The observers of the comet of 1577 paid great attention to detail and made frequent observations. Because of their continuity, the observations of the comet of 1577 by such men as Scultetus and Dasypodius, whose training in mathematics stood them in good stead, even though they did not accept the new advances in theory, were more valuable than the observations of comets in earlier times. Another sign of progress was that a great many were aware of the value of the method of parallax for determining a comet's distance, although the method was not successful in the hands of the unskilled, or of those whose preconceived notions made it imperative for them to fit their new data to their old theories.

The usual fears attendant upon the appearance of a comet still prevailed in 1577. The awe in which the comet of 1577 was

5 Regiomontanus or Peurbach. See chapter II, above, especially notes 42 and 77, for the suggestion that Regiomontanus was not the first to use the method of parallax on a comet.

held prompted theological leaders[6] to offer prayers. Many clerics found that the appearance of the comet furnished an opportune moment for preaching a sermon, pleading for the betterment of the human race. In those cases, the comet was definitely not the central theme and the resulting sermons consequently added little, if anything, to scientific knowledge. In a similar manner, anyone with a bit of learning was led to attempt a dissertation. Some physicians studied comets because they believed that those bodies influenced health.[7] The frequent single page and other short writings on the comet can be said to have filled the place now held by articles in newspapers. On the other hand, in the closing years of the sixteenth century several men attempted to combat superstition concerning comets. Among these[8] were Johannes Praetorius, Dudith, Erastus, and Squarcialupus, observers of the comet of 1577. At the same time the influence of Peter Apian, who died in 1552 but whose works were read long thereafter, was felt. Tycho gave the astrological implications of the comet no place in his Latin work; and though Maestlin did, it was somewhat as an afterthought. Superstition concerning comets, however, has been said to have "reached its highest development and received its sharpest attacks" at the time of the comet of 1680.[9]

6 Such as Chytraeus, Selneccer, and Heerbrand.

7 Smith (1917), 128.

8 See Janssen, VI, 440, note 3; Pingré, I, 73; White, 198-9 and note; appendix, below. Dudith's *Commentariolus* (see appendix, items 13, [34], [35], and 37 (5), is discussed by Scheibel, 112 ff. It is in the form of a letter to Crato, and deals with the meanings of comets. It was reprinted as late as 1665 (see appendix, item 13). A letter by Dudith to Hagecius, dated from Breslau, September 26, 1580, was printed by Scheibel, 161-182. See also item [36] of appendix, below, and Thorndike, V, 656-7, VI, 183-6. That Giovanni Ferrerio's work on the comet of 1531, which was published in 1540, was published in an Italian translation in 1577 shows that his attack on astrology had not been forgotten. See chapter II, above, especially note 216.

9 Robinson, vii. In 1683, in his *Kometographia*, with observations of the comets of 1680 and 1682 freshly in mind, Increase Mather, who has been said to be up to date on the subject of comets (Holmes, I, 318), portrayed the current attitude toward those bodies. He had previously written two sermons, basing his exhortations on the comets (of 1680 and 1682).

Besides the printed books dealing solely with the comet of 1577, there were many notices in diaries [10] and compendia, which, because they were not tracts on the comet, have not been here incorporated in the literature of that subject. However, since they did not embrace the new theory, they were a deterrent to its acceptance. Among them was Bodin's treatise on nature, a compendium of science by a learned man who could not quite break with Aristotelian tradition.[11] Other books,

He was fully aware that those bodies were in the "Starry Heaven" but he held to the belief that comets were signs of evil events (Holmes, I, 312-9). Even in the nineteenth century, belief in comets as signs of war persisted (Lauffer, 13). These superstitions were entirely separate from any theory of the motions of comets. See Lauffer, for an historical sketch of comet superstition. Indeed, such superstitions were still present, though to a lesser extent, when Halley's comet made its 1910 appearance. In this connection see Emerson's book. This work not only accepts the tales of disasters following comets in past history, but speaks of predictions from Halley's comet in its 1910 appearance and ends with a story (from Flammarion) of the end of the world because of a collision with a comet. In spite of his superstitious leanings, Emerson grasped the modern theory of comets.

10 Haton's memoirs provide an excellent example of a diary. They were first published in 1857. However, their author can be considered as portraying the point of view of a casual observer of the comet of 1577, and because Haton was a leader in his community, his opinions must have had considerable weight there. For the life of Haton, a French priest who was born in 1534 and died after 1605, see Haton, I, xx-xliv. For Haton's account of the comet of 1577, see Haton, II, 909-911. Haton's description of the comet, both of its appearance and of its position, is most vague, but he was sufficiently interested to include a crude drawing of it.

11 The political theories of Bodin have won him fame, but in the field of science he was not so advanced. See Bodin, 302-311. Although Bodin's work is not a part of the literature devoted specifically to the comet of 1577, it can be taken into account as reflecting the immediate reception of the newly acquired information. Using the form of a dialogue, Bodin was able to express several points of view concerning comets, but the result is confusing, and the reader is left with the impression that perhaps the ancient theories are the better. Bodin, or rather "Le Mystagogve", professing his ignorance of the subject, cited Aristotle's *Meteorologica*. However, he also pointed out that exhalations can rise only to a limited height and that some observers found no parallax for the "comet of 1573". He placed little value on the evidence of observations of parallax, but seems to have had confidence in the supposed meaning of comets, and cited Cicero

adopting the new knowledge, furthered its diffusion, and showed the influence of the nova of 1572 and the comet of 1577, so often discussed together.[12]

on that score. On the other hand, this same "Mystagogve" spoke of the inconveniences of the Aristotelian theory. Bodin stated the theory, which he assigned to Democritus, that comets are the souls of famous men and bring famine, pestilence, and war. He put this theory into the mouth of the interrogator, "Le Theoricien", and therefore it is not at all certain that Bodin himself held this opinion, although White, 178-9, seems to have taken such for granted. At the close of the passage it is stated that comets are not exhalations from the earth; but unfortunately the statement is unconvincingly presented. It is the confusion resulting from a perusal of Bodin's passage, rather than any positive statement upholding the Aristotelian doctrine, which may have acted as a deterrent to the acceptance of the new theory of comets.

12 That the comet of 1577 became fixed in the minds of seventeenth century astronomers as having furnished the data necessary for subsequent theorizing is evident. Already in 1605, an Englishman, Thomas Lydiat, in the third chapter of his text-book *Praelectio Astronomica* (Lydiat, 23-8), said that there was no essential difference between the aethereal and sublunary worlds. For, he said, it has been proved by observations by Cornelius Gemma, Tycho Brahe, and others, of the star of 1572 and the comet of 1577, which showed no parallax, that these bodies were in the region of the fixed stars. These distance measures which were accepted by Lydiat did not interfere with his belief in the Aristotelian theory of the generation of comets. In 1623 Galileo published *Il Saggiatore*, his last offering in a dispute which had started after the publication of his work on the three comets in 1618, the third of which he had found could not be sublunar (Galileo, VI, 31). Galileo harked back to Tycho's observations of the comet of 1577 (Galileo, VI, 229-233 or Galileo (1623), 21-6). The observations of that comet were also the basis of the arguments of Galileo's opponent, Sarsi or Grassi. (See Favaro, editor, VI, 143 or Galileo (1623), 122.) Galileo demonstrated how to measure the parallax of a comet from the observations made at two widely separated places, using the observations of Tycho and Hagecius. However, he did not agree with Tycho on the generation of comets but rather leaned toward the side of Claramontius, whose *Antitycho* had appeared in 1621. Johnson (1937), 276, said that John Swan's *Speculum Mundi*, which appeared in 1635, "shows that he is fully aware that the researches of astonomers had proved that the new star of 1572 and all subsequent novae and comets were far above the moon, and that the idea of solid orbs was therefore completely demolished..." An example of the partial acceptance of the change in theory is given by a work by William Gilbert, first published in 1651, which shows that at the close of the sixteenth century Gilbert still believed that some comets were sublunar (Gilbert, 242), although he was fully aware of the connotations to be derived from

Although the newly acquired information was applied to the observation of comets in the years following 1577, tracts about comets were still written which did not take this knowledge into account and which consequently slowed up its acceptance. Take as examples two tracts chosen at random, one from 1596,[13] the other from 1677.[14] These completely disregarded the astronomical data pertaining to comets. Thus, in a negative way, they detracted from the importance of the observations of the absence of parallax.

Not all astronomers immediately accepted the new findings. There were some open dissenters, but by the time another very bright comet appeared, these had diminished in number. Moreover, in spite of dissenters from or non-supporters of the new

the observations of Tycho, Maestlin and others of the nova of 1572 (Gilbert, 155, 236). See the discussion of Vögelin, chapter II, above. It becomes apparent that in the opening years of the seventeenth century, some lesser known writings tended to advance rather than hinder the acceptance of the new theories, and, furthermore, that they ascribed the changes in theory to the nova of 1572 and the comet of 1577.

13 Marlishusanus. The only information about the comet of 1596 which this tract contains is that the comet appeared the 7th, 8th, or 9th of July, with a pale, shining tail like a broom, and was very high and was visible from midnight to morning. Nothing was said of the comet's path or duration. According to the title-page, the authority for the expressed opinions about comets was Paracelsus. A comet was a prediction and its origin was not natural, for it was made of the spirits of the air. Mankind was warned to read the Scriptures.

14 Uranophilus. This is a catalogue of comets from 14 A. D. to 1677 A. D. A "famous mathematician" was mentioned as having observed the comet of 1472 and Milichius and Vögelin were said to have observed the one in 1532; Apian that of 1539; Maestlin that of 1580; Kepler that of 1582; Tycho that of 1590; and Longomontanus that of 1607. Almost all the comets were listed with the events which followed them. These included deaths of prominent persons, earthquakes, thunder-storms, pestilences, wars, religious and political changes. For some of the comets, their duration and the constellations through which they passed were briefly mentioned. The title-page has a picture of a comet, supposedly that of 1577, and that comet was said to have been observed by many important mathematicians, to have been visible from November 11th until January of the following year, moving from the middle of Capricorn through Aquarius to the middle of Pisces. But no mention was made of what the "important mathematicians" discovered from this comet.

theory, the position of the comet as a celestial phenomenon was definitely established by Tycho and a small group of his contemporaries. That ground never had to be retraced. For the most part, the authority of Tycho was taken as evidence of the truth. Although Tycho's book did not appear until 1588, and then in a small edition, the tracts by Maestlin, Roeslin, and Gemma were published shortly after the comet's appearance. The new theories became a part of the literature immediately and were not left to be rediscovered. Men were ready for the inclusion of comets in the Copernican-Keplerian system, an accomplishment which was left for Halley and others in the seventeenth century.

The outstanding dissenters were John Craig,[15] the Scot, and Scipio Claramontius,[16] the Italian, both of whom wrote against

15 See Dreyer, editor, IV, 416, 515-8, VIII, 454; Dreyer (1890), 208-9, 272, 305, 369; D.N.B., IV, 1372-3; Chalmers, XX, 243; Wood, V, part 1, 310. Craig was born in Scotland, entered the University of Wittenberg in October, 1570, taught mathematics and logic on the continent for some time and returned to England in 1584. He practised medicine in Edinburgh and was first physician to James VI of Scotland. He died in 1620. In 1588 Craig obtained a copy of Tycho's book on the comet of 1577 and wrote to Tycho, attempting to disprove the latter's conclusion concerning the supra-lunar position of the comet. Tycho prepared an *Apologia* for his book and sent it to Craig in 1589. Dreyer edited it in 1922 (Brahe, IV, 415-476, *Apologetica responsio ad Craigum Scotum de cometis*). It seems possible that this work was printed in Uraniborg in 1591, and that there is now no extant copy of that edition. Tycho had intended to add the refutation of Craig's ideas on parallax to his *Progymnasmata.* See appendix, items 15a and 15b. In 1591 Craig published a refutation of Tycho's book and violently attacked all who disagreed with Aristotle's theory of comets. This work was entitled *Capnuraniae restinctio seu cometarum in aethera sublimationis refutatio.* A fragment published by Dreyer in 1922 (Brahe IV, 477-488) shows that Tycho took notice of the 1591 attack, and he defended himself against it in January, 1595, in a letter to Rothmann. Kepler began a reply to Craig but never published it although it is included in the Frisch edition of his works. Longomontanus, likewise, intended to answer the attack against his friend and patron, but his refutation never appeared in print.

16 Claramontius (sometimes called Chiaramonti) was born in Cesena in 1565 and died there on October 3, 1652. The B.M. and B.N. catalogues have lists of Claramontius' writings. Riccardi, I, 347-350, listed Claramontius' mathematical and astronomical writings in chronological order. See also Favaro, editor, xx, 418.

the Tychonic theory of comets. Craig might have remained obscure in the history of science had Tycho ignored his first attack, but, as it was, their controversy was given considerable publicity. It sems to have hurt Tycho's pride. Fortunately, for the development of the theory of comets, however, Craig had few, if any, supporters and made little headway.

The opposition of Claramontius to Tycho and his followers must be weighed more carefully than that of Craig. Claramontius differed from Craig in that he would have held a place in the annals of science even if he had not voiced his disagreement with Tycho's theories. As professor of philosophy in Perugia and later in Pisa and as the author of innumerable scholarly tracts, both in Latin and in Italian, Claramontius commanded a wide audience. His attack on the theory of the supra-lunar position of comets did not begin until after Tycho's death, but it centered about the observations made by that great observer. The brunt of upholding the new theories fell upon Tycho's pupil and successor, Kepler.

Although Claramontius had written about the comet of 1618,[17] his direct attack upon Tycho began in 1621 with the publication of his *Antitycho,* in which he attempted to prove from Tycho's own observations that comets are below the moon.[18] This was answered by Kepler in 1625,[19] and Kepler in

17 Claramontius (1619).

18 Claramontius (1621). This is a lengthy work, setting forth the material in scholarly fashion. It contains many diagrams and mathematical calculations. It begins by giving the derivation and a definition of the word "parallax". Claramontius went into an elaborate discussion of different types of parallax, such as vertical and horizontal, and managed to confuse both himself and the reader. Besides using observations by Tycho, he used those by Maestlin, Cornelius Gemma and an observer named Dazlinus, all of whom thought the comet supra-lunar. He divided observers into two groups, those who believed comets sublunar and those who believed them supra-lunar, and mentioned a great many observers by name. See also Riccioli, II, 88, who told how Claramontius attempted to prove the sublunary nature of the comet of 1577 from the observations of Tycho, the Landgrave, Maestlin, Gemma, and Roeslin.

19 *Tychonis Brahei Dani Hyperaspistes, adversus Scipionis Claramontii ... Anti-Tychonem, in aciem productus à Ioanne Keplero ... quo libro doc-*

turn was answered by Claramontius in 1626.[20] In 1628 Claramontius published a tract about the new stars in 1572, 1600, and 1604, attempting to prove them sublunar.[21] In 1633, in keeping with his general attitude, Claramontius wrote against Galileo's great treatise on the systems of Ptolemy and Copernicus, at the same time defending his *Antitycho*.[22] The year 1636 saw the publication of Claramontius' *De sede sublunari cometarum,*[23] supplementing his *Antitycho,* and as late as 1648 he again wrote on the position of the comets.[24] Since Claramontius won few followers, he did not greatly hinder the development of the new theory, but undoubtedly its acceptance was somewhat retarded by his many writings.

In addition to their importance in the development of a theory of comets, the works on the comet of 1577 have considerable significance. Their authors took sides in the Copernican controversy. Tycho's rejection of Copernicus' theory was based on

trina praestantissima de parallaxibus deque novorum siderum in sublimi aethere discursionibus, repetitur, confirmatur, illustratur . . . , Frankfort, G. Tampachius, 1625. See Kepler, VII. The year before the appearance of Kepler's answer to Claramontius' book, the question of the distance of comets from the earth was discussed by another Italian writer, Gloriosi (see Gloriosi), who tried to set forth the pros and cons of the argument. As early as 1611, Santucci (probably the same as Santutius) had argued in favor of the more distant position of comets (see Santucci); but the lively discussion of the hypotheses nowhere else led to a controversy of the magnitude of that started by Claramontius.

20 Riccardi, I, 348.

21 *Idem.*

22 *Idem.* Dreyer (1906), 415, said that Galileo did not differ greatly from Claramontius on the nature of comets but that Galileo did believe novae to be celestial. Moreover, he found that one of the comets of 1618 could not be sublunar. See note 12, above.

23 Claramontius (1636). This work discusses the comets of 1577, 1582, 1585, 1597, 1607, and 1618. Observations by Antonius Santutius (probably the same as Antonius Santucci) of the comet of 1577 were used to add weight to Claramontius' contention that comets are sublunar. Observations of other comets by several men, including Rothmann and Kepler, and the Indian observations of the Jesuit, Jacob Rho, were similarly used.

24 Riccardi, I, 349.

a very sound inference from his failure to detect any annual motion of the fixed stars, an objection which was first overruled in the nineteenth century, although men had previously been able to meet it theoretically by assuming for those stars almost incredible distances from the earth. A survey of the literature on the comet of 1577 gives a nearly complete picture of astronomers and astronomy in the last quarter of the sixteenth century, showing an acceleration in the process of accumulating increasingly precise knowledge, such as is typical of "modern" times. It shows the interest of the observers in the improvement and use of astronomical instruments. And furthermore, it shows a growing tendency to distinguish between astronomers and non-astronomers. That is, it becomes apparent that those men who devoted themselves entirely or in greater part to the pursuit of astronomical learning were the ones who could grasp the situation and capitalize on it, bringing cool judgment to bear on the new data. Likewise, they could more easily shake themselves free from the weight of ancient authority. These men banded together, not physically, but by correspondence, and seem to have fostered a specialization which the learned societies founded in the seventeenth century carried still further, and which modern scientists consider a necessity.

When the chapters of this dissertation and the appendix are regarded as a continuous narrative, it becomes evident that in volume and accuracy of observation the year 1577 marks a tremendous leap forward. Furthermore, not only in the theory of comets, but also in the establishment of a general theory of the universe, the works on the comet of 1577 played a leading role.

APPENDIX

A BIBLIOGRAPHY OF TRACTS AND TREATISES ON THE COMET OF 1577

ALTHOUGH this bibliography has been made as complete as possible, it was confined to European observations and literature, because it was felt that Chinese and Arabic observations of comets had little effect on western thought in the sixteenth century. No search for manuscripts has been made, because, although they may throw light on sixteenth century beliefs, they could not have done much to mold the thought of that time, when most current material was circulated in printed, not in manuscript, form. A few manuscripts dealing with the comet of 1577 have been listed, but merely because they have been encountered. A few letters, while not included here, have been cited in the text. This bibliography deals only with the treatises devoted specifically to the comet of 1577. Nevertheless, it must be remembered that that comet was discussed in a great many general works appearing at the close of the sixteenth century and during the seventeenth. Only such general works as appeared before the comet of 1580 can be said to portray solely the influence of the comet of 1577.

Unless otherwise indicated, all the books in the bibliography have been examined. When it was impossible to see one of the items, the reference which bears witness to the existence of that item has been noted. Once a reference to a given book was found, no effort was made to find further references. However, they have been included when accidentally found. Thorough search for a copy of each book was made. In the case of books which have been examined, note has been made of the particular copy examined. In the instances where the copy examined does not belong to a large public or semi-public library, its distinguishing features have been described in order to insure its identification at some future time when the ownership may have changed. Once a copy of a tract was seen, no search was made for additional copies of the same book, but if more than one were seen, reference was made to all the copies examined. Books have been considered unlocated when no copy is known, or when the only known copy is in an inaccessible place, such as the Pulkova Observatory; but in such cases the existence of the so-called " inaccessible " copy was noted.

Title-pages have been copied exactly, although the original form of printing was not retained. That is, a capital letter has been used at the beginning of every word which begins with a capital letter on the original title-page, even when the whole word is capitalized or written in small capital letters on that title-page. No attempt has been made to show capitals appearing within a word, nor to indicate the ends of lines on the original pages, nor to reproduce italics from the originals. Diphthongs have been written out but abbreviations have usually been retained. If the printer's name or the date or place of publication was given in some part of the treatise in question other than on the title-page, that information has been quoted, and the source indicated. The method of citing titles also applies to quotations from other parts of the book. In all quotations from titles or text the original spelling is retained. Leaves have been referred to by number even in cases where the number was not printed on the leaf, as was usual with the fourth leaf of a quarto signature and not unusual in other instances.

Most of the bibliographical information given here was not contained in the original bibliography,[1] which was scarcely more than a list. However, in numbering the items, the numbers used in the first bibliography have been retained. This has necessitated the placing of square brackets around the numbers of items now seen not to belong in a bibliography of tracts dealing specifically with the comet of 1577. Retaining the old numbers has also made necessary the use of numbers with letters for tracts which have been added to the first bibliography, in order to retain the alphabetical arrangement, by author, which has been used.

In this bibliography, books have been described by the signatures, that is, by the way the printed sheets are folded. Thus a folio is a book in which each signature has two leaves; a quarto one where there are four leaves in a signature; an octavo one with eight, and so forth. This is the measure which can be uniformly applied to all books here described, whether they were seen in photostat or in actuality. However, because of the present tendency to describe a book by its size, and because a better picture of a book can thus be given, the height of each book has been noted

1 See Hellman.

where possible, although no attempt was made to collate these sizes with the current uses of the appellations quarto, octavo, etc. Measurement failed in the case of books which have been cut down. Those books of which only photostatic copies, without a scale, have been seen cannot be described by size. In the case of unlocated books it was necessary to use the title and description given in the source whence the title was taken. These sources are not always clear as to the meanings of terms such as quarto.

The abbreviations used for references are those which have been used throughout this dissertation and can be found in the index of references.

Andrea, Jacob

According to Janssen, VI, 441, citing Weber's *Anna von Sachsen,* (no date), the Elector August of Saxony, because of the "terrible sign of God's wrath", had Selneccer and Jacob Andrea write a church prayer and distribute it in all parishes. No record has been found of a separate prayer by Andrea although the prayer by Selneccer is listed below in this bibliography.

Anonymous

1 Astrologische Beschreibung des erschrecklichen langkschwantzigen und ungehörten Cometen im November des 77. Jares .. bei uns erschienen.

Not located. Houzeau, 5605

Description: The volume is a quarto printed in Wittenberg in 1578. Houzeau said that it had been wrongly attributed to Cardan. Possibly the work is the same as item 47, by Groplerus, for the following three reasons: 1. the part of the title of 1 which is given by Houzeau is part of the title of 47; 2. both works were printed in Wittenberg in 1578; 3. Cardan's influence is attested on the title-page of 47. See the title and description of 47.

1a Brief discours sur la signification véridique du [sic] comète apparu ... le 10e novembre 1577.

Not located. Delambre (1821), II, 537

Description: Delambre's sole comment concerning this work was "Fatras ridicule."

1b Declaration sur la comette qui a este veue en Allemagne en l'annee 1577 au moys de novembre, traduicte d'Alleman en François. A Lyon, par Benoist Rigaud, 1578.

Not located. Baudrier, III, 342

Description: This item is an octavo of eight pages and was printed in Lyons in 1578 by Benoist Rigaud. Baudrier referred to "Cat. Renard, *Lyon, Brun*, 1884. n° 879".

1c Ein Erschreckliches Wunderzeichen. Von eim grausamen Fewer, So am Himel gesehen ist worden, im Landt zu Preussen in der stat Dantzig, Vnd vmbher... (at end: Erstlich Gedruckt zu Königsberg in Preussen. 1577.)

Not consulted. Weller (1862-4), II, 436 no. 593, where mention is made of a copy in Berlin

Description: This tract is called an octavo of four leaves. It may or may not refer to the comet of 1577.

1d Ein kurtze erinnerung, von dem Cometen, so auff den 12. tag Novembris des 1577. Jars zu Augspurg erschinen, und erstmals gesehen worden. Zu Augspurg bey Bartholme Käppeler Brieffmaler, im kleinen Sachsen Gesslein.

Not located. Weller (1857-8), 324.

Description: This is a folio sheet with a colored woodcut showing the comet over a city. No date was given but Weller assigned the item to 1577.

2 Kurtze Beschreibung des grossen Kometen von 1577. Von zwei Liebhabern.

Not located. Carl, 53

Description: This item was printed in Nuremberg in 1577.

2a The Blazoning of a Comete or Blasing Star, &c.

Not located. Referred to (by Twyne ?) on the verso of A_{iii} of item 105a

Description: This pamphlet probably dealt with the comet of 1577, but the phraseology used by Twyne does not rule out the possibility that the pamphlet considered a comet in the abstract.

2b Newe Zeitung / Von dem Cometen / So jetzt im Nouember dieses 1577. Jares / vast in aller Welt erschienen / vnd sonderlich wie derselbig im Obernlande zu Aussburg vnd Nůrmburg ist gesehen worden / auch der ordts der Gelarten bedeůtung vnd beschreibung desselbigen / kurtz verfasset.

Crawford library; photostatic copy, C.U.L. B523.6 N44

Description: This pamphlet is probably a quarto of one signature. The second and third leaves are lettered A_{ii} and A_{iii} and the other two leaves are unlettered. As the available photostatic copy has no scale attached, the exact height of the book cannot be given here. However, the Crawford Library photostats usually retain the size of the original, and a page on the photostat measures 17.8 cm. The pages are unnumbered. The volume was printed in German, probably in 1577, when it would have been salable, but no date nor place of publication nor printer's name is given. A section of the book beginning on A_{ii} describes the comet seen in Nuremberg in November of the current year (1577) which would indicate that the book was printed in Nuremberg in 1577. On the title-page is a woodcut of a comet, labeled the comet of 1577, and beneath it is the legend " Et nunquam terris spectatum impunè Cometen.", which can be compared with the citations in items 12, 33, 80, and 91.

2c Neue zeyttung von dem Cometen, So jetzt im Nouember dises ain Tausent Fünff hundert Syben vnd Sybentzigk Jars erschinnen, vnd beschreibung der bedeüttung desselbigen. (last page: Gedruckt zu Landsshuet, durch Martinus Apianus.)

Not consulted. Weller (1872), 248; Schottenloher, IV, 378. There is a copy in Munich.

Description: Weller described the work as a quarto of two leaves, with a poor woodcut on the title-page and equally poor printing. He said that no date of publication is given; but both he and Schottenloher assigned the work to the year 1577.

2d Newe Zeytung von dem Cometen, So jetzt im Nouember dises 1577. Jars erschinen, vnd beschreibung der bedeütung desselbigen. (last page: Getruckt zu Augspurg, durch Valentin Schönigk, auff vnser Frawen Thor, vnd bey Hanns Schultes Brieffmaler vnd Formschneyder zufinden.)

Not consulted. Weller (1872), 248; Weller (1860), 77; Weller (1857-8), 359-360. There is a copy in Zurich.

Description: The title is taken from the first reference. Weller called this work another edition of 2c and assigned 2d to the year 1577 although he said that no date of publication was given. He described it as a folio sheet with a woodcut of a comet and under this a town.

2e O widzeniu Komety w tym ninieyszym przeszłym Roku Pańskim 1577 mieśiącá Listopadá y Grudniá, od wiela ludźi widźiánem ... Z Niemieckiego na Polskie przełożono.

Not consulted. Wierzbowski, 1534; Estreicher, XXXII, 436. There is a facsimile copy in the Imperial University library in Warsaw. The original seems to have been lost.

Description: The title is taken from Estreicher. The work is an octavo of eight leaves. Wierzbowski assigned it to the year 1577, although no date nor place of publication seems to have been given in the work itself. Estreicher said " without place supplied ... " but he gave Cracow as the place, Stan. Scharffenberg as the publisher or printer, and 1577 as the approximate date. The verso of the title-page and the last page are blank. A literal translation of the title reads: " On the appearance of a comet in this current year of our Lord, 1577, in the months of November and December, seen by many people ... Translated from German into Polish." Beneath the title and above the information about the translation is a quadrangular picture representing the comet with several stars around it; in the center, in the distance, there is a village; on the left side is an astrologer in cap and gown with a globe in his hand; on the right side two men are conversing. On the seventh leaf there is a picture which represents, on top, the sun, stars and moon; a man with a staff in his hands points to the sun; on the right side another man wrapped in a cloak and hood grasps the pointing hand with his own right hand and with his left hand holds a mask in front of him.

2f Verzaichnuss des Cometen, so im Novemb: in disem 77. jar zum ersten mal gesehen worden. Zu Nurmberg, bey Georg Macken, Illuministen beym Sonnenbad.

Not located. Weller (1857-8), 324

Description: Weller said that no date was given for this item but assigned it to 1577. He said that it was a folio sheet with a colored woodcut showing the comet with a long tail and beneath this four onlookers and the city of Nuremberg.

3 Vom Cometen / So jtzund in Latitudine Meridionali ascendente / de sůndlichen Welt zu einem zeichen Göttliches Zorns / vnd kůnfftiger Straffe sich ereugnet / kurtzer Bericht / an einen guten Freund schrifftlichen gethan / Durch einen Liebhaber der Astronomey vordeudscht / vnd in Druck vorfertiget. 1577.

R.A.S.

Description: The volume is a quarto with signatures A_1 to A_{1v}. Only a photostatic copy without a scale was available. The verso of A_1, the title-page, is blank. The pages are unnumbered. The book was printed in German, probably in Germany, in 1577. No place of publication nor printer's name is given. There is a woodcut of uncertain meaning on the title-page. The same cut appears on the title-page of item 28 and a similar one on the title-page of 79b. Across the top of the cut are the dates 1577 and 1578 and in the right hand corner is a comet. Beneath the date 1577 is a face completely filled and beneath the date 1578 is a similar one mostly in outline, filled in to the mouth. The lower part of the cut shows four figures, two fully clothed, one of which is on horseback, and two skeletons holding what might be a balance. Beneath the woodcut is the following quotation from the first chapter of Zephaniah: "Zur selbigen zeit wil ich heimsuchen die Leute / die auff jren Hefen ligen / vnd sprechen in jhren hertzen / Der Herr wird weder guts noch böses thun. Vnd sollen jre Gůter zum Raub werden / vnd jre Heuser zur Wůsten. Sie werden Heuser bawen / vnd nicht drinnen wonen. Sie werden Weinberge pflantzen / vnd keinen Wein dauon trincken. Denn des

Herrn grosser Tag ist nahe / Er ist nahe / vnd eilet sehr." On the last page, the book is dated St. Catherine's day, 1577.

3a Von dem Cometen / So jetzund in Latitudine Meridionale ascendente / der Sůndlichen Welt zu einem Zeichen Göttliches Zorns / vnd kůnfftiger Straffe sich ereugnet / kurtzer Bericht / an einen guten Freund Schrifftlichen gethan / Durch einen Liebhaber der Astronomey verdeutschet / vnd in Druck verfertiget. 1578.

Nationalbibliothek Wien, 72 t 145 (18)

Description: The volume is a quarto with signatures A_i to A_{iv}. Only a photostatic copy without a scale was available. The verso of A_i, the title-page, is blank. The pages are unnumbered. The book was printed in German, probably in Germany, in 1578. No place of publication nor printer's name is given. There is a woodcut on the title-page, which might represent the Last Judgment. It is not the same woodcut as that on the title-page of number 3. However, the same quotation from Zephaniah is given on both title-pages. The title is the same as that of number 3, except that the words " Von dem " appear in number 3a whereas the contraction " Vom " is used in item 3. The spelling and division of words is slightly different. The text of number 3a is the same as the text of item 3. Occasionally words are spelled differently in the two editions. The type used is very different. The lines in 3a are slightly longer, but there are fewer per page. Although there are the same number of pages in both, the pages start on different words. Because of the difference in type and arrangement and because of the use of a different woodcut on the title-page, it seems unlikely that the 1578 copy was printed by the printer who made the earlier one.

3b Von dem Cometen, welcher in Nouemb. 1577. erschinen observ. in Leipzig.

Not located. Weller (1857-8), 215

Description: Although no year was given, Weller or his source assigned the work to 1577, and called it a quarto with one woodcut, printed in Augsburg.

3c Warhafftige Beschreibung / Was sich zugetragen hat / vō Kriegen / Vnglůck / Wunderzeichen / zwischen der zeit / da der Comet / Anno 1577. erschienen / Vnd wie viel Cometen gestanden haben / Von der allgemeiner Sůndfluth an / biss auff die jetzt zween brennenden Cometen / Anno. M. D. LXXX ... Gedruckt Anno 1581. (at end: Getruckt zü Cöllen, Durch Nicolaum Schreiber.)

Not located. van Someren, no. 306 and plate

Description: On the title-page above the date of publication there is a woodcut, with a legend, of two comets in 1580. The pamphlet is a quarto of eight unnumbered pages printed in Gothic type. It is a moot point whether a tract with the above title should be included in a bibliography on the comet of 1577.

Archidamus

See Crespin

Arma, Giovanni Francesco (or Johannes Franciscus)

4 De significatione stellae crinitae.

Not located. Houzeau, 5606.

Description: According to Houzeau this is a quarto and was printed at Turin in 1578 in Latin and Italian.

B. I. T.

See 79b

Baldinus, Bernardinus

[5] De stellis, iisque quae in stellis et in numina conversi dicuntur homines.

Not consulted, BN. V 7744(3); Houzeau, 165; Lalande, 108-9; Scheibel, 111

Description: The title is given above as it appears in Houzeau. Schiebel wrote "qui" in place of "quae" in the above title. The volume was described by the B.N. catalogue as a quarto of fifty-two pages. Scheibel listed the work under publications for the year 1579 and gave the following information, which probably appears on the title-page: "Ad Io. Thom. Odescalcum et Galeat. Brugoram, Senatores regios. Venetiis, ex officin. Domin. et Ioh. Babt. Guerreorum." The sources seem to agree that the book was printed in Latin verse at Venice in 1579. Scheibel did not list it with comet writings and Houzeau listed it under the heading "Astrolatrie, Mythes, Images, Symboles" and not under works on comets. It probably does not deal with comets but with those persons in Greek mythology who, after death, became stars or gods. It is included in the present bibliography because it was included in the original bibliography published in *Isis* in 1934 (Hellman) and can be excluded only with this note of explanation.

Bariona, Laurentius

See [Johnson, Laurence]

Bazelius, Ant.

[6] Descriptio cometae qui die 14 nov. 1577. apparuit.

Not located. Carl, 53; Lalande, 107

Description: The book was printed in Antwerp in 1578. The title is given above as it was given by Carl. Since perusal of the usual references has yielded no information concerning Ant. Bazelius and the only clue to his identity is the title of [6] given by Carl and Lalande, who did not describe the book, it seems likely that they erroneously called Nicolaus Bazelius by the name Ant. and that number [6] is identical with number 7 and possibly number 10 of this bibliography. (See below.)

Bazelius, Nicolaus

7 Descriptionem Cometae, qui apparuit an. CIↃ. IↃ. LXXVII. XIV. Novemb. unà cum Prognosticis novis anni calamitosissimi CIↃ. IↃ. LXXVIII. Excudit Antver. Henr. Henricius, 1578.

Not located. Bib. Belg. Valerii Andreae, 678-9

Description: This work may very possibly be the same as item 10. The accusative ending of "Descriptionem" is due to the fact that the title is the object of the verb in the sentence Andreas used to tell what Bazelius wrote. The title is similar to that of item 10, except for order. This change in order might be due to Andreas' desire to put the most important part of the title first, adding the rest as an afterthought.

8 Een nieuwe Prognosticatie || vanden wonderlijcken ende || ellendighen Jare ons Heeren 1578. || Midsgaders de beschryuinghe vande Comete des voorleden Jaers. || Ghecalculeert by M. Niclaes Bazelius, Medecijn ende Chirurgijn ordinaris der stede || ende Casselrye van Sinte VVinox-Bergen in VVest-Vlaenderen. || (*Figure sur bois: la comète.*)

T'Antwerpen, || By Heyndrick Heyndricksen / op onser Vrouwen Kerchof / || inde Lelie-bloeme. || 1578. ||

Not located. Bib. Belg. Gand, series 2, v. 1, B 314; Petit, I, 33

Description: Nijhoff (Bib. Belg. Gand) described the volume and its contents as follows, comparing items 8 and 9 : " In-4°, sign. Aij-Cij [Ciiij], 12 ff. non chiffr. Car. goth.

" N'ayant pas eu sous les yeux la traduction flammande, quand nous avons décrit celles en français et en latin, nous avons supposé, à bon droit, que la traduction flamande pourrait bien différer des deux autres. Cette supposition se vérifie complétement en ce qui concerne le texte, qui est du tout au tout différent.

" Dans le prologue, l'auteur parle des éclipses solaires et lunaires des années 1556, 1558, 1560, 1565 et 1567. Celle de l'année courante (26 sept. 1577) dépassera, dit-il, dans ses conséquences terribles, toutes les autres. Il relate ensuite les conséquences qu'ont eues les comètes des années 1527, 1530, 1531, 1532, 1533, 1538, 1539, 1556, 1558, 1569 et 1572, pour s'occuper de celle de 1577 dans quatres chapitres, intitulés: 1°, *Vanden tydt, openbaringe, ghedaente, plaetse, loop ende beteeckenisse van deser Co– || mete ofte ghebaerde Sterre.* || ; 2°, *Vande cause materiale van deser Comete ende ooc van all andere* ... ; 3°, *Traiectiones & Crinitae (inquit Ptolomaeus) secundas partes in Iudicijs habent* ... ; 4°, *In wat Landen, Steden, ende subiecten, dese Comete haer hinderlijcke beteecke– || ninghen wtdeylende is.* || . Dans ce dernier chapitre, l'auteur dit notamment (f. Cij r°) : ... *ick vreese grootelijcx dat in dit alderellendichste iaer / seker onde* (sic, per *oude*) *Prophetie vol || com-*

men sal wesen / de welcke ouer vele iaeren aengaende den Staet van ... Vlaenderlandt / in een oudt tafereel || binnen de Stadt van Brugge inde Abdye vanden Eeeckhoutte in parckemente be= || schreuen, ende geschildert gheuonden gheweest is / het welcke langhe van te voren || vanden ... Heer Lubert Hau = || schilt ... ghemaect ende ghepropheteert was . Het welcke wy figuerlijck soo || wel als schriftuerlijck hebben willen by voegen ... (Voir notre article : Lubert Hautschilt, *imago Flandriae*).

"A la fin: *Typis Radaei.*, et l'approbation: *In dese Prognosticatie ... is niet begrepen dat teghen die Heylighe Catho= || lijcke ... Religie is, aengesien dat daer= || in nae de conste der Astrologie, ende loop des Hemels || neerstich gheprocedeert is, ende niet voor seker affir= || meert. ... Tot Antvverpen, den xvij. dach Februarij, int Iaer 1578, || H. Henrick Ziberts van Dunghen, || S. T. D. Lib. Cens.* ||

" Les gravures sur bois sont celles qui ont servi pour les éditions latine et française." (This implies that the Latin and French editions appeared before the Flemish.)

It was not unusual for a censor to commend a book as not offending the pious. Gemma's work (item 43) was similarly commended.

9 Prognostication || nouuelle, de cest An calamiteux 1578. || Auec description de la Comete veuë le 14. || de Nouembre en l'an passé. || Par M. Nicolas Bazel, Medecin & Chirurgien de || Bergues S. Winoch, en Flandres. || Auec vne Prophesie fort vieille, nagueres trouuée a || l'Abbaye vanden Eeeckhoutte a Bruges. (*Fig. sur bois : la comète.*)

A Anvers. || Chez Henry Heyndricx, au cemitiere nostre Dame, || à la fleur de Lis. 1578.||

Not located. Bib. Belg. Gand, series 1, v. 2, B 27; Petit, I, 33

Description: As in the case of item 8, above, the description is Nijhoff's (Bib. Belg. Gand). " In-4°, sans chiffres, sign. A2-C2 [C4], 12 ff., dont le dernier porte au v° : *Typis Radaei.* ||

"Cette plaquette très rare est illustrée de 4 planches : 1° la comète de 1577 (sur le titre); 2° les effets de la comète du 14 nov. 1577 (r° du 2e f.); 3° une planisphère céleste et 4 figg. astronomiques indiquant les mouvements de la comète de 1577 (v° du 2e et r° du 3e f.); 4° figure allégorique: la Flandre représentée par une femme nue allaitant deux loups et entourée des douze villes de Flandre figurées par autant de portes fortifiées (v° du 10e et r° du 11e f.). Cette dernière planche est probablement une réduction de *l'Imago Flandriae*, que Jean Otho, de Bruges, fit paraître vers 1575 ..."

10 Prognosticon nouum, Anni huius calamitosissimi 1578. Cvm descriptione Cometae visi 14. Nouembris anni elapsi. Autore D. Nicolao Bazelio, Bergensium D. Guinochi Medico Chirurgo. Antverpiae. Apud Henricum Henricium, ad coemiterium B. Mariae, sub Lilio. 1578.

B.N. rés. P. v. 49; B.M. T 1753 (12)

Description: The volume is a quarto with signatures A_I to C_{IV}. The B.N. copy is 196 mm. high, but the B.M. copy has been cut down. The pages are unnumbered. On the title-page is a woodcut of the comet pictured under the symbol of Saturn. The verso of A_I is blank. On the recto of A_{II} is a very elaborate symbolic picture with the caption "Vltionem capiam, & visitabo in virga, iniquitates eorum." It pictures a comet and its accompaniments, the vengeance of God, such as the death of many, probably the result of plague; the siege of a city; war and inundation. These effects were not considered necessarily due to the particular comet of 1577. However, Nijhoff (see item 9 above) seemed to think so. The verso of A_{II} and the recto of A_{III} have one large figure with four smaller ones near the corners. The large figure pictures the path of the comet on the celestial sphere from November 10, 1577 to January 6, 1578. The four smaller ones picture the comet's position when first seen on November 14, 1577, and six hours past the meridian on December 1, 1577, December 13, 1577 and January 4, 1578. On the verso of C_{II} and the recto of C_{III} is a circular figure, picturing in the center a woman nursing two wolves and about her, as a border, twelve cities of Flanders, represented by twelve fortified gates. This illustrates a prophecy found in the monastery of Eeckhoute at Bruges. Nijhoff (*Bib. Belg. Gand*, series 2, v. 1, B 32,) said that this Latin edition corresponded exactly to the French edition (item 9). He said that the figures are the same and on the same pages and that the prophecy, mentioned on the title-page of the French edition, is the same in both editions. The Latin edition is dated at the end "Bergis D. Guinochi 6. Id. Ian. 1578" and the consent at the end reads and is dated as follows: "In hoc Latino Prognostico, quod diligenter & astrologicè elaboratum est, nihil continetur, quod Sanctae Catholicae Rom. Ecclesiae Religioni contrarium est, & dignum est, & vtile, quod imprimatur & vendatur, & legatur, datum Antwerp, Die 17. Feb. Anno 1578." This, according to Nijhoff, is not the same in the French edition. The date of the consent and the name of the man who gave it are the same in the Flemish and Latin editions, although those two editions are not alike. The Latin edition, like the French, is marked at the end "Typis Radaei." Judging from the dates cited above and the state of the imprints from the plates, Nijhoff thought that the Latin edition was printed before the French one. Nève (Biographie Nationale... de Belgique, I, 742-3) said that Bazelius' tract on the comet of 1577 was written in French under the title, "Description de la Comète...", which may mean that Nève thought that the French edition was the first.

de Billy, Himbertus

11 Descrittione della Cometa, vista nel cielo alli Novembre 1577.

Not located. Carl, 54; Struve, I, 787

Description: The book was printed in Lyons in 1580. Carl and Struve gave the same title.

12 Description, Et Ample Discovrs prognostic du Comete, qui s'est monstré au ciel le douzieme iour de Nouembre, mil cinq cens septante sept, enuiron les cinq heures du soir: & est esuanouy, & disparu le trentieme iour

de Decembre: & commencera produire ses effects vers la feste des Roys, en l'an 1578. qui dureront longuement. Par M. Himbert de Billy, natif de Charlieu en Lyonnois, disciple de noble Corneille de Montfort, dict de Blockland, &c. A Lyon, Par Benoist Rigavd. 1578. Auec permission.

B.N. V 21083

Description: The volume is a quarto with signatures A_1 to D_4. It is 153 mm. high. The leaves from A_2 to D_2 are numbered 2 to 14. The versos of A_1, D_3 and D_4 and the recto of D_4 are blank. On the title-page is a woodcut of a comet with its head between the directions "Occidens" and "Septentrio" and its tail between "Oriens" and "Meridies". The same cut was used on the title-page of item 78b. On the verso of A_3 is a figure representing the sky at the hour of the comet's appearance, 5 o'clock on the evening of November 12, 1577. On D_3 there is a printer's device. The book was printed in French in 1578 by Benoist Rigaud at Lyons. It is dedicated to "Philibert de Charnoz, Seigneur de Fauerges, la Becherie, &c. Gentilhomme de la maison du Roy Catholique...", and expresses the author's desire to serve him. The dedication is dated from S. Amour, January 1, 1578 and quotes "les Poëtes" thus: Iamais impunement on n'a veu les Cometes", giving the keynote for the whole book. This quotation is from Pachymerēs, 149 (book 3, chapter 23). It was quoted also on the last page of Dasypodius' book on the comet of 1577 (item 33, below; see also chapter V, above), and on the title-page of Adelarius Praetorius' book (see item 80, below). It was quoted in Italian by Rocca (item 91, below; see also chapter VI, above).

Bosius, Jo. Andreas

13 De Significatv Cometarvm Dissertationes et Jvdicia Doctorvm Hominvm: collecta, emendata, & Cometomanticae nostri temporis opposita a Jo. Andrea Bosio. Singulorum nomina ac seriem Indiculus praesationi subjectus indicabit. Jenae, Typis & Sumtu Georgii Sengenvvaldi, An. Chr. CIↃ IↃ CLXV.

B.M. 532. e. 23

Description: The volume is a quarto with one unlettered signature followed by signatures A_1 to Z_3. At the time this description is being written no measurements of the height of the volume are available. The pages, beginning on A_2, are numbered from 1 to 180. This indicates, although the book cannot now be examined, that there are no signatures for three letters, probably J, U, and W. The book was printed in Latin in Jena in 1665 by Sengenwaldus. It is a compilation of previously printed works. Pages 1 to 21 are the same as part II (pages 51 to 68), of item 34 (see below). Pages 23 to 54 are the same as item 37(5). Pages 55 to 131 are the same as pages 27 to 97 of item 37, i. e. 37(3). The section on pages 132 to the end is entitled "De Significatv cometarum excerpta e scriptis Doctorvm Aliqvot Virorvm." It contains subheadings as follows: page 133 is headed "Jvlivs Caesar Scaliger Exercitationum Exotericarum de Subtilitate ad Hieron. Cardanum, anno CIↃ IↃ LVII. editarum, LXXIX, sectione II;"; page 134 is headed "Benedictvs Pererivs Valentinus Soc. Jesu Theologus Romanus libro II. Commentariorum & Disputationum in Genesin, sectione XCVI:". Beginning page 134 is a section entitled "E Simonis Grynaei, Medici & Mathematici Heidelbergensis, libro de natura & significationibus Cometarum. De fine & Cometarum significatione, Caput XIV.", which is the same as chapter 14 of commentary 2 of item 37(6), pages 47 to 49. Similarly on page 136 begins chapter 15 which corresponds to chapter 15 of the second commentary in item 37(6) or pages 49 to 52; on page 140 begins chapter 16 which occupies pages 52 (numbered 25) to 54 of 37(6); on page 142 begins chapter 17 corresponding to the same chapter on pages 54 to 56 of item 37(6); on page 144 begins chapter 18 corresponding to the same chapter on pages 56 to 58 of item 37(6). Chapters 19 and 20 are also given in both item 13 and item 37(6). On page 150 of item 13 begins a section entitled "Ex ejusdem Grynaei Judicio de Cometa anni CIↃ IↃ LXXVII.", which starts "Ex dictis constat,..." and is the same as the text from the middle of page 84 to the end, page 88, of item 37(6). The sections entitled as follows begin on the indicated pages: page 154, "E Philippi Mvlleri, Med. Lic. & Mathematum in Academia Lipsiensi Professoris, Commentatione Physicomathematica de Cometa anni CIϽ IϽ CXVIII. Caput XXII. De significatis Cometarum, & vi aspectuum Coelestium."; page 157, "Capvt XXIII. Concertatio cum astrologis super iudicio Cometarum."; page 159, "Thomas Fienvs, Medicinae Professor Louaniensis, in Dissertatione de Cometa anni CIϽ IϽ CXVIII. ad Libertum Fromondum, sub finem:"; page 173, "Jo. Baptista Ricciolvs, Societatis Jesv, Philosophiae, Theologiae & Astronomiae Professor Bon — ensis [poor printing makes this place name illegible. It should read "Bononiensis".], Almagesti noui, anno CIϽ CLIII. Bononiae editi, libro IIX. sectione I. cap. V. numero XI."; page 174, "Petrvs Gassendvs, Diniensis Ecclesiae Praepositus, & in Academia Parisiensi Matheseos Professor regius, Physicae sectione II. libro V. Capite III:"; page 177, "Jacobvs Primerosivs, Medicus Anglus, libro II. de Vulgi erroribus in Medicina capite XXXIV:".

Bovio, Zefiriele Thomaso

14 Dechiaratione a [sic] confutazione dell' eccel. M. A. Raimondo intorno all' apparitione della cometa apparsa alì 9 Novembre 1577.

Not located. Houzeau, 5595; Riccardi, II, 338
Description: Both Houzeau and Riccardi called this work a quarto and said that it was printed in Verona in 1578, Riccardi citing the printer, "frat. dalle Donne".

15 Trattato contra le sinistre opinione d'A. Raymondo et G. Mazaro in materia della cometa 1577.

Not located. Houzeau, 5596; Carl, 53; Lalande, 107
Description: The title is given above as it was given by Houzeau, who called the work a quarto. It was printed in Verona in 1578.

Brahe, Tycho

15a An Answer to the Letter of a certain Scotsman, concerning the Comet in the year 1577.

Not located. Watt, I, 145q

Description: This is very likely the same item as 15b, cited below with its Latin title.

15b Apologia Illustriss. Viri Domini Tychonis Brahe Ad Craigum Scotum De Cometis

Not located. Reprinted in 1922 (Brahe, IV, 415-476) with notes (Dreyer, editor, IV, 515-521)

Description: Dreyer, editor, said that the work was written in 1589 and possibly printed at Uraniborg in 1591. He knew of no copy of the printed work. There is an addition to this work which bears the title *Additamenta Ad Apologeticam Responsionem Ad Craigum Scotum De Cometis.* It was first published in 1927 by Dreyer (Brahe IX, 151-7), who took it from the manuscript Vindobonensi lat. 9737 f., and added notes (Dreyer, editor, IX, 320). For the correspondence between Craig and Tycho see Brahe, VII.

[16] Tychonis Brahe Astronomiae Instauratae Mechanica Wandesbvrgi Anno cIↃ. IↃ. IIC. Cum Caesaris & Regum quorundam Privilegiis (end of book: Impressvm Wandesbvrgi In Arce Ranzoviana Prope Hamburgum Sita, Propria Authoris Typographia Opera Philippi De Ohr Chalcographi Hamburgensis Ineunte Anno M. D. IIC.)

Not consulted. B.M. C 45 h 3; reprinted in 1923 (Brahe, V, 1-162) with notes (Dreyer, editor, V, 317-327

Description: The B.M. has the presentation copy to Thaddaeus Hagecius ab Hayck. The woodcut on the title-page is similar to that on the last page of item 20. The work does not deal with the comet of 1577 except with reference to the use of instruments. Thus, technically, it does not belong in this bibliography and is included solely because of its presence in the original bibliography (Hellman).

[17 and 18] Tychonis Brahe Astronomiae Instavratae Progymnasmata. Quorum haec Prima Pars De Restitvtione Motvvm Solis Et Lvnae Stellarvmqve Inerrantivm Tractat. Et Praetereá de admirandâ Nova Stella Anno 1572. exortâ luculenter agit. Typis Inchoata Vranibvrgi Daniae. Absolvta Pragae Bohemiae. M. DC. II.

Reprinted in 1915-6 (Brahe, II and III)

Description: The work is in three sections, the first and second appearing in volume II of the collected works, and the third in volume III. The comet of 1577 is mentioned only in passing (II, 379, III, 156, 228), and the work requires notice here only because of its presence in the original bibliography (Hellman). The *Progymnasmata* was not edited in Tycho's lifetime but was printed in Prague in 1602, some of the copies bearing the imprint 1603. Another edition appeared in 1610. The *Progymnasmata* was supposed to form volume I of Tycho's monumental trilogy, volume II of which, on the comet of 1577, was completed in Tycho's lifetime (item 20, below) and volume III of which, discussing the comet of 1580 and later comets, was never finished, although much material on the comet of 1585 was published many years later (1867). The first volume, *Astronomiae instauratae progymnasmata,* deals with the new star of 1572, but the additional researches necessary for reducing the observations of that phenomenon involved considerable labor, and volume I was not completed until after volume II.

[19] Tychonis Brahe Dani Epistolarvm Astronomicarvm Libri Quorum Primvs Hic Illvstriss: Et Lavdatiss: Principis Gvlielmi Hassiae Landtgravii ac ipsius Mathematici Literas, vnaq; Responsa ad singulas complectitur. Vranibvrgi Cvm Caesaris et Regvm Qvorvndam Privilegiis. Anno CIↃ IↃ XCVI.

Not consulted. BM. 50.c.24; reprinted in 1919 (Brahe, VI) with notes (Dreyer, editor, VI)

Description: The work was printed in Uraniborg in 1596. Although it mentions the comet of 1577 and observations thereof, it does not concern itself with them, but rather with observations of the comet of 1585 and star places, instruments and so forth. Its importance for a study of the comet of 1577 is due to the mention it gives of the transmission of observations of that comet to Tycho and statements as to who did and who did not observe it, but the book does not properly belong in a bibliography of works on the comet of 1577. Letters which bear much more closely upon observations of that comet can be found in Brahe, VII (1924).

20 Tychonis Brahe Dani De Mvndi Aetherei Recentioribvs Phaenomenis Liber Secvndvs Qvi Est De Illvstri Stella Cavdata ab elapso ferè triente Nouembris Anni 1577, vsq; in finem Ianuarij sequentis conspecta. Vranibvrgi Cvm Privilegio. (last page: Vranibvrgi In Insula Hellesponti Danici Hvenna imprimebat Authoris Typographus Christophorus Vveida. Anno Domini. M. D. LXXXVIII.)

H.C.L. Astr. 310.5*; reprinted in 1922 (Brahe, IV, 1-378) with notes (Dreyer, editor, IV, 489-511)

Description: The volume is a quarto with 8 unnumbered pages plus 465 numbered pages plus 3 unnumbered pages. The information taken from the close of the book and included above appears under a picture of a man writing on a celestial sphere held by a small boy, with the words, "Svspiciendo Despicio" spaced on either side of the picture. The first part of the work was written immediately after the disappearance of the comet and the book was completed in 1587. Several copies were printed in 1588 and widely circulated among Tycho's friends, but the book was not for sale until 1603 at which time the second edition appeared, with two additional prefaces. These have been reprinted, in Dreyer, editor, or Brahe, IV, 494-7. The work was to be part two of the trilogy. Across the top of each two pages of the *De Mvndi Aetherei . . . Phaenomenis* one can read the words "Tychonis Brahe Lib. II. De Cometa Anni 1577", bearing witness to Tycho's plan for his works. The *De Mvndi Aetherei . . . Phaenomenis* was printed in Tycho's own printing office and a title-page was finished before the book, but was not used. However, a facsimile of it is given by Dreyer, editor, IV, 491. It reads "Tychonis Dani De Novis Aetherei Mvndi Generationibvs Hoc Aevo Conspectis Liber Secvndvs. Qvi Est De Stella Cavdata ingenti, quae iuxta exactum trientem Nouembris Anni 1577. primùm apparuit, & circa finem Ianuarij anni proximè sequentis videri desijt."

The H.C.L. copy of this work, one of those published in 1588, contains a picture of Tycho at the age of 40. The picture does not seem to appear in the other copies and was no doubt added when the book was rebound. The volume was nicely printed and contains excellent illustrations of Tycho's instruments (pp. 460 and 463). These, as well as the title-page and last page have been reproduced by Dreyer. Dreyer also reproduced (chapter VIII of the *De Mvndi Aetherei . . . Phaenomenis*) the illustration of Tycho's system, clearly showing the intersecting spheres, and the illustration showing that part of the system in which the comet was found, namely the earth at the center with the orbits of the moon and of the sun around it, and about the sun the orbits of Mercury, Venus and the comet, in that order, the comet being shown in that part of its orbit which lies between the earth and the orbit of Venus. The book and its early editions were well described by Dreyer in his notes to his edition of it. He also discussed the dates at which it was written, deducing the evidence from the book itself.

20a De Cometa Anni 1577 (1578)

Brahe, IV, 379-396 with notes (Dreyer, editor, IV, 511-2)

Description: This work is in German. The title seems to have been furnished in 1922, when the work was first published, although the book was written in 1578. Dreyer used the manuscript Vindobonensis lat. 10689[33], with reference to manuscript 10689[32] from the same library (Vienna) written in Tycho's hand. The work is divided into ten sections the titles of which seem to be Tycho's own.

20b Tychonis Brahe Dani Observationes septem cometarum ex libris manuscriptis qui Havniae in magna bibliotheca regia adservantur nunc primum edidit F. R. Friis. Havniae, 1867.

Not consulted. H.C.L. Astr 3068.67; B.M. 8560.h.22; Crawford library; reprinted in 1926, (Brahe, XIII, 287-293)

Description: The comets discussed are those of 1577, 1580, 1582, 1585, 1590, 1593, and 1596.

Brunfels, Otho

20c Beschreibung des Cometen / so bey Vns in diesem D. M. LXXVII. Jar / den xj. Nouemb. Erschienen ist / des abens zwisschen Funffen vnd sechs vren / Mit anzeigung was darauff erfolgen sall. Sampt anzeigung etlicher Cometen so an vielen örtern gesehen für etlich hundert Jaren / vnd was nach einem jeglichen / für ein straff dar auff erfolget ist. Durch D. Otho Brunfels.

Karl W. Hiersemann / Leipzig, catalogue 647, number 268 (photostatic copy furnished by Hiersemann)

Description: The volume has four leaves, the verso of the title-page and the verso of the last leaf being blank. It is probably a quarto, but there are no signature marks, and since only a photostatic copy without a scale was available, no accurate size can be given. Hiersemann called the volume a quarto. On the title-page there is a woodcut of a comet and the following biblical quotation, "Lvce XXI. Es werden Zeichen geschehen / an der Sonnen vnd Mond / vnd Sternen / Vnd auff Erden wird den Leuten bange sein / rc." On the recto of the last leaf, at the close of the book are the words "Gedruckt nach dem Nörinbergischen Exemplar." They imply that there was a previous edition of the book, printed in Nuremberg. The date and place of publication of the edition under discussion are not given. Judging from the text, it is probable that it was written early in December 1577 and published soon thereafter.

20d Beschreibung des Cometen . . . etc.

Not located. A Nuremberg edition was implied at the close of 20c.

Bucci, Agostino

[20e] a manuscript letter

Preserved in the ducal archives of Anjou; Tiraboschi, VII[1], 433

Description: The letter was written in Turin on March 5, 1578 to Antonio Montecatino, who taught philosophy in Ferrara. This letter deals with the comet of 1577 and is included here because it might otherwise not be connected with the material relating to the comet.

Busch, Georgius

21 Beschreibung / von zugehörigen Eigenschafften / vnd natůrlicher Influentz / des grossen vnd erschrecklichen Cometen / welcher in diesem 1577. Jahre erschienen. Zu Ehren / vnd gnedigem Wolgefallen / Dem Wolgebornen / vnd Edlen Herrn / Herrn Wilhelmen / der vier Graffen des heiligen Römischen Reichs / Graffen zu Schwartzburg / Herrn zu Arnstadt / Sundershausen vnd Leuttenberg / Meinem Gnedigen Graffen vnd Herrn. 1577. Gestellet durch Georgium Busch / von Nůrenberg / Bůrgern in Erffurdt. Gedruckt zu Erffurdt / durch Esaiam Mechlern / zum bundten Lawen / bey S. Paul.

R.A.S.; Nationalbibliothek Wien, 72 T 145 (22)

Description: The book is a quarto with signatures A_1 to D_3, but, since only photostatic copies without scales have been available, it is impossible to give the size. On the title-page, A_1, is a figure showing the comet and its path on the celestial sphere. The verso of A_1 is blank. On the verso of B_1 is a chart of the heavens showing the comet where it was first observed. The verso of B_2 has a complicated chart of the sky portraying many of the different constellations and picturing the comet in many of its positions along its path. The book was printed in Erfurt (Doppelmayr, * * 3, says Frankfort) in German in December 1577 by Isaiah Mechler and dedicated to William, count of the Roman kingdom, with many other titles as stated on the title-page. The dedication is dated December 1, 1577. The R.A.S. copy has three pages of manuscript notes at the end.

22 Von dem Würckungen des Cometen

Not located. Carl, 53

Description: Carl gave Erfurt and 1577 as the place and date of publication.

Caesius, Georgius

23 Catalogvs, Nvnqvam Antea Visvs, Omnivm Cometarvm Secvndvm Seriem Annorvm A Dilvuio conspectorum, vsque ad hunc praesentem post Christi natiuitatem 1579 annum, cum portentis seu euentuum annotationibus, & de Cometarum in singulis Zodiaci signis, effectibus : ex quibus prudens lector posthac facilimè de quouis Cometa iudicare poterit, &c. ex multorum Historicorum, Philosophorum & Astronomorum, quorum praefatio mentionem facit, scriptis, memoriae causa, & propter alias multiplices vtilitates, plurimo labore & diligentissima inquisitione collectus, & dedicatus Amplissimo, Prvdentissimoqve Senatvi Inclytae Reipub. Noribergensis, à M. Georgio Caesio Pastore In Oppidvlo Leutershausen : Et eiusdem Iudicium de Cometa nuper in fine anni 77. elapsi viso. (last page: Noribergae Excudebat Valentinus Furmannus. Anno. M. D. LXXIX.)

B.M. 8560. aa. 32; C.D.H., incomplete copy completed by photostats from B.M.

Description: The volume is an octavo and is 157 mm. high (C.D.H. copy) or 153 mm. (B.M. copy). It has signatures A_i to K_{iiii}. There is no signature J. The pages are unnumbered. The versos of the title-page and of K_{iii} and the recto and verso of K_{iiii} are blank. K_{ii} is wrongly marked I_{ii}. The C.D.H. copy misses signature K, and, in that copy, towards the close of the book many of the key phrases and beginnings of paragraphs are underlined in ink. The Latin edition is substantially the same as the German edition discussed below. Both were printed in Nuremberg in 1579 although the preface of the Latin edition is dated almost a year earlier than the preface of the German edition.

24 Chronick / Oder ordentliche verzeichnuss vnnd beschreibung aller Cometen / von der algemeinen Sůndflut an / nach erschaffung der Welt 1656. biss auff dis gegenwertiges itztlauffends nach Christi vnsers Herrn vñ Seligmachers geburt 1579. Jar / vnd was darauff fůr zufell / straffen vnd verenderungen erfolget / von Kriegen / Theurung / Pestilentz / etc. Auch ein sonder-

liche erklerūg vnd Exempel / was der Cometstern durch alle 12. Himlische zeichen wirckung sey: Auss welchem der vernůnfftige Leser forthin von einem jeden Cometen leichtlich wird vrtheilen können / rc. Auss vilen Scribenten mit sonderm fleiss vnd bedencken / auch auff das kurtzest zusamen gezogen / Durch M. Georgium Caesium itzt zu Leutershausen. (last page: Gedrnckt zu Nůrnberg / bey Valeutin Fuhrman. Anno 1579.)

B.M. 8561. aa. 9; C.D.H.

Description: The volume is an octavo and is 151 mm. high (C.D.H. copy) or 145 mm. (B.M. copy). It has signatures A_i to T_{iiii}. There is no signature J. The verso of T_{iiii} is blank. In the C.D.H. copy the leaves have been numbered in pencil from 1 to 148 and the date 1579 written on the title-page in indelible pencil. The Crawford library catalogue, 93 and 71, listed two copies, one of which has part of the title printed in red. Although the verso of the title-page of the Latin edition is blank, the corresponding page in the German edition bears a quotation from Isaiah. The prefaces are similar in the two editions, both beginning by dedicating the work, the Latin edition to "Amplissimo, Prvdentissimoqve Senatvi Inclytae Reipub. Noribergensis" and the German edition to George Frederick, Margrave of Brandenburg, and others. The preface to the Latin edition is dated from Leutershausen, June 26, 1578, and signed " V. Ampl. & prud. Reuerenter colens M. Georgius Caesius." and the preface to the German edition from the same town, June 3, 1579 and signed " Vntertheniger / gehorsamer M. Georgius Caesius Pfarherr zu Leutershausen." In the German edition, the word " June " in the date is followed by the symbol for Mercury. The work is mainly a numbered list of comets and deals with the comet of 1577 scarcely more than with the earlier ones. However, it belongs to the literature on the comet of 1577 because it was a product of the furor created by that comet, and undoubtedly the hopes for the sale of the book were based on the interest aroused in the subject by the comet of 1577.

The first real difference between the two editions occurs after item 183 of the second section (comets after Christ), which in both editions deals with the comet of 1545. Number 184 of the German edition deals with eclipses in 1547, number 185 with eclipses in 1551, and number 186 with the comet of 1554. In the Latin edition, the information about eclipses is included at the end of section 183, and number 184 discusses the comet of 1554. Five more comets are listed in both editions, numbered consecutively with the comet of 1554. In the German edition, fiery signs, a planet conjunction, and so forth, not numbered in the Latin edition, are listed as item 192; then the comet of 1569 and the star of 1572 are numbered in both editions, and the chasms of 1575 are numbered in the German edition but unnumbered in the Latin edition. The comet of 1577 is numbered in both editions. Both editions include a short discussion of a comet in May, 1578, and cite, as authority, Helisaeus Roeslin, the title of item 93, below, being given. The differences seem to be merely in numbering, as the material appears in both editions. The last section is similar in both, discussing comets resembling the different planets and appearing in the various constellations. The Latin edition closes with two short non-scientific poems and the German edition with a quotation from Luther.

Camerarius, Ioachimus

[25] De Eorvm Qvi Cometae Appellantvr, Nominibvs, Natvra, caussis, significatione, Cvm Historiarvm Memorabilivm Illvstribus exemplis, Disputatio atque narratio Ioachimi Camerarii Papeperg. edita. 1578. Lipsiae. (last page: Lipsiae Imprimebat Iohannes Steinman. Anno M. D. LXXVIII.)

C.D.H.; B.N. V 21080

Description: The volume is an octavo and is 157 mm. high (C.D.H. copy). It has signatures A_1 to G_8 and the pages are numbered from 2 to 110 from the verso of A_2 to the verso of G_8. It was printed in Latin in Leipzig in 1578 by Johannes Steinman. It has sometimes been wrongly catalogued, as in Cat. Belg., 2584, under "Papeberg". In the B.N. copy of this work, opposite page 2 is "Figura I pag. 4" which belongs in Gemma's work, item 43, and opposite page 18 is "Figura 2. pag. 19", also from Gemma's work. These figures are not in the B.N. copy of Gemma's book although they are clearly indicated in the text. The figures do not appear in the C.D.H. copy of Camerarius' book but do appear in the H.C.L. and C.D.H. copies of Gemma's. Since the items by Camerarius and Gemma, together with other tracts, are bound in one volume in the B.N., it is reasonable to suppose that when the volume as it now stands was first made up, those two figures were wrongly placed.[2] The C.D.H. copy has no peculiarities except that the upper right hand corner of the title-page was once folded over and the lower right hand corner has a slight pencil mark.

The work does not deal with the comet of 1577 but with comets in general. It first appeared in 1558.[3] Its presence in the original bibliography necessitates its inclusion here. There was also an edition in 1559, a copy of which belongs to C.D.H., and another in 1582, copies of which can be found in the Crawford library, the B.N., and the B.M. The B.M. catalogue also listed an edition for 1561 (possibly 1661) printed in Braunschweig, and Scheibel, 34-8, listed a Strasburg edition for 1561 in a German translation. The book may have been reprinted in 1578 because of the interest in comets at that time, but nothing was done to bring it up to date or to include material on the comet of 1577, the later edition containing nothing not already in the 1559 edition. Probably all the editions are substantially the same. The book was written by Ioachimus Camerarius, the Elder, who died in

2 These statements concerning the B.N. copy were true in the summer of 1931, but the book may have been rebound since then.

3 Crawford library catalogue, 93 and 75; Scheibel, 27-9. These two sources said that the 1558 edition is an octavo of ninety-two pages. This can also be said of the 1559 edition (C.D.H. copy). Pingré, I, 209 mentioned the 1558 and 1578 editions and said that the book was written in 1558.

1574, and it is possible, although not certain, that the 1578 edition was prepared by his son of the same name.

Celichius, Andreas

26 Christliche, Notwendige, Nützliche vnd Theologische Erinnerung von dem newen Cometen.

Not located. Carl, 53; Friedrich, 33; Lalande, 107; Scheibel, 101

Description: The book was printed in Leipzig in 1578. The title is given above as it appeared in Carl. Scheibel described the book as a quarto of eight sheets and he and Lalande called it a new edition of the 1577 work (see 26a below).

26a Christliche notwendige nůtzliche und theologische Erinnerung von dem newen Cometen. Sampt einer Vorrede D. Nicol. Selnecceri Superint. zu Leipzig.

Not located. Scheibel, 96

Description: This work was printed in Leipzig in 1577. It may be the same as item 98.

27 Theological Reminder of the New Comet.

Not located. White, I, 190.

Description: The book was printed in Magdeburg in 1578. The title is given above as it was given by White, who may have translated it from the German before including it in the text of his book. This would make 27 and 28 alike.

28 Theologische erinnerung / von dem newen Cometen. Andreas Celichius Alt Merckischer Superintendens. Gedruckt zu Magdeburgk / durch Joachim Walden / Anno 1578.

Nationalbibliothek Wien 72. T. 145 (8)

Description: The volume is a quarto with signatures A_1 to E_{1v}. Its size cannot be given because only a photostatic copy without a scale was available. The verso of the title-page, A_1, is blank. The pages are unnumbered. On the title-page is a woodcut like that on the title-page of item 3, only larger, and similar to the cut on the title-page of 79b, showing the dates 1577 and 1578 and a comet, with the faces beneath the dates and the four figures, two of which are skeletons, holding what might be a balance. The dedication to " Dem Edlen / Gestrengen vnd Ehrnuesten Junckern / Dietericheu von der Schulenburck / etc." was dated " Stendal / am andern Sontag des Abuents. Anno 1577." and was signed by the author. There are references in the margins, largely to biblical citations.

Chytraeus, David

29 De Stella Invsitata Et Nova, Qvae Mense Novembri, anno 1572. conspici coepit. Et De Comato Sidere, Qvod Hoc Mense Nouembri Anno 1577. videmus. Commonefactiones in Schola propositae A Davide Chytraeo. Rostochii. Anno M. D. LXXVII.

B.M. 532. e. 55

Description: The volume is a quarto with signatures A_i to C_{iv}. It is 187 mm. high. The pages are unnumbered. The verso of the title-page (A_i) and the verso of C_{iv} are blank. The printing on the title-page is uneven. The book was printed in Rostock in Latin in 1577. It deals with the nova of 1572 and the comet of 1577. The part dealing with the comet of 1577 is dated December 2nd, 1577.

30 Vom Newen Stern / Welcher Anno M. D. LXXII. im Nouember erschienen. Vnd vom Cometen / Welchen wir im Nouember dieses lauffenden M. D. LXXVII. Jars / vnd noch jtzund sehen. Erinnerung D. Davidis Chytraei in Latein gethan. Verdeudscht durch M. Iacobvm Praetorivm Profess: zu Rostock. Cum Priuilegio. Gedruckt zu Rostock durch Jacobum Lucium. Anno 1577.

C.U.L. 523.6 Z v. 2 (43); C.D.H.

Description: The volume is a quarto with signatures A_i to D_{ii}. It is 192 mm. high (C.U.L. copy) or 181 mm. (C.D.H. copy). The pages are unnumbered The book was printed in Rostock in German by Jacob Lucius in 1577. It deals with the nova of 1572 and the comet of 1577. The preface is dated December 16, 1577. The book is a translation by Iacobus Praetorius of the book by Chytraeus (item 29) and also includes a tract by Simon Pauli on the comet of 1577. There is a woodcut of a comet and some stars on the title-page. The C.D.H. copy is a clean copy bound in boards with a back made of part of an old music manuscript. There are some handwritten notes in ink, probably in Latin, at the top of the title-page and there are some marginal notes in ink on the recto and verso of C_{iii}, the verso of C_{iv} and the verso of D_i. The author's name and the date and place of publication are written in pencil on the inside of the front cover.

Codicillus, Peter

30a Von einem Schrecklichen und Wunderbarlichen Cometen, so sich den Dienstag nach Martini dieses lauffenden M. D. Lxxvij. Jahrs, am Himmel erzeiget hat. (at end: Gedruckt in der Alten Stadt Prag durch Georgium Jacobum von Datschitz.)

Not located. Weller (1857-8), 323

Description: This work is a folio sheet with a colored wood-cut showing the comet in the night sky, with black-clothed men in the foreground guided by one man with a lantern. The sheet is signed " M. Peter Codicillus. Mit Ihrer Fürstlichen Gnaden Antonij Ertzbischoffs zu Prage, ubersehung und bewilligung." No date was given, but Weller assigned the work to 1577.

Creat, Johannes

31 Kurtze Beschreibunge des Cometen / welcher ist gesehen worden am Himmel Anno 1577. den 11. Nouembris / Auch von etlichen Wunderzeichen die vorher gegangen sein / zu trewer warnung an alle Christen geschrieben. Durch Johan Creat: T.

N.Y.P.L. OAI p.v. 105 no. 11

Description: The volume is a quarto with signatures A_i to B_{ii}. It is 191 mm. high. The pages are unnumbered. It was printed in German but no place nor date of publication nor printer's name was given. It can be surmised that the book appeared either in December 1577 or January 1578, because at a later date it would have had no value, due to the large number of similar tracts which appeared. On the title-page there is a woodcut representing the comet of 1577 and beneath the cut the following biblical quotation from Joel, "Joelis Cap. 2. Ich wil wunder geben / spricht Gott der Herr / am Himel / vnd auff Erden / nemlich Blut vnd Fewr / vnd Rauchdampff / die Sonne in Finsternis / vnd der Mond in Blut verwandelt werden / ehe denn der grosse Tag des Herren komme." The tract was listed by Weller (1857-8), 215 and by Koch (1907-1910), LXXXIII, 53, n.3, the latter citing a copy in the Zittau library.

Crespin, Antoine (called Archidamus or Nostradamus)

31a Av Roy. Epistre Et Avx Avthevrs De Dispvtation Sophistiqves De Ce Siecle sur la declaration du presage & effaictz de la Comette qui à esté commencée d'estre veuë dans l'Europe, x. de Nouembre à cinq heures du soir 1577. Assez veuë & cogneuë à tout le mõde. Per M. Crespin Archidamus Seigneur de haute ville Astrologue de Frãce docteur, & medecin Cõseiller ordinaire du Roy, & de Monsieur, son frere vnique Dediée à messieurs de la ville Cité & Vniuersité de Paris Ville cappitale de ce Royaulme. A Paris, Pour Gilles de S. Gilles, ruë S. Nicolas du Chardonneret à l'enseigne de la corne de Cerf, auec priuilege-general du Roy, suyuant la coppie de Poytiers. Auec permission de l'Auteur. 1577.

B.M. 1192.e.15; photostatic copy, C.U.L. B523.6 C86 Description: The volume has four leaves numbered to A_{iiij} and was probably printed as an octavo of which only the first eight pages were used. The pages are unnumbered and there are no blank ones. The book is 163 mm. high, but seems to have been cut. It was printed in French in Paris in 1577. There is a woodcut of a comet and two stars on the title-page. The work is dated (A_{iiij} v) from St. Denis, November 18, 1577.

31b Av Roy. Epistre Et Avx Avthevrs De Dispvtation Sophistiqve de ce siecle sur la declaration du presage & effaicts de la Comette qui a esté commencee d'estre veuë dans l'Europe 10. de Nouembre à cinq heures du soir 1577. Assez veuë & cogneuë à tout le monde. Par M. Crespin Archidamus Seigneur de haute ville, Astrologue de France, Docteur, Medecin & Conseiller ordinaire du Roy, & de Mõsieur son frere vnique. Dedié à messieurs de la ville Cité & Vniuersité de Paris, ville capitale de ce Royaume. A Lyon, Par Benoist Rigavd. 1578. Avec Permission.

B.N. Rés. p.V. 201; photostatic copy, C.U.L. B523.6 C861

Description: The volume is a quarto with signatures A_1 to B_3. Only a photostatic copy without a scale has been available in preparing this description. The pages are unnumbered. The verso of the title-page (A_1) and the verso of B_3 are blank. On the title-page there is a woodcut representing a globe, the sun, the moon and stars. On the recto and verso of A_2 is a notice to the reader. The text commences on the recto of A_3 and is addressed to the gentlemen of the city and university of Paris. On the verso of B_2 the author addressed himself to the farmers and workmen, and on the following page to the queen. The text is in prose with a bit of verse interspersed. With the exception of differences in spelling, printing and an occasional word, the text, including the title-page, corresponds to that of 31a. The spelling in 31b is more modern than that in 31a.

31c the first edition of 31a and 31b

Not located. Mentioned on the title-page of 31a

Description: It follows from the title-page of 31a that 31c was printed in Poitiers in 1577.

Crügerus, Petrus

32 Theoremata exegetica de Cometis tam in genere quam in specie de tribus aevi nostri insignioribus, anno nempe 1572, 1577 et praeterito 1604 conspectis.

Not located. Carl, 54; Struve, I, 550, 762 and 788

Description: The book was printed in Dantzic in 1605. The title is given above as it appeared in Carl.

Dasypodius, Cunradus

33 Brevis Doctrina De Cometis, & Cometarum effectibus. Per M. Cvnradvm Dasypodivm. Argentorati Excudebat N. Vvvriot M. D. LXXVIII.

H.C.L. 24281.1; N.Y.P.L. Reserve

Description: The volume is a quarto with signatures a_i to e_{iv} or 40 pages. It is 182.5 mm high (Harvard copy, the top margin of which probably has been trimmed since the tract is bound with four others) or 198 mm. high (N.Y.P.L. copy, bound in 1939). The pages are unnumbered. On the title-page there is a woodcut of an astronomical diagram for 6 P.M., November 11, 1577, showing the comet. The versos of the title-page and the last page are blank. On the verso of a_{iv} are three drawings representing three so-called varieties of comet, "Stella Comata", "Stella Barbata", and "Stella Caudata". The book deals with comets in general and with the comet of 1577 in particular. It was printed in Strasburg in Latin in 1578 by N. Vvvriot or Wyriot. The preface dedicates the book to "CL. V. Ioanni Sambvco, Caesareae Maiestatis Historico, atque Consiliario" and was dated from Strasburg, February 1, 1578.

33a Von Cometen, vnd jhrer würkung. durch M. Cunradum Dasypodium beschriben. Gedruckt zu Strassburg bey Niclauss Wyriot. 1578.

Not located. Blumhof, 23-4; Scheibel, 101; Schottenloher, IV, 378; Wolf, III, 54; Lalande, 106

Description: The title is cited from Blumhof. The volume is listed as a quarto with four and a half signatures, or two leaves less than the Latin edition. It seems to be merely a German edition of item 33. It was printed in German in Strasburg in the same year as the Latin edition, 1578, and by the same printer, N. Wyriot.

Dybvad (or Dybvadius or Dibaudius), Georg Christoph

33b En nyttige Vnderuisning / Om den Comet, som dette Aar 1577. in Nouembri, først sig haffuer ladet see. Bescreffuen ved Georgium Christophorum Dibuadium / Professor i den hellige Scrifft. Prentet i Kiøbenhaffn / aff Andrea Gutteruitz / Anno M. D. LXXVII.

Not consulted. Lund, University Library; Oslo, Deichm. Library; description and title from Nielsen, 553

Description: The volume is a quarto of twenty-two leaves, with signatures A to F_2. The printing measures 144x95 mm. The book was printed in Danish in Copenhagen in 1577 by Andreas Gutterwitz. There is a woodcut of an astronomical figure on the title-page. On the second and third leaves there is a dedication to Frederick II, dated December 21, 1577. The verso of the last leaf is blank.

33c En nyttig Vnderuissning / Om den Comet, som dette Aar 1577. in Nouembrj, først haffuer ladet sig see. Bescreffuen ved Georgivm Christophorum Dibuadium, Professor i den hellige Scrifft. Prentet i Kiøbenhaffn / aff Laurentz Benedicht. Anno M. D. LXXVIII.

Not consulted. Royal Library Copenhagen (2 copies); Skara, Stifts-o. Läroverks Library; title and description from Nielsen, 554. The work is also cited in Bruun, II, 72; Lalande, 107; Poggendorff, I, 636; Scheibel, 101-2; Stolpe, XI-XII.

Description: The volume is a quarto of sixteen leaves with signatures A_1 to D_4. The printing measures 152 or 155x97 mm. On the second and third leaves is the dedication to Frederick II, dated December 21, 1577. The text begins on the recto of A_4. The verso of the last leaf is blank. The book was printed in Danish in Copenhagen in 1578 by Laurentz Benedicht. There is a woodcut of an astronomical figure on the title-page.

33d Om Kometers Betydning Som Jaertegn I Fordums Dage Genoptryk Af Jørgen Christoffersen Dybvads Skrift Om Kometen 1577 Holstebro Trykt Hos Niels P. Thomsen MDCCCCXXII

John Crerar Library.

Description: This item was limited to 225 copies. There is a half-title-page, the title-page given above, and a title-page for the reprint. The reprint of the early edition has the title: *En nyttig Underuissning, Om den Comet, som dette Aar 1577. in Nouembrij, først haffuer ladet sig see. Bescreffuen ved Georgivm Christophorvm Dibuadivm Professor i den hellige Scrifft. Prentet i Kiøbenhaffn, aff Laurentz Benedicht. Anno M. D.*

LXXVIII. At the end is an essay of XIII pages, signed " Emil Selmar ", with the title: " Et Stykke Dansk Folkelitteratur Fra Det Sekstende Aarhundrede ". This essay contains a comment on the original work. The text appears to be a reprint but not a reproduction of the 1578 edition, item 33c. The lining off of the title-page and the format of the reprint do not agree with those described by Nielsen, 554, and Stolpe, XI-XII. It is stated on pages XII to XIII of Selmar's essay that because the format was changed, the typographical arrangement of the reprint does not agree with the original. Neither is the title-page an exact reproduction. However, the principles of sixteenth and seventeenth century typographers were followed and sixteenth century letters were used. Moreover, Benedicht's printer's mark was used on the title-page, and the first page of the dedication is framed in a border actually used at one time by Benedicht.

Dudith (Duditius), Andreas

[34] Andreae Dvditii Viri Clarissimi De Cometarvm Significatione Commentariolvs. In quo non minùs eleganter, quàm doctè & verè, Mathematicorum quorundam in ea re vanitas refutatur. Addidimus D. Thomae Erasti eadem de re sententiam. Basileae Ex Officina Petri Pernę. Anno 1579.

C.D.H.; B.M. 532. e. 12 (2)

Description: The volume is a quarto with signatures A_1 to I_2. It is 206 mm. high (C.D.H. copy). The pages from the recto of A_2 to the verso of I_2 are numbered from 3 to 68. There is a printer's device on the title-page, A_1, and the verso of A_1 is blank. The book was printed in Latin in Basle in 1579 by Peter Perna. The preface, by John Michael Brutus, occupies pages 3 to 12 and is dated " ex Arce Cracouiensi X. Kalend. Septemb. M. D. LXXIX." On pages 13 to 50 is the main body of the book, headed "Andreas Dvditivs Ioanni Cratoni S." and ending " Ex solitudine mea Pascouiana, apud Morauos, pridie Kal. Mart. M. D. LXXIIX." This section corresponds to item 37(5), that is, to pages 167 to 196 of 37, not word for word, but closely. The opening sentence of the section starts with a woodcut initial, which is a different cut although the same letter, in items [34] and 37. On pages 51 to 68 of item [34] is a section entitled "De Cometarvm Significationibvs Sententia Thomae Erasti Veris ac certis ex ipsa rei natura petitis argumentis probata." It is dated, at the close, January 6, 1578. This section corresponds to (37)1, that is, to pages 1 to 21 of item 37, but in it, too, whole phrases and sometimes the word order are different. The section in [34] is the second edition and that in 37 the third edition of that material (see Scheibel, 121 ff.).

The book, although written because of the comet of 1577, does not deal specifically with it, but is important in studying the state of cometary theory at the time of that comet. Technically it does not belong in this bibliography.

The C.D.H. copy is a clean copy and is bound in boards covered by an old Latin manuscript. On the inside of the front cover is pasted the book-plate of J. L. E. Dreyer.

[35] De Cometarum Significatione Cl. Virorum Andreae Duditii Commentariolus, & D. Thomas Erasti sententia. Elias Maior Vratislaviensis denuò edidit, & adjecit παραδοζον . . . Breslae, Typis Baumannianis. Impensis Davidis Mülleri, Bibliopolae Vratise. Anno 1619.

B.M. 8610. a. 26

Description: The volume is an octavo with signatures A_1 to K_8. It was printed in Breslau in 1619. The recto and verso of K_8 are blank. The pages are numbered from 1 to 140, from the recto of B_1 to the verso of K_8. Pages 1 to 95 correspond to item 34, thus forming the fourth edition of Erastus' writings on comets. This book, also, does not deal with the comet of 1577.

Emmen or Emmenius, Gallus

35a Beschreibung des Cometen, So im Abgelaufenen 1577. Jahr, den 12. Nov. gesehen ist worden, Beyneben seiner bedeutung vnd vermutlichen Wirkung, Aus wahrem grundt der Astronomiae genommen vnd mit fleis durch Gallum Emmen lutrebocensem Medicinae Doctorum gestellet.

Not located. Koch (1907-1910), LXXXIII, 79, n. 2
Description: The work was printed in Budissin (Bautzen) by Mich. Wolrab in 1578. It was dedicated to Christof and Hans Haugolden, brothers from Schleinitz.

Erastus, Thomas

[36] De Astrologia Divinatrice Epistolae D. Thomae Erasti, Iam Olim ab eodem ad diuersos scriptae, & in duos libros digestae, ac nunc demum in gratiam ueritatis studiosorum in lucem aeditae, opera & studio Ioannis Iacobi Grynaei. Basileae, Per Petrvm Pernam, M. D. LXXX.

C.D.H.; B.M. 718. f. 13 (2)
Description: The volume is a quarto with signatures α_1 to α_4 plus A_1 to Z_4 plus Aa_1 to Hh_4. It is 211 mm. high (C.D.H. copy) or 195 mm. (B.M. copy). The following quotation from Origen, headed "Origenes.", appears on the title-page above the place of printing, "Si quis uestrûm Mathematicorum deliramēta sectatur, in terra Chaldaeorum est. Si quis natiuitatis diem supputat, & uarijs horarum momentorumq́; rationibus credens hoc dogma suscipit, quia stellae taliter ac taliterfiguratae faciunt homines luxuriosos, adulteros, castos, aut certè quodcunq; eorum, in terra Chaldaeorum est. Iam quidam existimant ex astrorum cursibus Christianos fieri, &c. Hom. 3. in Ierem." B_2 is marked B. There are no signatures J, U or W. The "3" in Aa_3 is backwards. The verso of the title-page, α_1, and the recto and verso of Hh_4 are blank. The pages are numbered from 2 to 236 from the verso of A_1 to the verso of Gg_2. There are several errors in printing the numbers. The index occupies the pages from the recto of Gg_3 to the end. The book was printed in Basle in 1580 by Peter Perna, but the letters contained in it are all of a much earlier date. Joannes Jacobus Grynaeus was the editor. The first preface (recto of α_2 to verso of α_4) is dated 1580 but the second preface (recto of A_1 to verso of A_2) is dated 1564 and the letters

all seem to have been written before the latter date. There is the same printer's device on the title-page of [36] as on that of [34], and there are woodcut initials at the beginnings of the letters. The volume is divided into books one and two, book two starting on page 129, there being nine letters in book one and five in book two. These letters are not arranged chronologically. This work was included in the original bibliography but does not belong in this one because it does not deal with the comet of 1577. Indeed, only letter XIII, pages 190 to 201, is devoted to comets, and it was written in 1558.

The C.D.H. copy is clean and can be distinguished by " N. 9." written in black ink in the upper right hand corner of the title-page and by a blue pencil line underneath " Erasti " on the title-page.

37 De Cometis Dissertationes Novae Clariss. Virorvm Thom. Erasti, Andr. Dudithij, Marc. Squarcialupi, Symon. Grynaei. Ex Officina Leonardi Ostenij, sumptibus Petri Pernae. M. D. LXXX.

C.D.H.; B.N. V 8797 and 8798; B.M. 532. e. 12 (3)-(6)

Description: The volume is 200 mm. high (C.D.H. copy). It is a quarto with signatures $*_1$ to $*_4$ plus α_1 to α_4 plus β_1 to β_4 plus γ_1 to γ_4 plus A_1 to Y_4 plus a_1 to l_4. γ_2 is marked A_2 and γ_4 is marked A_4. There are no signatures J, U or W. Leaf a_3 is marked a_5. There is no signature j. The pages are numbered from 1 to 196 from the recto of α_2 to the verso of Y_3. There are no pages 7 and 8 although the text is continuous. Pages 28 and 29 are marked 38 and 39; 60 is marked 40; 67 is marked 77. Pages 98 to 104 are unmarked, 97 being the recto of K_1 and 105 the recto of L_1. Page 147 is marked 145. The verso and recto of S_4 are unnumbered, and 166 is the verso of S_3 and 167 is the recto of T_1. Page 190 is marked 192. The recto and verso of Y_4 are unnumbered and new numeration starts with the new signatures, 3 being on a_2. This numeration is continuous to the end, that is 88. Page 52 is marked 25. The recto and verso of $*_4$, the verso of K_1, the recto and verso of

S_4, the recto and verso of Y_4 and the verso of a_1 are blank. On the verso of the title-page are indicated the various parts of the book as follows,

"Iudicium Tho. Erasti de Cometis.
Andr. Dudithij Epistola ad Erastum de Squarcialupi sententia.
Squarcialupus de Cometis aduersus Erastum.
Erasti aduersus Squarcialupum defensio.
Dudithij de Cometis Epistola ad D. Ioan. Cratonem.
Simonis Grynaei Commentarij duo, vnus de Ignitis Meteoris: alter de Cometarum causis."

These parts have often been considered as separate books. In the B.M. copy the six parts are all bound together in the proper sequence, although the sixth part could very easily be mistaken for another and separate volume and the shelf marks imply four items. The B.N. copy is divided in two parts, bound separately, and apparently considered as two separate books. They have the catalogue numbers V8797 and V8798. The first volume contains the first four parts and ends with the leaf S_3. The letter to Crato, which should be on the leaves T_1 to Y_4, is not contained in either volume. The second volume contains the sixth part.

The book can be divided as follows:

37(1) De Cometarvm Significationibvs Ivdicivm Thomae Erasti Medicinae in Scola Heydelbergensi Professoris.

C.D.H.; B.M. 532. e. 12 (3); B.N. V 8797

Description: This occupies leaf α_1, plus pages 1 to 21. The verso of $*_3$ gives corrections. On the recto and verso of α_1 is a preface to the reader. The treatise is dated January 6, 1578. The edition 37(1) is the third and corresponds to the second part of [34].

37(2) Andreas. Dvdith. Caes. Consiliarivs, Thomae Erasto Philosopho, & Medico Clarissimo, professori in Academia Heidelbergensis. P.D.

C.D.H.; B.M. 532. e. 12 (3); B.N. V 8797

Description: This occupies pages 22 to 26 and is dated the first of February, 1579, from Pascov. ("Ex secessu meo Paschouiano ". Evidently Dudith purchased an estate, Paskov (or Pascow or Paskow), in the northeastern part of Moravia. See Rieger, II, 325; Zedler, VII, 1547; Zeiller, map opposite page 86. The town still exists.)

37(3) De Cometa In Vniversvm, Atque De Illo Qvi anno 1577, uisus est, opinio Marcelli Squarcialupi Plumbinensis: Ad Ampliss. & sapientem virum Andream Dvdithivm, Caesaris Consiliarium.

C. D.H.; B.M. 532. e. 12(3); B.N. V 8797

Description: This occupies pages 27 to 97 and is dated from Pascov in 1578. There is an index to this section on the recto of K_2 to the verso of K_3, corresponding to pages 99 to 102, but unnumbered. Lalande, 104, listed this item separately and said that it was printed in 1577. Lalande, 110, cited the whole of 37, so perhaps part three had been previously printed.

37(4) De Cometarvm Ortv, Natvra Et Cavsis Tractatvs: In Qvo Aristot. sententia explicatur, & contra D. Marcellum Squarcialupum Plumbinensem defenditur À Thoma Erasto Medicinae in Schola Heidelbergensi Professore.

C.D.H.; B.M. 532. e. 12(4); B.N. V 8797

Description: There is a bastard title-page on the recto of K_4, corresponding to page 103 but unnumbered. The verso of K_4 bears a table of contents, which is not completely followed, is repetitious, and applies partly to the other parts of 37. The page headings (two pages together) for this section read " Thomae Erasti Defensio." The section is followed by a blank leaf, S_4.

37 (5) Andreas Dvditivs Ioanni Cratoni S.

C.D.H.; B.M. 532. e. 12(5)

Description: This occupies pages 167 to 196 and is followed by a blank leaf, Y_4. It is dated " Ex solitudine mea Pascouiana, apud Morauos, pridie Kal. Mart. M. D. LXXIIX." The page headings (two pages together) read "Andr. Dvdit. De Comet. Signific." and the section is a later edition of the first part of item [34].

37(6) Commentarii Dvo, De Ignitis Meteoris Vnvs: Alter De Cometarvm Cavsis atque significationibus: Conscripti per Simonem Grynaevm Med. & Math. Accessit eiusdem Observatio Cometae, Qvi Anno superiore 77. & ab initio 78 fulsit. Et Dispvtatio De Invsitata magnitudine & figura veneris conspecta in fine anni 1578, & ad initium cIↄ Iↄ LXXIX.

C.D.H.; B.M. 532. e. 12(6); B.N. V 8798

Description: There is a bastard title-page on the recto of a_1, the verso of which is blank. The first part, containing a preface and twelve chapters, is on pages 3 (recto of a_2) to 26. The second part, containing twenty chapters, is on pages 27 to 61. It is followed, pages 62 to 70, by observations of Venus in December, 1578, by Simon Grynaeus and, pages 71 to 88, by his observations and opinion of the comet of 1577.

Not all the separate sections of the book deal with the comet of 1577, but that comet undoubtedly was the cause of the publication of the volume as a whole, and the third and last sections bear directly on the comet.

On the recto and verso of $*_2$ is a dedication to Dudith dated March, 1580 and signed by Erastus, and on the verso of $*_3$ a poem in Greek and in its Latin translation, praising Erastus.

In the C.D.H. copy, the parts are all bound together, in tooled vellum, with Archangelus Mercenarius' *De Pvtredine Dispvtatio Aduersus Thomam Erastvm . . .* (Basle, 1583) and Erastus' answer (Basle, 1583). The inside of the front cover bears the book plate of W. H. Cortield. The fly leaf has "Charles V Emperor" and "Frederick I Elector of Sax." written in pencil on it. The title-page has "Sum. Francisci Pyalthasar. á Lindern Medii Alumn. Ao. 1702." written in ink on it. There are several marginal notes in pencil in the first twenty-six pages and on the verso of $*_4$, which do not seem to bear on the text.

[38] Disputatio de auro potabile . . . Adjectum est ad calcem libri judicium ejusdem authoris de indicatione cometarum.

Not consulted. B.N. 8° Te ^{131}. 16

Description: This work was printed in Basle in 1578 by Peter Perna. According to Scheibel, 121 ff., this represents the first edition of Erastus' work on comets (that is

[34], 37 (1)). This work is probably what Lalande referred to, page 108, when he listed the book "De cometarum significationibus Judicium Thomae Erasti", an octavo printed in Basle in 1578. The work was referred to by Wolf, III, 29, note 53. The B.N. also catalogued a 1584 edition. As was stated above, under item [34], this book does not deal with the comet of 1577.

F., P. S. T. A.

38a Petit Traite De La Natvre, Cavses, Formes, Et effects des Cometes. Par P.S.T.A.F. A Paris, Pour Lucas Breyer, Marchant Libraire, tenant sa boutique au second pillier de la grand' salle du Palais. 1577.

B.M. 8561. aaa. 7; photostatic copy, C.U.L. B523.6 F1

Description: The volume is a quarto with signatures A_i to C_{iv}. It is 155 mm. high, but the upper margin seems to have been cut. The pages are unnumbered. The verso of the title-page (A_i) and the verso of C_{iv} are blank. On the title-page, under the author's initials is a decorative design. The book ends on the verso of C_{iii}, C_{iv} holding merely a decorative design. The book was printed in French in Paris in 1577, the author's initials being given in place of his name.

38b Traicte De La Natvre, Cavse, Formes, Et effects des Comettes, & de celle qui s'apparoist maintenant au Ciel. Par P.S.T.A.F. A Paris, Pour Lucas Breyer marchant Libraire, tenant sa boutique au second pillier de la grand' salle du Palais. M. D. LXXX.

B. M. 1192. e. 27; B.N. V 21101; photostatic copy, C.U.L. B 523.6 F11

Description: The volume is a quarto with signatures A_i to C_{iv}. It is 16 cm. high (B.M. copy), but the upper margin seems to have been cut. The pages are unnumbered. The verso of the title-page (A_i) and the verso of C_{iv} are blank. On the title-page, under the author's initials, is a decorative design, different from the one on the title-page of 38a. The book ends on the verso of C_{iii}, C_{iv} holding merely a decorative design, the same as in 38a. The book was printed in French in Paris in 1580, by the same printer who printed 38a.

Except for the title-page and a few minor differences in the fourth leaf of the first signature, items 38a and 38b are identical, even as far as the spacing of letters and as far as errors, such as " c " for " e " in the eighth line of B_1 v. The differences between the two books on the recto of A_{iv} are as follows:

Line	*1577 edition*	*1580 edition*
5th	comma after "vapeurs"	semi-colon after "vapeurs"
6th	" seche "	" seiche "
7th	" ramassée "	" ramassee "
8th	" aisée "	" aisee "
8th	" disposée "	" dissposee "
9th	" à "	" a "
15th	" engēdrées "	" engendrees "

The differences on the verso of A_{iv} are as follows:

Line	*1577 edition*	*1580 edition*
7th	" nō "	" non "
9th–10th	" destinées "	" destinees "
19th–20th	" vēteuse "	" venteuse "
20th	" seche "	" seiche "
23rd	" etherée "	" etheree "
24th	" suiette "	" suiecte "
28th	" suyte "	" suitte "

Except for the fact that " venteuse " on the nineteenth to the twentieth lines is divided at the point where the difference in abbreviation occurs, the lines all end on the same letter.

Fabricius, Paulus

See item 83

39 Ivdicivm De Cometa, qui anno Domini M. D. LXXVII. A 10. Die Novemb: Vsqve Ad 22. Diem Decemb: Viennae conspectus est. In quo varia de Cometarum natura & forma in genere breuiter tractantur. Ad Magnificum & Generosum Dominum, Dom: Hartmannum, Dom: à Liechtenstein etc. Autore Pavlo Fabricio Med: Doct: & Caesaris Mathematico. Cum gratia & priuilegio Sac:

Caesar: Maiest: Impressum Viennae Austriae apud Michaëlem Apffelium.

R.A.S.

Description: The volume is a quarto with signatures A_1 to B_5. That is, the first signature has four leaves. The second signature has five leaves and was probably folded as an octavo. In fact, B_4 is lettered, which is unusual in a quarto. Because the available photostat was without a scale, the height of the book cannot be given. The verso of the title-page, A_1, is blank. On the title-page there is a colored woodcut of a comet, beneath which is the following information, " Progressus est autem Cometa intrà hos 42.dies 57. grad. si ad eclipticam comparetur: In ambagibus verò sui motus 70.gradus superauit. Via eius fuit transuersè obliqua & sinuosa inter tropicos, super Zona torrida, vt pictura ostendit." The book was dedicated to Lord Hartmannus, Lord in Lichtenstein, Niclasburg etc. It was printed in Latin at Vienna by Michael Apffel probably in 1577, the date given for it by Carl, 53, Houzeau, 5590, and Lalande, 104. Weller (1857-8), 323 gave the date as 1578.

39a Cometae qui Anno 1577. à die 10. Novemb: ad 22. diem Decemb: conspectus est intra circulos, stellas & Asterismos ad singulos dies designatio, in qua & Lunae locus ad multos dies secundum longit: & latitudinem annotatus est. Autore Paulo Fabricio Med: Doct: Caesaris Mathem: Mit Röm: Kay: Mt: etc. Gnad und Privilegien. Gedruckt zu Wienn in Oesterreich, bey Michaël Apffel.

Not located. Weller (1857-8), 323

Description: No date was given but Weller assigned the work to 1578. He called it a folio with a woodcut showing the comet's path on a star map, followed by three and a half folio sheets. He said that Fabricius referred the reader who wanted further information to item 39.

Fernandez Raxo y Gómez, Francisco

See Raxus, Franciscus Fernandez

Fiornovelli, Giovanni Maria

40 Discorso sopra la cometa apparsa nell' anno presente 1577: con le osservationi degli effetti di molte altre comete apparse in diversi tempi antichi e moderni etc.

Not located. Houzeau, 5594; Riccardi, I, 462; Struve, II, 550

Description: The title is given above as it was given by Riccardi. According to both Houzeau and Riccardi the volume is a quarto and was printed in Italian in Ferrara in 1577. Riccardi gave the printer as Baldini. This was probably the Vittorio Baldini who printed the volume by Thurneysser, item 104, which contains the same figure as that on the title-page of the French edition of Fiornovelli's tract (item 41). Riccardi said that Riccioli (XXXVIII) and Ughi mentioned this work as being printed in Latin. Riccioli, I, XXXVIII said that Fiornovelli wrote an "cruditum opusculum de Cometis", which might indicate that reference was being made to item 42. Whether or not such a volume actually existed will be discussed below. Item 40 is undoubtedly the original Italian edition of item 41, the title of the latter being an exact translation of the title of the former. In 1880 there must have been a copy of item 40 in the Pulkowa observatory.

41 Discovrs Svr La Comette Aparve En L'An mille cinq cents septante sept, és terres de Ferrare, Auec l'obseruation des effets de plusieurs autres Comettes aparues en diuers temps antiques & modernes. Recueillis par M. Iean Maria Fiornouelli. A Lyon, Par Iean Patrasson. M. D. LXXVIII. Auec permission.

B.N. V 21092 bis; B.M. 531. e. 28 (1)

Description: The volume is a quarto with signatures A_1 to B_4. It is 163 mm. high (B.N. copy) or 147 mm. high (B.M. copy, which has been cut down). The pages are unnumbered. The versos of A_1 and B_4 are blank. The figure on the title-page shows a comet, stars and clouds and is the same as the one on the title-page of Marzari's tract (item 71) and in the tracts by Raymundus (item 85) and Thurneysser (item 104). The tract is in French and was printed at Lyons by Jean Patrasson in 1578. In the summer of 1931, the B.N. copy of this tract had the signature B of Marzari's tract in place of its own signature B and vice versa. Dreyer, editor, IV, 510, listed a copy of this work which was added to the Crawford Library after 1890, the date of the library's catalogue, and called it an octavo.

42 Opusculum de cometis.

Not located. Houzeau, 5598; Lalande, 107, citing Riccioli; Riccioli, I, XXXVIII.

Description: The title is given above as it was given by Houzeau. From the title one would gather that the work is in Latin. According to Houzeau it is a quarto and was printed in Ferrara in 1578. Riccioli gave neither date nor place of publication for it and merely said that on the occasion of the comet of 1577 Fiornovelli "scripsit eruditum opusculum de Cometis"; and "Opusculum de Cometis" is not necessarily the title of the work. Riccioli may have been merely describing the tract. Scheibel, 102, listed it among the books of the year 1578, but he got his information from Weidler's *Bibligraphia Astronomica* which in turn took it from Riccioli. The way Riccardi, listing only one title for Fiornovelli, made the statement that Riccioli and Ughi listed the *Discorso* . . . as printed in Latin, would indicate that Riccardi did not know of nor believe in the existence of a Latin edition. Riccioli's sentence can be interpreted as saying nothing about the title or language of the tract. Perhaps there never was a Latin edition. The determining argument, although not conclusive, is that Houzeau listed the Latin work and gave the date 1578, a year later than the Italian edition which he also listed.

Gemma, Cornelius

43 De Prodigiosa Specie, Natvraq. Cometae, Qvi Nobis Effvlsit Altior Lvnae sedibus, insolita prorsus figura, ac magnitudine, anno 1577. plus septimanis 10. Apodeixis tum Physica tum Mathematica. Adivncta His Explicatio Dvorum Chasmaton anni 1575. nec non ex Cometarum

plurium Phaenomenis epilogistica quaedam assertio de communi illorum natura, generationum causis atque decretis supra quàm hactenus à Peripateticis annotatum est. Per D. Cornelivm Gemmam, Louaniensem, Ordin. ac Regium Medicinae professorem. Antverpiae, Ex officina Christophori Plantini, Architypographi Regij. M. D. LXXVIII.

C.D.H.; H.C.L. 24281.12; B.N. V 21081; C.U.L. 523.6 G 28

Description: The volume is an octavo with signatures A_1 to E_8. It is 158 mm. high (C.D.H. copy, in which the upper margin seems to have been cut) or 148 mm. high (C.U.L. copy, in which both the upper and lower margins seem to have been cut). The verso of the title-page, A_1, the recto of E_2, the verso of E_7, and the recto and verso of E_8 are blank. The pages from the recto of A_2 to the verso of E_1 are numbered from 3 to 66 leaving fourteen unnumbered pages, the last three of which as well as the first being entirely blank. The book is in Latin and was printed in 1578 in Antwerp at the famous Plantin press. On the title-page is a woodcut design. Van Ortroy (1920), 392, called this design Plantin's typographical mark, number 14. Opposite page 4 there is an extra leaf, " Figura I.", on which is pictured the path of the comet on the celestial sphere. Opposite page 19 is figure 2 which gives a detailed picture of the comet's path through the constellations. On page 26 is a small illustration of the comet as it appeared at the end of November. On the verso of E_2, unnumbered, is a picture of " Belgica " weeping while the city behind her burns and Fate points to the comet in the sky. This is described by Dreyer (1890), 68. On the recto of E_7 appear the words, " Descriptio haec [sic] Cometae nihil habet quod pium Lectorem offendere possit. *Walterus vander Steeghen, S.T. Licentiatus, Canonicus Antuerpiensis* ", showing a mark of censorship of the press which in the sixteenth century was political as well as ecclesiastical.

The figures 1 and 2 do not appear in the B.N. copy of this work but in the work by Camerarius, item 25, as is described above. The C.D.H. copy is a clean copy, bound in boards, with no distinguishing features other than the cut upper margin and the placing of figure 1 opposite page 3.

Giuntini, Francesco

See Iunctinus, Franciscus

Glisenti, Antonio

43a Dialogo Del Gobbo da Rialto et Marocco dalle pipone dalle colonne di S. Marco, sopra la Cometa alli giorni passati apparsa su nel cielo. Di M. Antonio Glisente Bresciano.

Not located. Cicogna, 758; Riccardi, I, 612, and part II, 198

Description: The title is given above as it appears in Cicogna, where it is stated that the work has six pages and that no date nor place of publication nor printer's name is given but that the work deals with the comet seen in November 1577.

Graminaeus, Theodorus (or Dietrich)

44 Cometae Anni Domini 1580. Physica Explicatio, Et Eivsdem Cvm Eo, Qvi Anno &c. 77. apparuit, Analogica collatio, in gratiam & obsequium Reuerendiss. ac Illustriss. Principis ac Domini, Domini Ioannis Gvilielmi Dvcis Ivliacensis, Clivensis, Ac Montensis, &c. Postulati ac Administratoris Episcopatus Monasteriensis, Domini sui clementissimi, descripta, & eius Celsitud. ad initium anni Domini 1581. per Theodorum Graminaeum Iuris Licentiatum, ac Philosophiae Doctorem, &c. exhibita. Coloniae Agrippinae, Anno M. D. LXXXI.

Bibliothèque Nationale et de Université, Prague 14 B 22

Description: The volume has twelve unnumbered pages. The available photostatic copy has no scale on it for determining the size of the book. Cat. Belg. called it a folio, which probably referred to the size. There are forty-one lines of printing on one of the pages. The second and third leaves have the signature marks a_2 and a_3 respectively. The title-page has a woodcut framing the title. Above the date and place of publication the following quotation is given, " Psal. 18. Coeli enarrant gloriam Dei, & opera manuum eius annunciat firmamentum." The verso of the title-page is completely covered by a woodcut as a decoration and symbolizing the weighing of truth. The next page has a short section entitled " Qvid Astrologico Ivdicio Sit Tribvendvm." The text starts on the verso of this page. The first letter is framed in a woodcut. There is also a woodcut of Saturn in a chariot, drawn by a monster, and bearing a scythe. The verso of the third leaf has three woodcuts, one symbolizing Pegasus, the same cut as on page 75 of item 45, which was printed before 44, one symbolizing an arrow with stars on it, probably that of Jove, and one seeming to symbolize a fish. The next page has the same woodcut of Ophiuchus, the snake holder, as is found on page 73 of item 45. The eighth page has a woodcut which possibly portrays Pisces. On the ninth page is a woodcut representing Aquarius and on the tenth page two woodcuts, one representing Capricornus, the same as on page 64 of item 45, and the other representing Sagittarius, like the cut on page 74 of item 45. The volume was printed in Cologne in Latin in 1581. The printer's name is not given.

45 Weltspiegel odor / Algemeiner widerwertigkeit / dess fůnfften Kirchen Alters / kůrtze verzeignuss. Darinnen Dess Cometen / oder aussgereckter Růthen / so im Jar Christi 1577. den 11. Nouembris, am hohen Himmel vernomen / stand / lauff / vnd bedrewung zuersehen / so Physicè, Astrologicè, Metaphysicè, oder aber Formaliter erklert vnd aussgelagt wirt. Durch Theodorum Graminaeum, LL. L. Amplissimi Senatus Colonienis, Mathematicarum Ordinarium. Gedruckt zu Cölln / Durch Ludouicum Alectorium, Vnd die Erben Jacob Soters. Im Jar M. D. LXXVIII.

B.M. 8610. bbb. 14; photostatic copy, C.U.L. B523.6 G76

Description: The volume is a quarto with signatures A to M_{iv} and one signature before A marked by arabic numerals. The book is 198 mm. high. Leaf I is marked K. I_{ii} and I_{iii} are marked J_{ii} and J_{iii} and the fourth leaf in that signature as in the other signatures is unmarked. The following signature is marked with K, K_{ii}, and K_{iii}. Thus there is but one signature labeled I or J, which is not uncommon, and the first leaf thereof is wrongly marked K. The pages, beginning on A, are numbered from 1 to 96. Numbers 32 and 33 are repeated but there are no numbers 34 and 35. The text is continuous over those numbers. Page 44 is numbered 34. 70 is numbered 60, 82 is numbered 78, and 96 is numbered 84. The book was printed in German in Cologne in 1578. On the title-page there is a woodcut of a celestial sphere, showing the sun, moon, and a comet, enclosed in a circle, around which is the legend "Der Warer Spigel Diser Welt Dar In Ein Ieder Sich Gefelt". This circle is in turn enclosed in a square, the corners of which are ornamented. There is a table of contents on the verso of the title-page. The preface occupies the next five pages. It was addressed to the Burgomaster of Cologne, dated from that city on March 6, 1578 and signed by Graminaeus. On the verso of the last leaf of the first signature is a woodcut said to be a picture of the comet which appeared November 11, 1577. Below the picture is a quotation from the seventh psalm. The picture shows the comet and some stars and four wise observers. Below these are references to parts of the Bible.

The first chapter starts on the recto of A, page 1. On page 5 is a diagram of the world according to John Stadius. On page 7 is a celestial sphere showing the path of the comet from November 11th to December 27th, from Capricorn to Pisces. On page 13 there is an astronomical diagram of the world creation, on page 35, numbered 33, another showing the position of the planets on September 26, 1563, on page 53 a third such diagram representing the planets on November 11, 1577, and, on page 59, a fourth showing the supposed position of the comet on November 9th in relation to the planets. On page 64 is a picture of a goat (Capricornus) with the stars of that constellation and on page 67 a similar one of Aquarius, on page 73 one of Ophiuchus, the snake holder, on page 74 one of Sagittarius, page 75 of Pegasus and on page 76 of Cassiopeia.

46 Weltspiegel; speculum mundi de minitante cometa anni 1577.

Not located. Houzeau, 5608.

Description: The volume is a quarto printed in Cologne in 1578.

Groplerus, Ioachimus

47 Astrologische Beschreibung Des erschrecklichen / langkschwentzigen vnd vngehewren Cometen / so im Nouembre des 77. Jares / in dodecatimorio Capricorni, mit dem Newen Liecht des Christmonats / bey vns erstlich erschienen / vnd fast drey gantzer Monschein gewehret / vnd auch das dritte dodecatimorion Zodiaci erreicht hat. Nach Anweisung vnd Lehr des hochgelarten vnd weitberůmbten Medici vnd Astrologi D. Hieronymi Cardani M. Gestellet vnd Verfasset / durch M. Ioachimvm Groplervm Brandenburgensem, der Mathematum vnd Physicae Studiosum. Gedruckt zu· Wittenberg durch Clement Schleich vnd Antonium Schön / Im Jar M. D. LXXVIII.

C.D.H.

Description: The volume is a quarto with signatures A_1 to F_4. It is 189 mm. high. The pages are unnumbered. It was printed in German in Wittenberg in 1578 by Clement Schleich and Antonius Schön. The verso of F_4 is blank. On the title-page there is an astronomical diagram showing the comet and constellations. On the verso of the title-page there is a verse entitled "Rosa Inter Spinas Veprarvm." The book is divided into ten numbered sections, preceded by an unnumbered section on the use of the book. The C.D.H. copy is clean and has no distinguishing features although the upper margin may have been cut down (it is not straight and at certain points measures only 6 mm.).

Grynaeus, Ioannes Iacobus

See item 36

Grynaeus, Simon

See item 37

Guilielmus IV

See William IV

Hagecius ab Hayck, Thaddaeus

48 Descriptio Cometae, qui apparuit Anno Domini M. D. LXXVII. à IX. die Nouembris usque ad XIII. diem Ianuarij, Anni &c. LXXVIII. Adiecta est Spongia contra rimosas & fatuas Cucurbitulas Hannibalis Raymundi, Veronae sub monte Baldo nati, in larua Zanini Petoloti à monte Tonali. Avtore Thaddaeo Hagecio ab Hayck. Pragae Excusum typis Georgii Melantrichi ab Auentino, Anno à Christo nato 1578.

B.N. V 7793

Description: The volume is a quarto, 185 mm. high, with signatures A_1 to E_4. The pages, beginning on the recto of A_3, are numbered from 1 to 34, the recto and verso of C_2 being numbered 15 and 16 and the recto and verso of C_3 being likewise numbered 15 and 16. On the verso of the title-page is a Latin poem to the pious Christian reader, signed by Procopius Lupacius. On the recto and verso of A_2 is the dedication to Augustus, Elector, Duke of Saxony, Landgrave of Thuringia, and so forth, dated from Prague, February 24, 1578, and signed by Hagecius. The text begins on the recto of A_3. The book ends with a prayer in German. On the recto of B_4 is a map of the path of the comet through the constellations. The book was printed in Latin in Prague in 1578 by the press of George Melantrichus. The B.N. copy does not seem to have the appendix against Raymundus. However, the B.N. catalogue listed a " Spongia " with the title given on the title-page of the *Descriptio Cometae* . . . , but with no date nor place of publication given. The B.M. catalogue mentioned the appendix in listing the *Descriptio Cometae* . . .

49 Thaddaei Hagecii Ab Hayck Epistola Ad Martinum Mylium. In qua examinatur sententia Michaelis Moestlini et Helisaei Roeslin de Cometa Anni 1577. Ac simul etiam piè asseritur contra profanas et Epicureas quorundam opiniones, qui Cometas nihil significare contendunt. Gorlicii Ambrosius Fritsch excudebat. Anno M. D. Lxxx.

Bibliothèque Nationale et de Université, Prague 14 F 289

Description: The volume is a quarto with signatures A_1 to G_2. Only a photostatic copy, without a scale, was available. There are no signatures E or F. The pages are unnumbered. On the title-page is the following poem:

" Quem tam multa virûm cecinerunt scripta Cometen,
Quaeris, cur redeat rursus in ora virúm:
Ille quidem coelo deflagrans, desijt esse:
Non quam portendit desijt ira Dei.
Nam caue nil rerum credas hoc crine notari:
Hoc ipsum, esse hominum corda profana, notat."

The verso of the title-page, A_1, is blank. The first section ends on the recto of D_4, and on the verso of that page begins a letter written in 1578 by Mylius. The verso of G_2 is blank. The book was printed in Latin at Görlitz in 1580.

[50] Litterae Hagecii ad Mylium 22. Sept. 1578 Pragis datae.

Not consulted. Struve, I, 787, a manuscript

Since item [50] is a manuscript, it does not belong in this bibliography.

Hamel, L. (probably Lud. du. See Zedler, XII, 365)

50a Theologischer Bericht von dem erschrecklichen Cometen und seinen Effecten, 1577. In Deutsche Reym verfasset. Frankfurt a. M. 1578.

Not located. Janssen, VI, 441; Weller (1862-4), I, 246; Weller (1857-8), 215

Description: The title is from Weller (1857-8). Weller called the book a quarto.

Heerbrand, Jacob

[51] Ein Erndt vnd Herbst Predig. Auss dem 26. Capittel / des fůnfften Buch Mosis. Zur Dancksagūg fůr die reiche Ernd vñ Herbst / dises gegenwertigen Jars. Geschehen den 9. Wintermonats zů Tübingen. Durch Jacob Heerband / Doctor vnd Professorn der H. Schrifft daselbst. Getruckt zů Tübingen / bey Alexander Hock / an der Burgsteyg / Anno 1578

B.M. C 68. h. 8 (12)

Description: The volume is a quarto, 200 mm. high, with signatures A_1 to D_4. The verso of the title page, A_1, is blank. The leaves are numbered from 1 to 13 from A_3 to D_3 inclusive. D_4 is unnumbered and its verso is blank. The recto and verso of A_2 contain a section headed " Das Sechs vnd zwensigst Capittel des fůnfften Bůchs / des heyligen Propheten Mosis." The book was printed in German in Tübingen by Alexander Hock in 1578. Although the book mentions the comet of 1577 once (verso of A_3) and bears witness to the tremendous interest of the people in the comet and to the awe in which it was held, it is not a book about the comet and is included in this bibliography only because of its presence in the original one.

52 Ein Predig von dem erschrockenlichen Wunderzeichen.

Not located. Carl, 53; Struve, I, 787

Description: The title is given above as it was given by Carl. The book was printed in Tübingen in 1578.

53 Ein trewe Warnnung vnd gutthertzige Vermanung zur Bůss / an die allgemeine Christenheit / vnd sonderlich Noch Teutschland / vber das schröckliche Wunderzeichen / den Cometen / oder Pfawenschwantz / der jetzt eine gutte zeitlang am Himmel ist gesehen worden. Auss der Christlichen vnd Eyfferigen Predig des Ehrwůrdigen vnd Hochgelehrten Herrn D. Jacob Heerbrands / gehalten zu Tůbingen / den 17. Nouemb. Anno 1577. Gestelt durch Vitalem Kreidweiss / diser zeit Schulmeister zu Leonberg. Getruckt zu Tůbingen / durch Alexander Hock / an der Burcksteig. 1578.

B.M. 11521. ee. 30 (4)

Description: The volume is a quarto, 190 mm. high, with signatures A_1 to C_3. The pages are unnumbered. The verso of the title-page, A_1, is blank. There is a leaf between C_1 and C_2 and there is no C_4. It is evident that the leaves were wrongly numbered. The alternative, that the leaves were bound out of order, does not apply, because the poem ends on the recto of C_3. The verso of C_3 is blank. B_3 is wrongly marked "A_3". The book is a poem by Vitalis Kreidweiss taken from a sermon by Heerbrand. Janssen, VI, 440 ff., said that it was taken from item 54, and the similarity of the two is so great that this seems likely. The book is listed here under Heerbrand's name because the ideas were his. Carl, 53, and Bassaeus, II, 171, listed it under Kreidweiss' name. Scheibel, 104, listed the tract under both names. He did not have the full title of the tract and was of the opinion that it was wrongly attributed to Heerbrand. The B.M catalogue listed the book under both names, but (in the supplement) referred the reader from Heerbrand to Kreidweiss and not vice versa and gave the title a bit more fully under the latter. However, that might be due to the relative positions of the names in the alphabet. Janssen, VI, 440 ff., cited the book under Kreidweiss' name. The book was printed in German in Tübingen by Alexander Hock in 1578.

54 Ein Predig / Von dem erschrockenlichen Wunderzeichen am Himmel / dem newen Cometen / oder Pfawenschwantz / Gehalten zu Tübingen den 24. Sontag nach Trinitatis / wölcher ist der 17. Wintermonats / Durch Jacob Heerbrand / der heiligen Schrifft Doctorn vnd Professorn daselbsten. Getruckt zu Tübingen / durch Georg Gruppenbach / 1577.

R.A.S.

Description: The volume is a quarto with signatures A_1 to C_2. The available photostatic copy has no scale to provide a measure of the height of the book. The versos of A_1 and C_2 are blank. Beginning with A_2 the pages are numbered from 1 to 17. The book was printed in German in Tübingen in 1577 by George Gruppenbach.

Weller (1857-8), 215, listed a book with a similar title but printed in Heidelberg in 1577. Weller or his source of information, F. Heerdegen's catalogue 228, may have erred.

Hellbach, Wendelin (der Pfarrer zu Eckartshausen)

54a Eigentliche und warhafftige beschreibung, der dreyen erschrecklichen Commeten, welche zu Cascha in Ungerland, auch viel andern orten mehr gesehen worden, dero deutungen etc. In Reimenweiss fleissig verfasst, und aussgelegt, etc. Gedruckt bey Anthony Corthois zu Franckfurt am Mayn, im Jar 1580.

Not located. Weller (1857-8), 360

Description: Weller called this a folio sheet with a woodcut and 278 lines of verse beginning " Christus der Herr ins him̃els thron etc." The comet of 1577 is probably one of the three comets mentioned in the title.

Henischius, Georgius

55 Ivdicivm De Pogonia Ad Finem Anni M. D. LXXVII. Conspecto Georgii Henischii, Medici & Mathematici Augustani. Avgvstae Excudebat Valentinus Schönigk Anno 1578.

C.D.H.; Biblithèque Nationale et d'Université, Prague 14 F 287

Description: The volume is a quarto with signatures A_1 to C_4. It is 190 mm. high (C.D.H. copy, where the lower margin seems to have been cut). The verso of the title-page, A_1, and the recto and verso of C_4 are blank. On the title-page there is a woodcut of an astronomical diagram, showing the constellations and the comet on the 9th of November. The book was dedicated to Jacobus Villingerus, Baron in Schönenberg, and so forth. It was printed in Latin in Augsburg in 1578 by Valentin Schönig. The C.D.H. copy is a clean copy with no distinguishing features except that at the foot of the title-page something written in ink was mostly cut off when the book was cut down.

Hooker, John (alias Vowell)

56 The Events of Comets or blazing Stars, made upon the

sight of the Comet Pagonia, which appeared in the month of Nov. and Dec. 1577.

Not located. Wood, I, 713; Watt, I, 511u; Houzeau, 5601

Description: The title is taken from Wood. All three sources agree that the book was printed in London. Wood gave no printer's date; Houzeau and Watt said 1577. Houzeau called it a quarto and Wood and Watt called it an octavo.

Huernius (or Heurne or Heurnius), Ioannes

57 De Historie / Natuere / ende Beduidenisse der erschrickelicke Comeet / die geopenbaert is int Jaer ons Heren 1577. Dorch D. Ioannem Hvernivm Vltraiectinum Medicum & Astronomum. Des Heeren Dach compt gruwelick / toornich / ende grimmende / om dat Landt te verdestruĕren / ende de Hundaers daer wt te roeyen. Want de Sternen aenden Hemel / ende synen Orion en schynt niet claer. Esa. 13. Gedrůckt zů Cölln / durch Nicläs Bohmbargen. Men verköstse by Willem Iansen Briefmaler, by der Godte Genade.

B.M. T 1753 (11)

Description: The volume is a quarto with signatures A_1 to B_4. It is 167 mm. high. The pages are unnumbered and there are no blank ones. This item is the original edition of which item 59 is the exact translation. Even the quotation from Isaiah, the woodcut and the printer's device on the title-page, and the dedication to the burghers of Utrecht are the same. The B.M. catalogue gave 1577 as the probable date for item 57, which was printed in Cologne.

58 De natura et praesagio horrendi cometae qui anno MDLXXVII orbem terrarum terruit.

Not located. Houzeau, 5592; Freher, 1307-8

Description: The title is given above as it was given by Houzeau, who said that the book is a quarto printed in Cologne in 1577. Freher also gave the date 1577, but, as Freher's reference appears in the text, the Latin title, almost identical with Houzeau's, might be merely Freher's reference to item 57.

59 Die Histori / Natur / vnnd Bedeutnuss des erschröcklichen Cometen / welcher gesehen ist im Jar vnsers Herrn 1577. Durch D. Ioannem Hvernivm der Statt Vtrecht Medicum vnd Astronomum. Des Herrē Tag kompt greuwlich / zornig / vñ grimmend / vmb das Landt zů verderben / vnd die Sůnder darauss zu reutten / dann die Sternen am Himmel / vnd sein Orion geben ihren schein nit klar. Esa. 13. Auss den Niderteutschen transferiert. Zů · Cölln / durch Nicolas Bohmbargen. Anno. 1578.

Nationalbibliothek Wien 72. 7. 145 (11); photostatic copy, C.U.L. B523.6 H87

Description: The volume is a quarto, with signatures A and B. Only a photostatic copy, without a scale, has been available in preparing this description. The pages are unnumbered. On the title-page there is a woodcut representing a comet. Also on the title-page is the printer's device with the legend "Perficit Qui Perseverat." Carl, 53, listed this work with a slightly different title and gave 1577 as its date of publication. Lalande, 104, also gave the date 1577. Item 59 is the German translation of 57. The same printer printed both.

Ioachimicus, Iohannes Praetorius

See Praetorius, Johannes

[Johnson, Laurence] (using the pseudonym of Laurentius Bariona)

59a Cometographia quaedam Lampadis aeriae quę 10. die Nouemb. apparuit, Anno a Virgineo partu. 1577. Iris vt est sigunm [sic], terras perijsse sub vndis: Ignibus est ignis cuncta casura suis. Londini excudebat Robertus Walley. Anno Domini. 1578.

H.C.L. 24281.11*; microfilm in C.U.L.

Description: The volume is a folio with a title-page and its blank verso plus signatures *ij plus Bj to Lij. It is 17.75 cm. high. There is no signature J, and Lij has no signature mark. The pages are unnumbered. There is a woodcut of a comet on the title-page below the date 1577. The dedication to the "Reverendissimo Patri, ac viro ornatissimo Edmundo Episcopo Peterburgensi." starts on the recto of *ij. The dedication is signed, on the verso of *ij, " Vale Ketteringa Ianuarij. 20. 1578. Tuae amplitudinis studiosiss. Laur. Bariona." The text starts, with a woodcut initial, on the recto of Bj and continues to the end of the book. There is a decorative design on the verso of *ij and on the verso of Lij. The volume was printed in London in Latin in 1578 by Robert Walley.

Jones, Richard (printer)

59b A christian coniecture of the newe blasinge starre
Not located. Arber, II, 145 (licensed January 16, 1578)
Description: A ballad

59c Certen notable effectes of the comet.
Not located. Hazlitt (1876-1903), 2nd. ser., 136
Description: Hazlitt said that this work was licensed to R. Jones, 24 March, 1578-9.

Ireneus, Christophor

60 and 61 Prognosticon Aus Gottes Wort nötige Erinnerung / Vnd Christliche Busspredigt zu dieser letzten bösen Zeit An hohe vnd nider Standes Deutsches Landes: Auff den Cometen / so von Martini des 1577. Jars / biss zum Eyngang des 1578. Jars gesehen. Sampt Erzehlung vieler Cometen vnd anderer schrecklicher Zeichen / vnd was allwegen darauff erfolget. M. Christoph. Ireneus. Anno M. D. LXXVIII.
C.D.H.
Description: The volume is a quarto with signatures A_i to Dd_{iv}. There is no I or U or W. It is 189 mm. high. The pages are unnumbered. The verso of the title-page, A_1, and the recto and verso of Dd_4 are blank. The C.D.H. copy is a clean copy except that the edges are slightly water marked. Its only distinguishing features are the number 308293 written in pencil in the upper right hand corner of the title-page, probably an old catalogue number, and a paragraph crossed out in ink on the recto of K_{ii}, the ink having blotted onto the verso of K_i. The upper margin may have been cut as it measures only 8 mm. The book was printed in 1578 in German, possibly at Alford, although no place of publication is given. The

work has been listed by Bassaeus, II, 282, Carl, 53, Houzeau, 5610, Janssen, VI, 440-1, Lalande, 108, and Scheibel, 104, with varying amounts of information. Bassaeus, Houzeau and Scheibel called the book a quarto, the latter describing it as "4. 1 Alph. 4 Bog." (probably meaning a quarto of one alphabet plus four signatures). Lalande wrote, *"Alphor. in-4⁰"*, indicating a place name. Carl gave the place of publication as "Alphor.", and Houzeau said that the work was printed in Alphordiae (Alford), although Scheibel and Janssen said that no place of publication is given. Bassaeus, Houzeau, Janssen and Scheibel said that it was printed in 1578 and Carl that it was printed in 1588 (probably a misprint). Probably there is but one treatise, which has been variously catalogued, and not two, as was presumed in the original bibliography in *Isis* (Hellman). No record has been found of a copy bearing the imprint "Alphordiae". Houzeau may have taken his entry from a catalogue and may have interpreted the notice "Alph.", referring to the collation, as the notice of a place name "Alph.", referring to "Alphordiae". Houzeau knew of the existence of Scheibel's work (*cf.* Houzeau, II, col. 308).

Iunctinus (or Junctinus or Giuntini), Franciscus

62 Discovrs Svr Ce Qve Menace Devoir Advenir La Comete, apparüe à Lyon le 12. de ce mois de Nouembre 1577. laquelle se voit encores à present. Par M. François Iunctini grand Astrologue & Mathematicien. A Paris, Chez Geruais Mallot, rue S. Iaques à l'enseigne de l'Aigle d'Or. 1577. Iovxte La Copie De Lyon.

B.N. V 21093; C.D.H.

Description: The volume is a quarto with signatures A_i to B_{iv}. It is 164 mm. high (B.N. copy). There is a woodcut of a comet on the title-page, A_i. The verso of the title-page is blank. On the verso of B_{iv} the book is dated from Lyons, November 13, 1577 and signed by the author. The pages are numbered from 3 to 16 beginning on the recto of A_{ii}. The book is addressed to Monseigneur de la Mante. It was printed in Paris in 1577. The C.D.H. copy is bound in paper and has been considerably cut down. It is 147 mm. high and the line " Iovxte La Copie De Lyon." and the lower half of the line above, " 1577 ", do not appear. There is an unmended tear in the lower right hand corner of the title-page.

63 Discours sur ce que menace devoir advenir la comète apparue à Lyon le 12 Nov. 1577 etc.

Not located. Houzeau 5599; Poggendorff, I, 1211; Baudrier, IV, 91.

Description: The title is from Poggendorff. This item is an octavo published in Lyons in 1578. Du Verdier, 404, gave the title of the work as "Ample Discours sur ce que la Comete apparue au mois de Nouembre 1577. menasse deuoir aduenir à plusieurs princes, païs & peuples de la Chrestienté " and said that it is an octavo printed in Lyons in 1578 by François Didier. Baudrier gave the title as " Discours sur ce que la Comete apparue au mois de Novembre 1577, menace devoir advenir à plusieurs Princes, Pays & Peoples de la Chrétiente, par François Junctini, Florentin." He also said that it is an octavo printed in Lyons by François Didier in 1578. He does not seem to have taken the title from a copy of the book, but from Du Verdier.

63a Discovrs Svr || Ce Qve Menace D'E || uoir aduenir la Comete apparue le 12. de ce || present mois de Nouembre 1577. laquelle || se voit encore auiourd'huy à Lyon, & au– || tres lieux. || Dedié A Monsievr || de la Mante, Cheualier de l'Ordre du Roy, Capitaine || de trois cens cinquante hommes de guerre a pied || François, Colomnel des legionnaires au Marquisat || de Saluces, Gouuerneur pour sa Maiesté en la Cita– || delle de Lyon, commandant en ceste ville en l'absen– || ce de Monseigneur de Mandelot. || A Lyon, || Par François Didier à l'enseigne du Fenix. || Auec Permission.

Not consulted. Baudrier, IV, 84. There is a copy in Lyons.

Description: The volume is a quarto by signature or an octavo by size. It has 8 leaves. The pages are unnumbered. The emblem of Iunctinus is on the title-page above the place of publication, and on the verso of the title-page is a figure of a comet. The book was printed in Lyons by Didier but the date of printing is not given. This item may explain the words " Iovxte La Copie De Lyon." on the title-page of 62.

[64] Tractatio vtilis & lectu digna de Cometarvm Cavsis, Effectibvs, Differentiis, Et eorundem proprietatibus. Cvm Plana Et Expedita Declaratione Eventvvm, quos diuersos pro diuersitate Planetarvm & Signorvm Zodiaci sortiuntur, tam ad communem vitae vsum necessaria, quàm ad natiuitatum figuras dextrè dijudicandas idonea, ex Francisci Iunctini Florentini voluminibus excerpta. 1580. Lipsiae. (at end: Lipsiae, Imprimebat Ioannes Steinman. Anno M. D. LXXX.)

C.D.H.

Description: The volume is an octavo with signatures A_1 to C_7. It is 148 mm. high. There is a printer's device on the title-page, A_1, and another on the recto of C_7. The versos of A_1, C_6 and C_7 are blank. The pages, beginning on the recto of A_2 and ending on the recto of C_6, are numbered from 1 to 41. The C.D.H. copy is bound in paper. The number " 222 " is written in pencil on the title-page. On page 3, the first four letters of " Cometa " at the beginning of the second paragraph have been crossed out in ink.

The book is about comets in general and does not deal with the comet of 1577. It was erroneously included in the original bibliography (Hellman). The heading of the pages, two pages together, is "Annotationes De Cometis." by which title the book was cited by Riccioli, II, 117.

Kreidweiss, Vitalis

See item 53

Launer, Martin

65 Eine kurtze Erinnerung vnnd Erklerung, von dem Erschrecklichen Cometstern, so im vorlauffenen 1577. Jar, den 10. Tag Nouemb. biss auff den 7. Januarij des jetztlauffenden 1578. Jares, erschienen und gesehen ist worden, gestellt durch Martinum Launer vom Stolz, diese zeit Pfarrer zu Lauterbach im Mittelwäldischen. (at end: Gedruckt zu Neyss, bey Johann Creutziger.)

Not located. Carl, 53; Scheibel, 104-5; Lalande, 147.
Description: The title is given above as it was given by Scheibel, who called the volume an octavo of two signatures. It was printed in Neisse in 1578.

Lemovicensis, Peter (Peter of Limoges)

[66] Judicium de stella Cometa [1298] a Magistro Petro Lemovicensi canonico Ebroicensi.
Not consulted. Cambridge University library, manuscripts, III, 406, V, 599 (shelf mark: Ii. III. 3. f. 283)
Description: This item is a manuscript by a thirteenth century author, undoubtedly the well known Peter of Limoges. It was erroneously included in the original bibliography because of the reference to it by Houzeau, 5591, where, however, the author's name was given as Lemoniensis.

Liberati, François

67 Description De L'Estrange Et Prodigievse Comete, Apparve Le Vnziesme iour de Nouembre, à six heures du soir. Auec la figure du lieu de sa scituation, en la huictiesme maison du Ciel, & la prediction & intelligence de ses effroyables effets. Plus l'approbation tant par Historiens Ecclesiastiques que Prophanes que lesdites Cometes n'apparurent oncques pour neant. Par tres-docte & excellent Astrologue M Françoys Liberati, de Rome. A Paris. Pour Iean de Lastre, libraire, demourant en la rue S. Iean de Latran, pres le College de Cambray. M. D. LXXVII. Avec Privilege Dv Roy.
C.D.H.
Description: The volume is a quarto with signatures A_1 to B_4. It is 146 mm. high, but the lower margin may have been cut. The pages are unnumbered. The book was printed in Paris in 1577 and is in French prose with the exception of some Latin, Italian, and French verse at the end. The verso of the title-page, A_1, and the verso of B_4 are blank. A figure on the title-page pictures stars, clouds and a comet with its tail pointing in the direction of a crescent moon. The figure is enclosed by a rectangle the sides of which are marked north, south, east and west. The verso of A_3 contains an astronomical figure

showing the comet and the planets for November 11, 1577 at 6. P.M. B_4 contains a decorative design. The work is not dedicated to anyone but acknowledges the graciousness of " Seigneur H. PG.A." with verse addressed to him as "D.H.P.A." Except for spelling and punctuation, the opening words, one word in a quotation from the prophet David, the use of " en ce que " in place of " par ce que ", the omission of one line of prophecy and a difference in closing paragraphs, this work corresponds word for word with items 67b and 67c. The C.D.H. copy is bound in paper. In the upper right hand corner of the title-page " 287177 " is written in pencil, and below those numbers is written " 36.–", also in pencil. In the upper left hand corner of the title-page is written in pencil " 8 Watt." There is a spot, like rust, in the middle of the outer margin on both sides of A_4, but darker on the recto.

67a Discovrs || de la comete || commencee a appa– || roir sur Paris le XI. iour de || Nouembre, mil cinq || cens septāte – sept, || a six heures || du soir. || Auec la declaration de ses presages & effets. || Par excellent Astrologue M. Francoys || Liberati, de Rome. || A. Lyon, || par Benoist Rigaud, || 1577. || Auec Permission.

Not located. Baudrier, III, 334; Rothschild, III, 367-8 Description: The title is taken from Baudrier. The volume is an octavo of eight unnumbered leaves. On the title-page, above the place and date of publication, is a woodcut representing the comet and the moon. The work is addressed to M. d'Habin. The recto of the last leaf has an astronomical figure and the verso is blank. Probably, counting by signatures rather than size, the volume is a quarto, and thus probably corresponds to item 67c described below, which however, was printed a year later.

67b Discovrs De La Comete Commencee A Apparoir Svr Paris Le XI, iour de Nouembre, mil cinq cens septante-sept, à six heures du soir. Auec la declaration de ses presages & effets. Par excellent Astrologue M. Françoys Liberati, de Rome. A Paris. Pour Iean de Lastre, libraire, demouran en la ruë S. Iean de Latran, pres le College de Cambray, M. D. LXXVII.

B. M. 1192. e. 17; photostatic copy, C.U.L. B. 523.6 L61

Description: The volume is a quarto with signatures A_1 to B_3. It is 16 cm. high but may have been cut. The pages are unnumbered. A_2 is wrongly marked A. The versos of the title-page and of B_3 are blank. B_3 contains the same decorative design as appears on the recto of B_4 of item 67. On the title-page is the figure of a comet with its tail pointing in the direction of a crescent moon. A_4 contains the same astronomical figure as appears on the verso of A_3 of item 67, showing the comet and the planets for November 11, 1577 at 6 P. M. The work is addressed to Monsieur d'Habin, Knight of the order of the King and His Majesty's Ambassador to Rome. It was dated from Paris, November 12, 1577, and was printed in French prose in Paris in 1577.

67c Discovrs De La Comete Commencee A Apparoir sur Paris le XI. iour de Nouembre 1577. à six heures du soir. Auec la declaration de ses presages & effets. Par excellent Astrologue M. François Liberati de Rome. A Lyon, Par Benoist Rigavd. 1578. Avec Permission.

B.N. Rés. p. V. 200; photostatic copy, C.U.L. B523.6 L611

Description: The volume is a quarto with two signatures, A_1 to B_3. Only a photostatic copy, without a scale, has been available in preparing this description. The pages are unnumbered. The verso of the title-page, A_1, and the versos of B_2 and B_3 are blank. On the title-page is the same figure as on the title-page of 67b. On B_3 is a circular diagram showing the points of the compass and the corresponding winds. This edition, like 67a and 67b, was addressed to Monsieur d'Habin and dated from Paris, November 12, 1577. It was printed in Lyons by Rigaud in 1578. Allowing for slight differences in spelling and punctuation, the text of 67b and 67c is the same, except that the text of the earlier one indicates the astronomical figure described above.

Longus, Io. Bernardinus

68 Io: Bernardini Longi, Phylosophi Neapolitani. De Cometis Disputatio. Ad Illvstrissimvm, Et Eccellen-

tissimum Marchionem Mondeyar, in Neapolitano Regno Proregem. Neapoli Apud Horatium Saluianum. 1578. (at end: Imprimatur, Gaspar Silingardus Vic. Gene. Neap. Andreas Sarnus. P. Regius. f. 5. Neapoli Apud Horatium Saluianum. 1578.)

C.D.H.

Description: The volume is a quarto with signatures A_1 to I_4. It is 187 mm. high. It was printed in Latin in Naples in 1578 at the press of Horatius Salvianus. On the title-page, A_1, is a printer's device. The verso of the title-page is blank. On the recto and verso of A_2 and the recto of A_3 is the dedication to Inaco Lopez de Mendoza Marchioni Mondeyar. This is followed by a preface on the verso of A_3 and the recto and verso of A_4. The text is divided into nine chapters, which occupy the remainder of the book. Beginning with B_1, the beginning of the text, the leaves are numbered from 1 to 32. The page headings, two pages together, are "Io: Bernardini Longi De Cometis Dispvtatio." The C.D.H. copy has been cut down, but is clean and complete. It is bound with three other tracts, namely, Tramontano's *Discorso Filosofico,* Naples, 1619; Thomas di Ruggiero's *Brieve E Compendioso Discorso,* Naples, 1619; Sordi's *Discorso Sopra Le Comete,* Parma, 1578 (item 101). The volume bears a book-plate with no name but with the phrase "Comme Fus Je". The binding is vellum and the back has "Delle Comete" stamped on it.

69 Traicté des Cometes du Seigneur I. Bernard Longve Philosophe & Medecin . . . mis en Français par Charles Nepvev, Chirurgien du Roy . . . Plus vn Recueil de la peste tres approuué & experimenté par ledict Nepvev Paris, Claude de Monstroeii, 1596

B.N. V 21091

Description: The volume is a quarto with signatures a_i to a_{iv} plus A_i to L_{ii} (one letter, probably J, not being used). Information for determining the size of this item is not now available. (The book was seen in 1931.) The verso of L_{ii} is blank. The leaves, from A_i to L_{ii} are numbered from 1 to 42. When the material for this description was gathered, it was not ascertained whether or not the work deals with the comet of 1577, but it is probable that item 69 is the French translation of 68.

L. T.

See T., L.

Maestlin, Michael

See item 49

70 Obseruatio & demonstratio Cometae Aetherei, Qvi Anno 1577. Et 1578. Constitvtvs In Sphaera Veneris, Apparvit, Cvm Admirandis eius passionibus, varietate scilicet motus, loco, orbe, distantia à terrae centro, &c. adhibitis demonstrationibus Geometricis & calculo Arithmetico, eiusmodi de alio quoquam Cometa nunquam visa est. Avtore M. Michaele Maestlino Goeppingensi. Tubingae, excudebat Georgius Gruppenbachius, 1578.

B.N. V 7918

Description: The volume is a quarto with signatures A_1 to I_2. It is 187 mm. high. The verso of A_1, the title-page, and the verso of I_2 are blank. The pages A_2 to the verso of A_4 inclusive contain the dedication, which is printed in italics. The book was dedicated to Louis, Duke of Wurtemberg, where Backnang was, and Teck, Count of Montbélard and so forth, and was signed by Michael Maestlin, " Ecclesiae, quae est in Backnang, Diaconus ", and dated from Backnang in 1578. There is a map of the path of the comet of 1577 on the title-page. The book was printed in Latin in Tübingen in 1578. The text of the book begins on B_1 and the pages are numbered from 1 to 59 inclusive, from the recto of B_1 to the recto of I_2 inclusive. There are ten chapters and there are mathematical diagrams on pages 9, 15, 23, 26, 28, 35, 39, 41, 43, 44 and 47 and tables on pages 36, 37, 41, 42, and 52 to 53.

Major, Elias

See item 35

Marzari, Giacomo (or Iacopo)

71 Notable Discours de M. Iacques Marzari, Vicentino, Tovchant la Comette apparue au moys de Nouembre 1577. Auquel est traité des occasions d'icelle, de ce qu'elle peut predire & signifier: & de la reigle de viure

en ce temps, pour obuier à la maligne affection d'icelle. Traduit de nouueau, d'Italien en langue Françoise. A Lyon, Par Iean Patrasson. M. D. LXXVIII. Auec permission.

B.M. 531. e. 28 (3); B.N. V 21092

Description: The tract is a quarto with signatures A_1 to B_4. It is 148 mm. high. The pages are unnumbered. It was printed in French in Lyons in 1578 by Jean Patrasson, and is a translation of the Italian edition. The verso of A_1 and the verso of B_4 are blank. The woodcut on the title-page is the same as that on the title-page of Fiornovelli's *Discours* . . . , item 41, and on the verso of B_4 of the tract by Raymundus, item 85. All three of the tracts were printed by Patrasson. The same figure also appears on the last page of the tract by Thurneysser, which was printed by Vittorio Baldini in Ferrari, item [104]. In the summer of 1931, the leaf B_1 of the B.N. copy of item 71 had been replaced by B_1 of Fiornovelli's work, item 41.

72 Discorso intorno alla cometa apparsa il mese di Novembre 1577 etc.

Not located. Riccardi, Part I, v. II, 131; Angiolgabriello, V, CCXXIII

Description: This is a quarto and was printed in Venice in 1577 by Domenico Niccolini.

72a Discorso intorno alla cometa apparsa il mese di Novembre 1577 etc.

Not consulted. Riccardi, Part I, v. II, 131; Haym, IV, 89 (9); Angiolgabriello, V, CCXXIII

Description: This is a quarto and was printed in Perugia by Bresciano in 1578. Riccardi said that there was a copy in the Marciana library. Angiolgabriello gave the date as 1588.

Mater, Richard

See item 88

de Maudin, Dauid

72b a book in French

Not located. Referred to by Twyne, item 105a, on the verso of A_{iii}.

Meister, Joachim

72c Cometes Qvi Apparvit Anno Christi cIↄ. Iↄ. Lxxvii. Mense Novembri, Carmine Descript. à Ioachimo Meistero Cogn: Gorl: Gorlicii Excudebat Ambrosius Fritsch. Anno M. D. LXXVIII.

Bibliothèque Nationale et de Université, Prague, 49 C 107

Description: The volume is a quarto with signatures A_1 to D_4. The verso of D_4 is blank. The pages are unnumbered. There is a woodcut, probably the printer's device, on the title-page. The verso of the title-page contains the dedication to Elias Melcerus. As is indicated on the title-page, the volume is in Latin verse. The verse begins on the recto of A_2 and continues to the middle of the verso of D_3 where Greek verse begins. The latter ends in the middle of the following page. The copy in the university library in Praha (Prague) is a presentation copy from the author to Thomas Mitis.[4]

Meyne, (or Mayne, or Menius) Matthias

73 Von aller geschlecht der Cometen / jeder zeit / wan die erscheinē zugebrauchen / vnd von dessen wirckungen / der vns zu Dantzigk den 12. Nouembris dieses 1577. Jar erschienen ist. Durch M. Matthiam Meyne Dantiscanum Mathematum studiosum. Gedruckt zu Dantzigk bey Jacobo Rhodo. M. D. LXXVIII.

Nationalbibliothek Wien 72.T. 145 (21)

Description: Only a photostatic copy, without a scale, has been available for this description. The volume is a quarto with signatures A to D_{iv}. The verso of D_{iv} is blank. On the title-page is a woodcut of a comet and beneath it an astronomical diagram for new-moon, at one in the morning of November 10, 1577. The book was printed in German at Dantzic in 1578 by Jacob Rhodo. It was dedicated to Scheppeu, who among other positions was the head of Dantzic. The dedication occupies the verso of A_i and the recto and verso of A_{ii}. It was dated January 6, 1578 and signed by Meyne. At the close of

4 Thomas Mitis of Naumburg wrote verse which was printed in Prague. See Jöcher, III, 558 and Zedler, XXI, 549.

the work, on the recto of D_{iv}, it is stated in Latin that the work was written and finished on the 20th, 21st, 22nd and 23rd of December, 1577. Estreicher, XX, 318, attributed the work to "Andreas" Meyne.

Micó, José

73a Diario y juicio del cometa que apareció á los 8 de Diciembre de 1577 en Barcelona.

Not located. Picatoste y Rodriguez, 195

Description: This work was printed in Barcelona in 1578.

73b Diario y juicio del cometa que apareció a los 8 días de Noviembre de 1577.

Not located. Marí, in *Archeion*, XXV (1943): 210-211

Description: According to Marí, this work was first printed in 1798. It may be the same as item 73a.

Misocacus, Wilhelm

74 Observationes astronomicae pertinentes ad novum cometam qui visus est jam anno 1578.

Not located. Carl, 53; Struve, I, 787

Description: The title is given above as it was given by Struve. Carl wrote "jam anno 1577" in place of "jam anno 1578", but this is probably wrong, as is seen from the meaning of the phrase. The tract was printed in Dantzic in 1578. The B.M. catalogue considered Misocacus a pseudonym for Johann Rasch, but Rasch criticized certain of Misocacus' writings. Misocacus was a real person, born in 1511, a doctor and astronomer. (See Hellmann (1883), 334; Jöcher, III, 555; Zedler, XXI, 468.)

Montelli, Ascanio

74a Lettres Dv Seignevr Ascanio Montelli Medecin, Et Mathematicien Napolitain, traduit d'Italien en François. Contenant la prediction des affaires de Flandres, & d'autres guerres des Chrestiens, selon la signification d'vn Comette apparue n'agueres: les Eclipses de la Lune: la conionction de Saturne & Mars, & plusieurs natiuitez des Princes. Auec vn discours de merueilleux presages & accidents menacez par la grande cōionction des planettes de l'an 1583. & des troubles qui doyuent suruenir à aucuns empires, monarchies & estats, selon l'opinion de plusieurs sçauans hommes Astrologues. A Paris, Pour Iean de Lastre, libraire: ruë saint Iean de Latran, pres le college de Cambray. Avec Privilege Dv Roy.

Crawford library; photostatic copy, C.U.L. B523.6 M76 Description: The volume is a quarto with signatures A_i to E_{iv}. It is 165 mm. high. The versos of A_i and E_{iv} are blank. The pages are unnumbered. The book was printed in Paris for Jean de Lastre, probably in October 1578, since the "Privilege" (E_{iv} r) was dated October 18, 1578. On the title-page is the same woodcut figure as on the title-page of item 67. The text starts on the recto of A_{ii} under the heading "Lettres Dv Seignevr Ascanio Montelli, Escrite Av Viceroy de Naples, sur la signification de la Comette, & autres constellations de l'an 77. & 78. & 79. escrite le 20. Aoust 1578." A section entitled "Prediction Dv Seignevr Ascanio Montelli tres-excellent Medecin & Mathematicien, sur les affaires de Flandres & autres guerres des Chrestiens." starts on the recto of B_{iv}. This section contains astrological diagrams on B_{iv} r and v, C_i r and v, and tables of astronomical values on C_i v and C_{ii} r, all for the horoscopes of the kings of France and Spain, of the Duke of Anjou, and Don Juan of Austria. A section entitled "Continvation Des Predictions De M. Ascanio Montelli, sur l'an 1579." begins on the recto of E_{ii}. This was dated from Palermo, August 28, 1578. On the recto of E_{iv} is an extract from the statement of the privilege to print and distribute the book.

Müller, Samuel

75 Astronomische Beschreibung dess Cometen / so zů ende dises verloffenen 1577. Jars erschinen / sampt seiner bedeutung. Zů Ehren vnd Wolfart / Dem Hochwirdigen Fürsten vnnd Herren / Herren Eberharten / Appte des Ehrwirdigen Fürstlichen Stiffts zů Kempten / seinem gnedigen Herren. Durch Samuelem Müllerum Physicum Campidonensem. (last page: Vetruckt zů Augspurg / durch Valentin Schönigk / auff vnser Frawen Thor.)

C.D.H.; photostat belonging to G.S.

Description: The book is a quarto with signatures A_{I} to B_{III}. It is 195 mm. high (C.D.H. copy). A_{III} is wrongly marked B_{III}. B_{III} is the page giving the name and city of the printer. The verso of B_{III} is blank. On the title-page, A_{I}, and its verso are pictures of two shields or coats of arms. The book was printed in German at Augsburg by Valentin Schönig, probably in 1578. It is dedicated to the Abbot of Kempten, as is stated both on the title-page and in the dedication. The dedication or preface is in Latin, and humbly offers the book. The C.D.H. copy is bound in boards. There is a dark greenish mark about 5 mm. broad on the upper inside corner of the verso of the title-page, which mark has penetrated through to the tile-page. The photostatic copy belonging to G.S. was probably made from the C.D.H. copy, judging from a similar mark on the title-page of the photostat.

Muñoz (or Mugnoz or Munosius), Jerome

See 88

75a Summa del Pronostico || del Cometa: y de la Ecclipse de la Luna, que fue a los || 26. de Setiembre del año. 1577. a las. 12. horas. 11. minu || tos: el qual Cometa ha sido causado por la || dicha Ecclipse. || Compuesto por el Maestro Hieronymo Muñoz || Valenciano. Cathedratico de Mathematicas || y de Hebreo d' la Vniversidad d' Valencia . . . || . (at end: Impressa en Valencia : || en casa de Joan Navarro, || Año. 1578. ||)

Not located. Picatoste y Rodriguez, 207, # 521; Ortroy (1920), 132-3

Description: The title is given above as it was given by van Ortroy. The book was printed in Valencia in 1578 at the press of Joan Navarro by Gabriel Rivas. It is a quarto and has, on the title-page, a picture of a comet between Aquila and Saturn. Van Ortroy mentioned a copy in the national library in Madrid, no. 3180, but even if that copy still exists, it is not available.

Mylius, Martin

See item 49

Mylius seems to have written about the comet of 1577 in his *Chronik*. (See Koch (1907-1910), LXXXIV, 72.)

Nagelius, Paulus (or Nagel, Paul)

76 Ander Theil Des in 1618. Jare erschienen vnd verschienenen Cometen / so an diesem Orte zur Proba begreifft / eine kurtze warhafftige deutung vnd interpretation Des neweñ wuñder - Sterns 1572. vnd des Cometen 1577. erschienen / wird auch angezeigt wie der Comet 1577. mit den Cometen 1618. in einer Harmonia stehe / was jhre praeludia gewesen / vnd was noch fůr Wunder in kůrtzen ja jetzt werden offenbar werden. Autore M. Paulo Nagelio. Im Jahr M DC XIX.

C.D.H.; B.M. 8563. aaa. 33 (16); C.U.L. 523.6 Z v. 2 (35) and 523.6 Z1 (2)

Description: The volume is a quarto with signatures A_{i} to G_{iv}. It is 184 mm. high (C.D.H. copy) or 180 mm. (B.M. copy) or 193 mm. (C.U.L. copy one) or 186 mm. (C.U.L. copy two). The pages are unnumbered. The leaf F_{ii} is marked by the Roman letter "F" and the Arabic numeral "2" whereas the other signature marks except "C" use German or Black letters and small Roman numerals. The verso of the title-page, A_{i}, is blank. On the title-page, beneath the name of the author, is a woodcut of a comet and two stars. The title-page has a woodcut design as a border. The preface dedicated the book to Jacob Schultz and Sigismund Deuerlin and was signed, on the verso of A_{iii}, by Nagel and dated "Argeliae 2. April. 1619." The text begins on the recto of A_{iv} and continues to the end. It is divided into two chapters, the first dealing with the nova of 1572 and the second, starting on the verso of E_{iii}, with the comet of 1577. The book was printed in German in 1619, but no place of printing nor printer's name was given. However, this item, although a separate tract with its own title-page and a new series of signatures, can be considered as part two of a larger work and consequently as being printed in the place where the first part was printed, namely at the author's own press, presumably in Arten (Argelia). The title-page of the first part reads: "Stellae Prodigiosae Seu Cometae per oculum triplicem observatio & explicatio. Das ist: Des newen Cometen vnd Wunder

Sterns im October / November vnd December 1618. erschienen / warhafftige Deutung vnd Auszlegung per Magiam insignem, dergleichen zuvor .nicht gesehen: Allen Menschen auff Erden zur guten Nachrichtung vnd Warnung fürgestellet Durch M. Paulvm Nageli'm L. M. Tehologum vnd Astronomum, &c. In Verlegung desz Autoris. Im Jahr 1619." This title-page is copied from the B.M. copy. The C.U.L. copy has the author's last name written out as " Nagelivm ". It is possible that what seems on the photostat of the B.M. copy to be an apostrophe before the "m" might be a blot covering most of the "v". In both copies " November " is divided between two lines but only in the C.U.L. copy is the sign " - " used. On the title-page there is a woodcut showing a comet, some stars labeled " Bootes ", "Arctur.", and so forth, the symbols for Libra and Leo, and " Mene Mene Tekel Uphirsin." Part of the title-page, at least in the C.U.L. copy, is printed in red. In the B.M. the two parts are bound together and catalogued only under the title of the first. The C.D.H. copy does not have the first part. The C.U.L. catalogued only the title of the first part and said that two parts are in one (C.U.L. 523.6 Z v. 2 (35)) and catalogued the second copy on the same card saying that only part two is there, all of which is true. The Crawford library catalogue, 119 and 317, listed the two books separately and recorded two editions of the first part, probably corresponding to the C.U.L. and B.M. copies. There seems to be no variance between the copies of part two.

Nepvev, Charles

See item 69

Neri, Giovanni de'

77 Pronostico e discorso di Giovanni de' Neri veronese, sopra la cometa apparsa il mese di Novembre l'anno 1577. Con il giudicio dell' ecclisse etc.

Not consulted. Houzeau, 5613; Riccardi, Part I, v. II, 195.

Description: The title is given above as it was given by Riccardi. The volume is a quarto printed in Mantua in 1578, by Giacomo Ruffinello. Riccardi said that there was a copy in the Marciana library.

Nicolaus, Johannes

See Nielson, Jens

Nielsen, Jens

77a De Portentoso Cometa, Qvi Anno M. D. LXXVII. apparuit, terrifico magna ex parte sidere, & non leuiter piato, vt loquitur Plinius. Carmen Iohannis Nicolai Diaecesium Asloïensis, & Hammarensis in Noruuegia Superintendentis. Rostochii Impressit Stephanus Myliander. Anno M. D. LXXVII.

Not consulted. Royal Library, Copenhagen (2 copies); description and title from Nielsen, 1186. The work is also cited in Bruun, II, 72.

Description: The volume is a quarto of one signature, A. The printing measures 138 x 98 mm. There is a vignette on the title-page. The book was printed in Latin verse in Rostock in 1577 by Stephan Möllmann (Myliander). The verso of the title-page is blank. On the recto of A_2 it is stated that the comet was first observed in Oslo, Norway on the tenth of November, 1577, between the fourth and fifth hour after sunset and that it had recently passed sideways through Cygnus.

77b De Portentoso Cometa, Qvi Anno Praeterito Apparvit, Terrifico magna ex parte sidere, & non leuiter piato, vt loquitur Plinius, Carmen Iohannis Nicolai Asloiensis. Hafniae Impressit Andreas Gutterwitz. Anno 1578.

Not consulted. Royal Library, Copenhagen; Lund University Library; description and title from Nielsen, 1187. The work is also cited in Bruun, II, 72.

Description: The volume is a quarto of one signature, A. The printing measures 139 x 108 mm. There is a vignette on the title-page. The book was printed in Latin verse in Copenhagen in 1578 by Andreas Gutterwitz. The verso of the title-page is blank. On the next page it is stated that the comet was first observed in Oslo on the tenth of November of the previous year, 1577, between four and five hours after sunset and that it had recently passed sideways through Cygnus. At the back of the book there is a plate measuring 288 x 210 mm. with a figure and the following inscription, " Schema Coeli, Positvm Exhibens Sydervm, Ad Tempvs Conivnctionis Solis Et Lvnae in Scorpio, quod Cometam grandem & horribili specie produxit, Mense Nouembri, Anno &c. 1577."

Nolthius, Andreas

78 Observatio, Vnd Beschreibung des Cometen / welcher im Nouembri vnd Decembri / des 77. vnd noch im Januario / dieses 78. Jharsz erschienen / Geschehen vnd gestellet / durch Andream Nolthium, Mathematicum. (at end: Bedruckt zu Erffurdt durch Georgium Bawman / wonhafftig auffm Fischmarckt.)

Crawford library; photostatic copy, C.U.L. B523.6 N72

Description: The volume is a quarto with signatures A_i to E_{iv}. It is 187 mm. high. The pages are unnumbered. The book was printed in German in Erfurt, probably in 1578, the date given by Schottenloher, IV, 378. The verso of the title-page, A_i, and the verso and recto of E_{iv} are blank. On the title-page there is a woodcut of a celestial sphere, showing the comet in the sign of Capricorn. On the recto and verso of A_{ii} and the recto of A_{iii} there is a preface, dated from Einbeck, February 2, 1578, dedicating the book to Philip Duke of Brunswick and Lüneburg, and signed by the author. The book is divided into seven chapters, each of which begins with a woodcut initial.

Nolthius, Valentinus

Squarcialupus, on the verso of C_3 of 37(3) mentioned a German tract on the comet of 1577 by Valentinus Nolthius, but it seems likely that he erred concerning the Christian name.

Nostradamus

See Crespin

Paduanus, Joannes (or Padovani, Giovanni)

78a a book on the comet of 1577

Not located. Riccioli, II, 13

Description: Riccioli, II, 2, also spoke of a book on the comet of 1577 by Antonius Paduanus, and, II, 117, of a "tractatus de Cometis" by Io. Paduanus, and Struve, II, 535, listed among general books on comets, a work by J. Paduanius, entitled "Opus de stellis crinitis", printed in Verona in 1578.

Papeberg

See Camerarius, Ioachimus

[Paradinus, Gulielmus]

78b Sommaire Discovrs Svr La Vision & presage du Comete, qui premierement s'apparut enuiron le cõmencement du moys de Nouembre, mil cinq cens soixante & dixsept, que l'on voit encores à present. Ignea flammantis picta est hic forma Cometae, Cuius fata notat pagina nostra sequens. A Lyon, Par Benoist Rigavd. 1577. Avec Permission.

B.N. Rés. p. V. 202; photostatic copy, C.U.L. B 523.6 P21

Description: The volume is a quarto with signatures A_1 to B_4. Only a photostatic copy, without a scale, has been available in preparing this description. The pages are unnumbered. The verso of the title-page, A_1, is blank. On the title-page is a woodcut representing a comet with its head between the directions "Occidens" and "Septentrio" and its tail between "Meridies" and "Oriens". This is like the cut on the title-page of item 12, which was printed by the same printer. On A_2 and its verso is a dedication, in verse, to Peter à Nagutio, signed "Gulielmus Paradinus faciebat", which last phrase, however, may apply to the whole book. The book was not listed by Nicéron, XXXIII, 165-9, in a bibliography of works by Paradin. A_3 and its verso contain an unsigned dedication in prose to "Monsievr Le Prothonotaire Vetus, Prieur & Seigneur de Chatonay & S. Laurens, &c." The *Sommaire Discovrs* begins on the recto of A_4 and ends on the verso of B_3 on which page there is also a woodcut showing the sun, stars, a comet, the moon, and a man holding a globe, with the legend "Astra Regvnt Orbem Sapiens Dominabitvr Astris." The recto of B_4 contains a poem about the then visible comet by Claudius Morellus of Wales and a short prediction. On the verso of that page there is a woodcut of a celestial globe.

Pauli, Simon

See item 30

Peter of Limoges

See item [66]

Philomathesius, R. P.

See item 84. The number is retained from the original bibliography, in which only the initials, not the author's name, were given.

Pianero, Giovanni

See Planerius

Planerius, or Planerus, Ioannes

79 Ioannis Planerii Qvintiani Brixiensis Artivm Et Medicinae Doctoris Varia Opuscula:

Epistolae morales.
Patriae descriptio : in qua de animorum immortalitate.
Henrici Regis ad Vrbem Venetam aduentus.
De Comete, 1577.
De Lacte.

Cvm Privilegio. Venetiis, Apud Franciscum Zilettum. 1584.

C.D.H.

Description: The volume is a quarto with signatures a_1 to a_4 plus A_1 to H_4 plus a_1 to f_4. It is 230 mm. high. The leaves are numbered from 2 to 32 from A_2 to H_4 and from 3 to 24 from a_3 to f_4. The work was printed in Latin in Venice in 1584 by Franciscus Zilettus. The versos of the title-page, a_1, and of the second title-page, a_1, and of f_4 are blank. There is the same printer's device on both title-pages. The work is divided into two sections, the first containing the moral letters and occupying the leaves through H_4, the second containing the other sections listed on the title-page. On the recto and verso of the first a_2 and the recto of a_3 is a dedication to Cardinal Ferdinand Medici dated from Venice in 1582. On the verso of the first a_3 and the recto of a_4 is a dedication to Alphonso Martinengus dated from Brescia in 1583.

On the verso of the first a_4 is an index of the letters. The second section has a new title-page which reads "Ioannis Planerii Qvintiani Brixiensis Artivm Et Medicinae Doctoris: Breuis Patriae suae descriptio: Et Illustrium Virorum eiusdem Patriae enumeratio: In qua de animorum immortalitate disseritur. Venetiis, Apud Franciscum Zilettum, M D LXXXIIII." The second section is dedicated to the brothers Jerome and Johannes Fugger. The treatise on the comet of 1577, headed "Ioannis Planerii Qvintiani Brixiensis Philosophiae, Et Medicinae Doctoris. Tractatus breuis de Comete à duodecima die Nouembris 1577. per duos menses mortalibus omnibus apparente.", occupies the verso of d_2 to the recto of e_2 inclusive. There is a poem to Planerius on the recto of f_4. The C.D.H. copy is bound in boards with a leather back upon which is stamped "Planer Varia Opuscu". On the inside of the front cover is pasted a long narrow strip of paper with "Schiavini." printed in ink on it. Beneath this are written in pencil the following notes, "18/6", "K_4", "25/287". The title-page has an "R" written in pencil and a figure like the lower three quarters of the number 8 written in ink in the lower right hand corner. Some pencil writing has been erased from the inside of the back cover.

Portantius, Johannes

79a Beschrijuinghe der nieuwer Cometen / Met aenwijsinge ende vermaninge wat zy bedieden ende voortbrenghen sal. (at end: T'Anwerpen by François van Ravelenghien, naest onser L. Vrouwen kerck-deure aende Noort-zijde.)

Plantin Museum, Antwerp

Description: This item is a single sheet, printed in Flemish, in Antwerp probably in 1577. The available photostatic copy, furnished by the courtesy of the Plantin Museum, does not have a scale on it to furnish the size of the sheet, but it is probably the size of the usual broad-side. The last line, giving the name of the printer, may read "T'Antwerpen . . ." instead of "T'Anwerpen" as given above, a fold in the sheet possibly concealing the letter " t " when the photostat was made. The text is signed by the author and gives his occupation as " Physicus & Mathematicus ". Beneath the title is a woodcut occupying almost as much space as the text. It shows a town at the edge of a body of water with a comet in the sky as well as the symbol for Capricorn and a half-moon and some stars (the constellation Capricorn).

79b Kurtze Erklerung / von den eigenschafften desz grossen / im M. D. LXXVII. jars erschienen vnd noch brinnenden Cometen / Sam̃t seiner bedeutung / Auch an was vhrt vnd stell / Himlischen Hausz / Planeten vnd Zeichen / er auff vnd vnter gangen / Durch der Astronomiae Liebhabern / erfahrne vnd Gelehrte / wie nachuolgt / vndterschiedlich beschrieben. (at end: Gedruckt zu Nůrmberg / durch Leonhard Heuszler)

B.M. C.29.h.4.

Description: The volume is a quarto of two signatures, A to B. It is 186 mm. high. The pages are unnumbered. It was printed on linen in Nuremberg by Leonard Heussler, probably in 1577. On the title-page is a woodcut closely resembling that on the title-pages of items 3 and 28. Below the woodcut is the following quotation, supposedly from the 20th chapter, really from the 21st, of the Gospel according to Luke, " Es werden Zeichen geschehen / an der Sonnen / Mond vnd Sternen / vnd auff Erden wird den Leuthen bange seyn / vnd werden zagen / vnd das Meer vnd die Wasserwogen werden brausen / vnd die Menschen werden verschmachten fůr forcht / vnd fůr wartten der ding / die da kom̃en sollen auff Erden / dass auch der Himel krefften sich bewegen werden / rc." The verso of the title-page, A_i, and the recto and verso of B_{iv} are blank. On the recto of A_{ii} is a woodcut of a comet and a description of the comet of 1577. The book is in two sections, the first of which is signed by Portantius and is headed " Beschreibung des newen Cometen / mit anweissung vnd vermanung / was er zubedeuten." It is an almost exact translation into German of the last two paragraphs of item 79a. It occupies the verso of A_{ii}, the recto and verso of A_{iii} and the recto of A_{iv}. The second section of the book occupies the verso of A_{iv}, and the rectos and versos of B_{i}, B_{ii} and B_{iii}. It is entitled " Ivdicivm Von dem grossen Cometen / so den XII. Nouemb. im M. D. LXXVII. Jar / am Himel erschienen.", is dated December 19, 1577, and is signed by "B.I.T."

Praetorius, Adelarius

80 Schade Wacht. Nothwendige Warnung vnd Vermanung an alle busfertige / glaubige Christen gethan / von dem erschrecklichen Comet / vnd Zornstern / der dieses 77. Jar / am Himel gesehen wird. Gedruckt zu Erffurt / Durch Johann Beck. Anno D.M. LXXVII.

Nationalbibliothek Wien 72. T. 145 (15)

Description: The volume seems to be a quarto of one signature, although the only signature marks are those of A_2 and B_3 on the recto of the second and of the third leaf respectively, and undoubtedly " B_3 " is a misprint for "A_3 ". Scheibel, 97, called the book a quarto. It was printed in German prose in 1577, but the verso of the title-page and the top of the following page contain a prayer in German verse, headed " M. Adelarius Praetorius Diener des Göttlichen Worts in Erffurt." This is the only mention of Praetorius in the book, but, no doubt, he was the author of it. The title-page contains a circular woodcut picturing three men and a comet to which one of the men is pointing. The inference from the woodcut is that the author, or printer, saw an analogy between the Star of Bethlehem and the comet of 1577. Notice of the astrological character of the book is given by the legend beneath the woodcut: " Georgius Pachymerius Historicus. 3 lib. Hist. ἀδέις κομήτης ösις ä φύσει κακόσ. Non est Cometa, quin natura sit malus." See Pachymerēs, 149 (Book 3, Chapter 23), and the description of item 12, above.

Praetorius, (or Richter) Johannes

See Thorndike, VI, 84

81 De Cometis, Qvi Antea Visi Svnt, Et De Eo, Qvi Novissime Mense Novembri Apparvit, narratio, scripta ad Amplissimvm Prvdentissimvmqve Reipvb. Noribergensis Senatvm, A Iohanne Praetorio Ioachimico, Reip. Noribergensis Astronomo, & Mathematum Professore in schola Altorfiana. Cum gratia & priuilegio Caesareae Maiestatis. Noribergae. cIↄ. Iↄ. Lxxviii. (at end: Imprimebatur Noribergae, in Officina Typographica: Catharinae Gerlachin, & Haeredum Iohannis Montani. M. D. LXXVIII.

C.D.H.; G.S.; N.Y.P.L. *KB 1578

Description: The volume is a quarto with signatures A_1 to C_4. It is 187 mm. high (C.D.H. copy, which has been cut down) and 197 mm. high (N.Y.P.L. copy). The verso of the title-page, A_1, and the recto and verso of C_4 are blank. A printer's design is on the title-page and another on the verso of C_3. The book was printed in Latin in 1578 in Nuremberg at the press of Catherine Gerlach and the heirs of Johannes Montanus. The C.D.H. copy is bound in boards with the title printed by hand in ink on a rectangular piece of paper pasted vertically on the front cover, the inside of which has the bookplate of J.L.E. Dreyer. The title-page has " 232 " written in pencil in the upper left hand corner, " 932 " written in ink along the outer edge, and " C – 1. 4." written in ink in the lower right hand corner. There is a pencil line in the outer margin on the recto of C_1

82 Narratio oder Historische erzehlung dern Cometen, so vor diser Zeit sind gesehen worden, und dann auch dessen, so jüngst im Monat November erschienen ist durch Joh. Prätorium. (at end: Gedruckt zu Nürnberg durch Kathar. Gerlachin und Joh. vom Berg Erben.)

Not located. Bassaeus, II, 303; Carl, 54; Schottenloher, IV, 378; Struve, I, 787; Weller (1857-8), 360-1; Zinner (1934), 97

Description: No date of printing was given in the book, but the work seems to have been printed in 1578. Weller called it a quarto, and Schottenloher said that it has twelve leaves.

Pribicerus, Iacobus

83 Tractatvs De Cometa Qvi Svb Finem Anni A Nato Christo 1577. conspectus est. Continens simul breuem eamque generalem expositionem de causis Cometarum, Conscriptus A Iacobo Pribicero. Novisolii. Excvsvm In Oficina Christophori Scvlteti. Anno M. D. LXXVIII.

Nationalbibliothek Wien 72. T. 145 (4)

Description: The volume is a quarto with signatures A_{i} to D_{ii}. The available photostatic copy has no scale for determining the height of the tract. The pages are unnumbered. The verso of the title-page, A_{i}, is blank. The book was printed in Latin in 1578 at Neusohl at the press of Christophor Scultetus. There is a woodcut of a comet enclosed in a rectangle on the title-page. On the recto and verso of A_{ii} and the recto of A_{iii} is the dedication by Pribicerus to the heads of Schemnitz, dated from Neusohl on the last day of January 1578. On A_{iii} and its verso is a short preface by Paul Fabricius addressed to Bartholomaeus Chrysaeus. This is followed, on the verso of A_{iii} and the recto and verso of A_{iv}, by two poems, the first by Bartholomaeus Chrysaeus, the second by Iohannes Leuchamerus, the latter being dated December 5, 1577. The text, which occupies the rest of the volume, and the two poems are printed in italics.

Philomathesius, R. P.

84 Kurtzer Ausszug aller Cometen, sò von Christi Geburth an bis dahero gesehen worden, nemlich aber, in was Jar dieselbige erschienen, und dann was auf einen jeden erfolgt sey? Aus allen fůrnembsten alten und newen Authoren, so bis aufs gegenwärtige 1578 jar von Cometen insonderheit geschrieben haben, zusammengetragen.

Not located. Houzeau: Vade-Mecum, 2764; Lalande, 107; Scheibel, 109; Struve, I, 761

Description: The volume is a quarto and was printed in Frankfort-on-the-Main in 1578. According to Scheibel it has three signatures. The title is given as it was given by Scheibel, who gave R. P. Philomathesius of Frankfort as the author, as did also Lalande. Struve gave only the initials " R.P." as author and Houzeau gave " Philomathesis ". The real name of the author is not known. Squarcialupus, on the verso of C_8 of 37(3), referred to a work by R. P. Philomathesius on the comet of 1577, undoubtedly item 84. The name " Philomathesius " has been applied to Valentin Steinmetz, as was seen in the discussion of his work in chapter V, above, but he was probably not the author of item 84. Professor Thorndike (Thorndike, VI, 184) seems to be in error in considering " R.P." and " Philomathesius " as two people.

R., I.

84a Discours de la Comete apparue à Lausanne le 8. iour de Nouẽbre 1577. fait en vers François par I. R. de Digne en Prouence.

Not located. Bassaeus, III, 43-4; Du Verdier, 243-4

Description: The title is given above as it was given by Du Verdier, who said that the book is a quarto and was printed in Lausanne by Franç. le Preux in 1578. Bassaeus gave the same format and place and date of publication and the information that the author was from Digne.

Raymundus (or Raimondo), Hannibal

See item 48

See items 14 and 15

85 Discovrs Sur la noble Comette apparue A Venise Av Mois De Nouembre 1577. plus notable, gracieuse, & beneuole que l'on ait veu de nostre temps. Auec l'arc qui la precedoit à l'heure de son apparition, contenant les grands effets d'icelle. Fait en Italien par M. Hannibal Raimondo de Veronne, & depuis traduit en langue Françoise. A Lyon, Par Iean Patrasson. M. D. LXXVIII. Auec permission.

B.N. V 21091 bis; B.M. 531. e. 28 (2)

Description: The volume is a quarto with signatures A_1 to C_{1v}. It is 161 mm. high. The pages are unnumbered. The versos of the title-page, A_1, and C_{1v} are blank. On the title-page above the place of publication is a woodcut showing six paths of a comet in relation to the horizon, that is, different points of setting. On the verso of B_{1v} is the same woodcut of a comet, clouds and stars as appears on the title-pages of items 41 and 71 and on the last page of item 104. The book was printed in French in Lyons in 1578 by Jean Patrasson and is a translation of item 86.

86 Discorso D'Annibal Raimondo Veronese, Sopra La Nobilissima Cometa, che cominciò apparire il Nouembre 1577. non mai piu veduta à ricordo de viuenti in questa nostra Regione, vna tanto nobile, & con tanta lunga, & larga coda. Indrizzato Alla Givstitia, Alla Carità, all' Abondanza, alla Sanità, & à tutti gli amanti del prossimo. In Venetia, M D LXXVII.

Crawford library; photostatic copy, C.U.L. B523.6 R14 Description: The volume is a quarto of two signatures. It is 22 cm. high. The verso of the title-page is blank. On the title-page is a woodcut with the legend "Dvcat Prvdentia Negocivm, Non Fontvna." The pages are unnumbered. The text begins with a woodcut initial. On recto of B_1 is a representation of the daily path of the comet, showing the daily differences in the place of setting and the visible space traversed. On the last page, the verso of B_4, there is a woodcut of Cassiopeia, seated on a throne. On the last page it is stated that the book was finished in Venice on the last day of November, 1577. It was printed in Venice in the same year, in Italian. Riccardi, I, 338, stated that the work was printed by Gratioso Perchacino, but there seems to be no statement to that effect in the book itself.

Rasch, Johann

87 Cometen Buech. Von dem newen Stern des 73. vnnd von den Cometen des 77. vnnd 81. Jars / auch von allen anderen Cometen vnnd newen Stern erscheinungen / geschicklicher erforschung zu vrtheilen / wie die beschreibungen derselben / biedurch zuuerstehen sey / fůr den gemainen Mann so auff frembde wort vnd art von Astronomischem circelwerck zu reden / nit geůbt / allen froñen Christen zum trost vnd dieser zeit sehr

notwendig zu wissen. Durch Johann Rasch / Burger zu Wien / Getruckt zu Mûnchen / bey Adam Berg. Mit Röm: Rey: May: Freyheit nit nachzutrucken .

B.M. C. 71. h. 14. (9); C.D.H.

Description: The tract is a quarto with signatures A_1 to H_1. It is 181 mm. high. The pages are unnumbered. The book is in German and was printed in Munich by Adam Berg. Since the dedication is dated the 6th of January 1582, it is safe to say that the book was printed in 1582 although the 1588 edition of Rasch's *Gegenpractic* listed the *Cometen buech* as printed in 1577. There are figures of three comets on the title-page. The dedication is to " Dem Durchleuchtigen / Hochgebornen Fûrsten vnnd Herren / Herren Wilhelm [5] / Pfaltzgraue bey Rhein / Hertzog in Obern vnd Nidern Bayrn / rc. meinem gnedigisten Herren." The title given above and the size of the volume were taken from the B.M. copy. On the title-page of the C.D.H. copy the words " vnd " and " circelwerck " in the phrase " auff frembde wort vnd art von Astronomischem circelwerck zu reden" are spelled "vn̄d" and " circkelwerck ". The C.D.H. copy is 186 mm. high. Otherwise the same description fits the two copies. The signatures E and F of the C.D.H. copy were interchanged in binding. The other distinguishing marks on that copy are the date " 1577 " written in pencil below the last line of printing on the title-page and two slight tears at the top of A_{ii}. The copy is bound in mottled brown boards with purple and green end papers.

Raxus (or Raxo), Franciscus Fernandez (or Fernandez Raxo y Gómez, Francisco)

88 De Cometis, Et prodigiosis eorum portentis, libri quatuor. Ad Illustrissimum. D. D. Bernardum Martinez de Bolea, Philippi, Hispaniarum Regis, secundi, vicecancellarium, atq́ue Consiliarium: Francisco Fernandez Raxo, Aragoniensi Medico, autore. Cvm Privilegio Excudebat Madriti Guillelmus Drouy, Typographus. Anno, 1579.

5 Wilhelm V.

C.D.H.

Description: The volume is an octavo with a signature of 10 leaves, the 2nd, 3rd, 4th, 5th, and 6th of which are signed ¶2, ¶3, ¶4, ¶5, ¶6, plus A_1 to M_8 plus a 12° signature N, of 12 leaves. There is no signature J. There are four blank leaves, two in front and two at the end, but they do not form part of the first or last signature. The book is 203 mm. high. The leaves are numbered from 1 to 96 from A_1 to M_8. The signature N is devoted to an index. The versos of the title-page and of the final leaf of the first signature and of M_8 are blank. The book was printed in Latin by William Drouy in Madrid in 1579. In the C.D.H. copy the date 1579 has been changed in ink to 1578. This is true also of a copy offered for sale by E. P. Goldschmidt, catalogue 45, no. 61, and catalogue 70, no. 31, and of the Morante copy (see Palau y Dulcet, III, 216). The recto of the first leaf after the title gives the approbation, in Spanish, of Doctor Heredia, dated May 22, 1578. The privilege is given, in Spanish, on the verso of that leaf and the recto of the next leaf, is signed by Antonio de Erasso and is dated May 8, 1578 from St. Martin de la Vega and is for a period of ten years. On the next three pages follow the privilege, in Spanish, for printing, and the copyright, for Aragon for the same period of time. This is dated from St. Lorenzo el Real, May 25, 1578. Then comes the preface, in Latin, which closes by outlining the book. The book is divided into four books bearing the titles " De Cometarvm Cavsis ", " De generalibus Cometarum effectibus ", " De specialibus Cometarum effectibus " and " De Cometarum iudicijs ". After the third book is a poem of twenty-six lines entitled " Carmina Cometae anni. 1577.", written in Rome by an Englishman, Richard Mater. Following this, and preceding the fourth book, is a preface to the fourth book dedicated to Bustus de Villegas and citing observations of the comet of 1577 by Jerome Mugnoz and Brother Ioannes de Victoria. It is in the fourth book that the comet of 1577 is discussed. There are a woodcut printer's insignia on the title-page, numerous woodcut initials, and, on the verso of I_5 and the recto of I_6, two woodcut astronomical diagrams, one for the eclipse in September, the other for the comet in November, 1577. The description of item 88 given by Perez Pastor, 68-9, corresponds to the C.D.H. copy, except that Perez Pastor counted eleven leaves preceding the signature A and spoke of a table of errors after the preface. He also said that this list of errors was printed separately. Thus it is an extra leaf present in the copy described by Perez Pastor but missing in the C.D.H. copy. Perez Pastor called the volume a quarto. Palau y Dulcet, III, 216, also called the work a quarto and mentioned eleven leaves before the numbered ones probably counting a table of errors. Sánchez Pérez, 111, described the work as printed in 1578, called it an octavo, and said that it had no figures. Picatoste y Rodriquez, 99-100, also gave the date of publication as 1578, called the book a quarto, and counted ten leaves before the numbered ones. He did not mention any astronomical diagrams in the text although he described the printer's insignia on the title-page. He made no mention of the poem by Richard Mater, although the omission may be due to considering the poem as part of book three. The descriptions of the book as quarto or octavo may not point to differences in the copies described but may be due to different standards of measurement. But the differences in the date of printing and in the presence or absence of figures cannot be so explained. In all probability there were two editions of the book by the same printer and closely resembling each other, one in 1578 without the astronomical diagrams and one in 1579 with the astronomical diagrams. This belief is consistent with action on the part of former owners of several of the copies of the 1579 edition in changing the date on the title-page to 1578. The C.D.H. copy is bound in limp vellum with the author's name and the title in ink on the back as well as the catalogue mark " Sc. s4. N4 ". This last mark also appears in ink on the upper right hand corner of the title-page and has blotted off onto the blank page opposite. The title-page has a cross in ink above the title, and, beneath the date of publication, changed in ink to 1578, is the following phrase written in ink "Collegii Societ. JESV Caesaraug.", which also has blotted onto the previous page. In addition to the two marks mentioned above, the page preceding the title-page has "100 pesetas" written in ink on it, no doubt the price paid for the book by some former owner. A great many phrases in the text of the book have been underlined in ink. The lower outer corners of D_8, F_7 and K_6 and the lower margin of G_7 have pieces out of them, but the defect looks more like a defect in the making of the paper than a tear in the finished product.

Reuchlin, Ernest

[89] Epistel oder Sendebrieff von des 1577. Jares nach unsers wahren Messiae Jesu Christi geburt, Catastasi, Constitution oder Witterung, und daraus erfolgenden Kranckheiten.

Not located. Bruun, II, 84; Houzeau, 4930

Description: The book was printed in Lübeck in 1577. The title is given above as it was given by Houzeau. He listed the book among general works on astrology and Bruun listed it among astrological calendars. Neither man listed it among books on comets, and it seems highly probable that the tract has nothing whatsoever to do with comets. It was included in the original bibliography on the comet of 1577 by error.

Richter,

See Praetorius, Johannes

La Rivière, R. L. de (probably Rochus le Bailly or Baillif)

90 Discours sur la signification de la comète apparue en Occident, au signe du Sagittaire, le 10 novembre.

Not located. Houzeau, 5600

Description: The volume is a quarto printed in Rennes in 1577.

Rocca, Angelo

91 Discorso Filosofico, Et Teologico Intorno Alle Comete. Oue si scuoprono molti secreti della Natura, con la dichiaratione de' Miracoli, & Portenti occorsi al Mondo. Del R. P. F. Angelo Rocca da Camerino Dott. Teologo dell' Ord. Agostiniano.

B.N. V 7952

Description: It is difficult to determine the format of this book. There are ten leaves, the first five marked with signatures to A_5, and the last five unmarked. This would indicate a 10°, but it is hard to see how such a signature would be folded for binding. Probably there were originally twelve leaves, the last two of which, being entirely blank, became separated in rebinding. Riccardi does not specify the format. The pages are unnumbered. The book is 197 mm. high. No date of publication nor printer's name nor place of printing is given. The volume was printed in Italian and probably appeared first in Venice. In all probability the book was not printed until 1578, since it says that the comet lasted until January 6, 1578 and it mentions a drought lasting until "this day", January 14th. The verso of the title-page has figures of three comets, similar to those in Dasypodius' book (see item 33). These illustrate the three "types" of comets still distinguished in the sixteenth century. Each has a starlike nucleus. One has a long narrow tail. One has nearly parallel streamers spread out to a width about equal to the diameter of the head. The third has streamers of about equal length going out in all directions from the head. The verso of the tenth leaf is blank. The book is addressed to Giovanni Pisani, "Al Clariss. S. Giovan Pisani. Fv Del Clarissimo Sign. Marco Ant."

92 Discorso filosofico, teologico intorno alle comete, ove si scuoprono molti secreti della natura con la dichiaratione de' miracoli et portenti occorsi al mondo, nel occasione del cometa del 1577.

Not located. Houzeau, 5593

Description: Houzeau called this volume a quarto and said that it was printed in Venice in 1577. If he reproduced the title correctly and if he was correct about the date and place of publication, this volume is not the same as number 91, the title of which contains no mention of the comet of 1577 and in which no date nor place of publication is indicated and which was probably not printed until 1578. The signatures of 91 do not indicate a quarto, although that term might be used by some to describe the size. Scheibel, 97, listed a *Discorso filosofico theologico delle Comete*, and assigned it to the year 1577, and said that it was printed in Venice. The book described by Riccardi, Correzioni ed Aggiunte Serię IIa, 146, is item 91, not 92. It is possible that items 91 and 92 are the same, but it seems preferable to list 92 as a separate item. Possibly Rocca's *Commentarius philosophicus ac theologicus de cometis* (see chapter VI, note 38, above), which appeared in Venice in 1577, has been confused with either item 91 or 92.

Roeslin, Helisaeus

93 Theoria Nova Coelestivm ΜΕΤΕΩΡΩΝ, In Qva Ex Plvrivm Cometarvm Phoenomenis Epilogisticῶs quaedam afferuntur, de novis tertiae cuiusdam Miraculorum Sphaerae Circulis, Polis & Axi: Svper Qvibvs Cometa Anni M. D. LXXVII. nouo motu & regularissimo ad superioribus annis conspectam Stellam, tanquam ad Cynosuram progressus, Harmoniam singularem vndiq; ad Mundi Cardines habuit, maximè verò medium Europae, & exactè Germaniae Horizontem non sine numine certo respexit. Authore Helisaeo Roeslin, Medico Tabernis Alsatiae. Argentorati. Excudebat Bernhardus Iobinus. M. D. LXXVIII.

C.D.H.; B.N. V 7960; B.M. 532. e. 12 (1)

Description: The volume is a quarto with signatures A_1 to H_5 (sic). It is 200 mm. high (C.D.H. copy). Evidently, the last signature is not in the form of a quarto. In fact, there is a blank leaf which would correspond to H_6 and is a continuation of H_1. The book was printed in Strasburg in 1578 by Bernard Iobin. The verso of the title-page is blank. There is a woodcut or engraving on the title-page representing a classical bust with the caption " Sapientia Constans ", probably the printer's insignia. The C.D.H. copy is bound in paper. It misses leaf G_4, which has been supplied by a photostat from the B.N. copy. On the title-page the number 1578 has been written in ink in Arabic numerals below the Roman numerals. Above and to the right of the Arabic numbers, " I C " has been written and crossed out in ink. The verso of the blank leaf at the end of the C.D.H. copy has a design, in ink, composed of arcs of circles.

Rothmannus, Christophor

See item 19

Sali, Valerio

93a Commento Sopra Alcvni Versi Della Cometa Dell' Anno M. D. LXXVII. Dove Anco Si Dimostra La Nobilta, E La Vera Pronontia Della Lingva Italiana. In Venetia, M. D. LXXIX.

Crawford library; photostatic copy, C.U.L. B523.6 Sa 3 Description: The volume is a quarto with signatures A_1 to D_3. Only a photostatic copy, without a scale, was available. In that copy a page is 20.3 cm. high. Photostatic copies made at the Crawford library usually retain the size of the original. The pages are numbered from 3 to 30 from the recto of A_2 to the verso of D_3. Pages 10 and 11 are numbered 8 and 5, page 14 is numbered 4 and page 15 is unnumbered. A brief note to the reader on the verso of the title-page says that the work was completed at the home of Monsignor Camillo Cautio (translating " a caso da " as though it were " a casa da ") and is signed by Valerio Sali. The text starts on page 3 and is addressed to Signor Orsato Giustiniano. The work was printed in Italian in Venice in 1579, but no printer's name was given.

Sanchez (or Sanctius), Francisco

93b Francisci Sanchez philosophi et medici doctoris, carmen de cometa anni 1577.

H. C. L. Baudrier, VIII, 374; Bassaeus, I, 372, 572, and II, 284; Lalande, 108; Ludendorff, 501-6;

Picatoste y Rodriquez, 286, # 736; Pingré, I, 100, n. (a); Scheibel, 107

Description: The title is given above as it was given by Baudrier. The volume is an octavo and was printed in Lyons in 1578 by Antonius Gryphius. If Baudrier and Bassaeus were correct, there were two tracts by Sanchez, printed in the same place and year by the same printer, one being called *Carmen de cometa anni 1577* and the other *Carmen de cometa anni 1578*. Picatoste y Rodriguez mentioned only one tract and gave it the title *Super cometa anno M. DLXXVII apparente*. Pingré, without giving a title, said that the tract was written in 1578 about the comet of the preceding year. Ludendorff, also, gave no title. He said that Sanchez had a Latin poem printed in Lyons in 1578. It is possible that there were two printings of the same poem in the same year, the first calling the comet by the date 1577 and the second calling it by the date 1578, although it is also possible that Sanchez thought of the phenomenon as two separate phenomena. Recently (1944) the Harvard library acquired a copy of this item. (See Picatoste y Rodriguez, 285-7; Smith (1917), 139; H. Friedenwald in the *Annals of Medical History*, ser. 3, v. II, n. 6: 521; G. Sarton in *Isis*, XXX (3): 437; Thorndike, VI, 560 ff.; and Iriarte-Ag.'s *Kartesischer oder Sanchezischer Zweifel?* . . . Bottrop i. W., 1935, 83 ff.)

Santutius, Antonius

94 on the comet of 1577

Not located. Implied by Claramontius (1636) and by Riccioli, II, 2 and 88, although Santutius' remarks on the comet of 1577 might be contained in a general work on comets written in the seventeenth century, i.e. if Santutius is the same as Santucci.

Saravezza, (or Seravezza, see Riccardi, pt. 1, v. II, 437) Bartolomeo

94a Il Breve Discorso Del Padre Saravezza, Sopra La Cometa Apparsa Alli XII. Di Novembre. M. D. LXXVII Detta La Scapigliata. Alla Molto Ill. Sig. Et Padrona sua osseruandiss. la Sig. Contessa Angela Bianca Christiana Beccaria.

Crawford library; photostatic copy, C.U.L. B523.6 Sa 7
Description: The volume is composed of four leaves, without signature marks, and might be called a quarto. It is 21.4 cm. high. The pages are unnumbered. On the title-page there is a woodcut representing the sun, moon and a figure, possibly Father Time, with wings. Between the title and the cut there is an Italian poem headed "Nunquam coelo spectatus impune Cometa." The poem reads "Se la Cometa giamai non fu uistae Senza portarci fame, querra, ò peste: O altri strani casi, oue fa uista, Farassi il ballo rondo senza feste." On the last page the book is signed by the author and beneath his name is the information that the work was seen by the papal inquisitor. The preface, on the verso of the title-page, is dated St. Catherine's day, 1577. According to Riccardi, pt. 1, v. II, 437, the following information is given at the end of the book, "In Piacenza, appresso Giouanni Bazzacho, & Anteo Conti compagni. 1577." The photostats from the Crawford library do not show this and Dreyer, editor, IV, 510, describing the Crawford library copy, wrote "s. l. & a., .. ". Thus there may be two printings of this work.

Schännis, Joh. Kaspar

94b Theses physicae de Cometis
Not located. Wolf, IV, 64 note 21.
Description: The book is a quarto printed in "Tig." (Zurich) in 1619, and speaks of the star of 1572 and the comets of 1577 and 1604.

Schinbain, Johann Görgen (possibly: Tibianus, Johann Georg)

[95] Sternen oder Cometen Bůch / In welchem die fürnem̃sten Cometan / deren bey 180. so hin vnd her / vor vnd nach Christi Geburt / an dem Firmament erschienen / sampt andern Meteorologicis / so sich in Läfften zůgetragen: Was auch gleich in jedem Jar besunder / für Effect oder würckung darauff gefolget: Beschriben vnd in Teutsche Rhytmos gestellet / Durch Johann Görgen Schinbain von Freyburg im Breyszgaw / diser

zeit Lateinischen vnd Catholischen / desz löblichen Römischen Reichs Statt Byberach / Schůlmaistern. Getruckt zů Ingolstatt / bey Dauid Sartorio. Anno cIↄ. Iↄ. Lxxviii.

Crawford library

Description: Only a photostatic copy of the title-page has been available. On the title-page there is a woodcut showing a town beneath a comet and the figure of Sagittarius. The Crawford library catalogue called the volume a quarto of 76 leaves, with many woodcuts and a colophon with the shins of two legs, illustrating the author's name. As is readily seen from the complete title-page, the work is not about the comet of 1577, although that comet may be mentioned. Consequently, item [95] does not belong in this bibliography.

Scultetus, Bartholemaeus

96 Cometae, Anno Hvmanitatis I.C. cIↄ. Iↄ. Lxx. Vii. à 10. Viiiibris per Xbrem in 13. Ianuarij sequentis anni, continuis Lx. & V, D. in sublunari regione adparentis, Descriptio. De illius motu Visibili et vero, adiectis cognitu dignioribus calculi, tabularum & demonstrationum ocularium fundamentis. Denique de huius meteoricae impressionis Significatione, ex praemissa descriptione, concepta. Authore Bartolemaeo Scvlteto Philomathe. (at end: Gorlicii Excudebat Ambrosius Fritsch, Anno 1578.)

B.M. 8610. b. 60

Description: The volume is a quarto with signatures A_1 to N_4. It is 182 mm. high. There is no signature J. The book was printed in Latin in Görlitz by Ambrosius Fritsch in 1578. On the title-page there is a woodcut, printed in red, of a quadrant with lines, some of them curved, drawn to a point not at the corner, and probably representing the different positions of a comet's tail. This figure also appears on the recto of F_4. The verso of the title-page is headed "Epigramma In Vrbis Gorlicii Insignia" and has a woodcut and a verse. There is an introductory chapter followed by a Latin poem to Scultetus by "Ioach: Maist: Gorl: f." (see item 73a) and another to "Magistro Bartholemaeo Suldeto [sic]" signed "Laurentius Ludouicus Leoberg:". The text starts on the recto of B_1, where there is a woodcut diagram of the comet's motion. There are numerous astronomical diagrams as well as astronomical tables, some of them full page, throughout the text, which is divided into three parts, which in turn are divided into chapters.

97 Des grossen und wunderbaren Cometen, so nach der Menschlichen Geburth Jhesu Christi, im 1577. Jahr, von dem 10. tag Novembris, durch den gantzen Decembrem, biss in den 13. Januarij des folgenden Jahrs, gantzer 65 tag, unter des Monden Sphär uber der Wolcken Region, gesehen worden: Astronomische und natürlicher Beschreibung: Von seiner sonderlichen Bedeutung und gewaltigen Wirckung, anfahend auff den negsten Augustum, vom 1578. biss uber fünff gantzer Jahr, in den eingang des 1583. wehrend. Welches Innhalt zu end der Vorreden mit seinen capp. zufinden. Durch Bartolemaeum Scultetum Gorl. Philomathem. (at end: Gedruckt zu Görlitz, durch Ambrosium Fritsch. Im Jahr, CIↃ. IↃ. LXXVIII.)

Not located. Bassaeus, II, 278; Carl, 54; Koch (1907-1910), LXIII, 79; Schottenloher, IV, 378; Struve, I, 787; Weller (1857-8), 361

Description: The title is given above as it was given by Weller. Carl and Struve gave the same title but not all of it. Bassaeus gave the title *Vom Cometen welcher Anno 1577. erschienen.* Weller said that the tract was a quarto of 52 leaves with a woodcut on the title-page. Schottenloher also called it a quarto of 52 leaves. Koch said that the German edition does not contain calculations and diagrams.

Selneccerus, Nicolaus

See item 26a

98 Nothwendiger Christlicher Bericht von dem Cometen.

Not located. Carl, 53.

Description: This item was printed in Leipzig in 1577. It may be the same book as 26a.

98a Ein Christlich Gebet in jetzigem elenden zustand / darin Gott selbs vns seine fewrige Rute / vnd seinen gerechten Zorn am Himmel zeiget / vnd vns mit allerley drawungen vnd zůchtigungen heimsuchet / zuthun / vnd sicht damit fůr Gott demůtiglich anzugeben / vnd vmb Gnad zu bitten. Der Christlichen Gemein zu Leipzig fůrgeschrieben / Durch D. Nicolaum Selneccerum, Pfarerrn vnd Superintendenten daselbst. Cum gratia & priuilegio. M. D. LXXVIII. (at end: Gedruckt zu Leipzig / Durch Jacob Berwaldts Erben.)

Nationalbibliothek Wien 72. T. 145 (9)

Description: The volume is a quarto with signatures A_i to B_{iv}. The available photostatic copy has no scale for use in determining the height of the book. The pages are unnumbered. The verso of the title-page, A_i, and the verso of B_{iv} are blank. The book contains two prayers, the second starting on the recto of B_{iii}. There is a small design on the title-page and another on the recto of B_{iv}. The book was printed in German in Leipzig in 1578 by the heirs of Jacob Berwaldt.

99 Erwartung an den Christianen nöthig über den Cometen.

Not located. Houzeau, 5603; Lalande, 104

Description: The title is given above as it was given by Houzeau, who described the book as a quarto, printed in Leipzig in 1577.

Seravezza

See Saravezza

Slovacius, Peter

99a Praktyka o komecie widzianey Roku Pańskiego 1577. dnia dziesiątego Listopada . . . przez Mistrza Piotra Słowacyusa ze Zdakowa sławney akademiey krakowskiey Astrologa.

Not located. Estreicher, XXVIII, 256 (There seems to have been (1930) a copy in Warsaw.)

Description: The volume seems to be an octavo, without date or place of printing. There are four maps or plates. The end is missing. A rough translation of the title is: " Practica of the comet seen in the year of our Lord 1577 on the tenth day of November, by Master Peter Slovacius of Zdacova, astrologer at the renowned academy at

Cracow." On the maps there are still visible remnants of a woodcut representing the comet. The lower part of the title-page is missing. On the verso there is a remnant with a coat of arms with a swan over which there are two verses to the priest, Peter Dunis, chancellor. Then follows the dedication to his excellency Peter Volsk and the chancellor K... the mighty. It is signed by Peter Slovacius under the protection of... In the preface to the reader the author said: "...It seemed to me not out of the way, as if I were the first to come upon it, that I describe how such stars emerge upon the sky. If anyone would want further information concerning it, we have published it widely in the Latin script."

99b Latin edition of 99a.

Not located. A Latin edition was implied in the preface to 99a.

Snell, Willebrord

See item [108] .

Sorboli, Girolamo (Jerome)

100 Dialogo In Materia Delle Comete, Del Sig. Girolamo Sorboli Dottore di Filosofia, et Medicina. In Ferrara, M. D. LXXVIII. Per Vittorio Baldini, Con licenza de i Superiori. (at end: In Ferrara, nella Stamperia di Vittorio Baldini. 1578.)

C.D.H.; B.N. V 7957

Description: The volume is a quarto with six leaves (B.N. copy) or four leaves (C.D.H. copy) plus A_i to H_{iv}. It is 196 mm. high (B.N. copy) or 183 mm. high (C.D.H. copy) The fifth and sixth leaves are present in the copy described by Riccardi, 467. The versos of the title-page and of the fourth leaf are blank. The pages are numbered from 1 to 58 from A_i to the verso of H_i. The leaves H_{ii}, H_{iii} and H_{iv} have an alphabetical index and, at the end, a list of printer's errors. There is a woodcut of a printer's device on the title-page and a small woodcut on the last page. The book was printed in Italian in Ferrara in 1578 by Vittorio Baldini. The dedication, to Giovanni Paolo Porti, was dated from Bagnacavallo, December 5, 1577 and was signed by

Giovanni Antonio Vandali. It occupies the second and third leaves and the recto of the fourth leaf. The fifth and sixth leaves (B.N. copy) contain verse. The text starts, with a woodcut initial, on the recto of A_i and continues to the verso of H_i. The C.D.H. copy is bound in stiff red paper, with a design of leaves in blue and yellow, and has a blank leaf before and behind. Besides the missing fifth and sixth leaves, the copy is distinguished by marginal notes in ink on the rectos of B_{ii} and F_i. There are some notes in pencil and some in ink on the inside of the front cover and on the recto of the extra blank leaf at the front. These seem to give previous catalogue numbers.

Sordi, Peter

101 Discorso Sopra Le Comete, Di Pietro Sordi. Alla Illma. Sig^{ra}. La Sig^{ra}. Barbara Sanseverina, Contessa Di Sala. In Parma, Appresso Seth Vioto. 1578. Con licenza de' Superiori.

C.D.H.; B.N. V 7968

Description: The volume is a quarto with three leaves plus A_i to P_{iv}. There is no signature J. The book is 183 mm. high (B.N. copy) or 187 mm. high (C.D.H. copy). The verso of the title-page is blank. The leaves are numbered from 1 to 60 from A_i to P_{iv}, 29 being marked 31 and 31 being marked 33. The dedication, on the second and third leaves, is to Barbara Sanseverina, was signed by the author and was dated from Parma, March 1, 1578. The text starts on the recto of A_i and continues to the verso of P_{iv}. It is divided into two parts, the second of which starts on the verso of I_{iv}. At the close of the text is a list of corrections or printer's errors. Both parts, as well as the dedication, start with woodcut initials. There is a woodcut on the title page, probably the printer's insignia. On the recto of A_{iii} and on the recto of C_{ii} there are mathematical diagrams, and on the verso of L_{iv} and the recto of M_i there are astronomical diagrams. The book was printed in Italian in Parma in 1578 by Seth Vioto. The C.D.H. copy is bound in vellum with other tracts (see item 68) and has no distinguishing feature except a water-mark on all the lower inside corners, fainter at the beginning of the book.

Squarcialupus, Marcellus

See item 37

Steinmetz, Valentinus

102 Von dem Cometen welcher im Nouember des 1577. Jars erstlich erschinen / vnd noch am Himmel zusehen ist / wie er von Abend vnd Mittag / gegen Morgen vnd Mitternacht zu / seinen fortgang gehabt / obseruirt vnd beschriben in Leiptzig. (last page: Gedruckt zu Augspurg / durch Michael Manger.)

C.D.H.

Description: The volume is a quarto with signatures A_I to C_{III}. It is 192 mm. high. The pages are unnumbered. The verso of the title-page, A_I, is blank. The book was printed in German prose in Augsburg by Michael Manger, probably in 1577 since the observations of the comet do not extend beyond December 7th and the dedication was dated December 10, 1577. The book was dedicated to Valentin Meder, the author's lord and relative. The dedication was signed "E.U.W. Vetter M. Valen. Steinmetz / Gersbach:". Latin phrases used in the book were printed in Roman letters. On the title-page is a woodcut enclosed in a rectangle the sides of which read "Mitternacht.", "Auffgang.", "Nidergang." and "Mittag." The cut represents a group of houses, including a church, on one end of which is a cross. The church has two steeples. Above these houses is the upper part of a celestial sphere showing the path of the comet through the constellations, which are pictured on the sphere. The C.D.H. copy is bound in boards with "Augsburg ca. 1578 M. Steinmetz Von den Cometen" written in black ink in printed letters, vertically, from bottom to top on the back. It is a clean copy and has no distinguishing marks except a smudge, probably from a thumb, in the upper left hand corner of the last page.

103 Von dem Cometen welcher im Nouember des 1577. Jars erstlich erschienen / vnd noch am Himmel zusehen ist / wie er von Abend vnd Mittag / gegen Morgen vnd

Mitternacht zu / seinen fortgang gehabt / obseruirt vnd beschrieben in Leipzig. (last page: Bedruckt zu Leiptzig / bey Nickel Nerlich Formschneider.)

C.D.H.; N.Y.P.L. *KB 1577; B.M. 8560.cc.1

Description: The volume is a quarto with signatures A_i to C_{iiii}. It is 189 mm. high (C.D.H. copy). The pages are unnumbered. The verso of the title-page, A_i, is blank. The book represents a different edition of 102 and was printed in Leipzig by Nickel Nerlich Formschneider, probably in 1577. The dedication, to Valentin Meder, was signed " E.U.W. Vetter M. Valen. Steinmetz / Gersbach.", and dated December 10, 1577. Latin phrases used in the book were printed in italicized Roman letters. The figure on the title-page is substantially the same as that of the Augsburg edition but all the lines in the figures do not represent a one to one correspondence, and the word "Auffgang" in the Augsburg edition was spelled "Auffgeng" in the Leipzig edition. This latter is the edition cited by Dreyer, editor, IV, 510. The C.D.H. copy is bound in an old vellum music manuscript. The upper right hand corner of the title-page has been patched, from the verso, by a rectangular piece of paper approximately 60 by 75 mm. in size. The edges of the book are water-marked and there are a few marginal notes in pencil on the recto and verso of C_{iii} and on the verso of C_{iiii}.

103a Von dem Cometen welcher in Nouember des 1577. Jars erstlich erschienen / vnd noch am Himmel zusehen ist / wie er von abend vnd Mittag / gegen Morgen vnd Mitternacht zu / seinen fortgang gehabt / obseruirt vnd beschrieben in Leipzig. (last page: Gedruckt zu Magdeburg / durch Joachim Walden.)

Crawford library; photostatic copy, C.U.L. B523.6 St3

Description: The volume is a quarto with signatures A_i to C_{iv}. Only a photostatic copy, without a scale, has been available in preparing this description. The pages are unnumbered. The verso of the title-page, A_i, is blank. There is a figure on the title-page similar to the figure on the title-pages of 102 and 103, but in 103a it does not have the directions " Mitternacht ", "Auffgang " etc. on the outer border. In 103a the church also has two steeples, but a cross is on the lower one. Most Latin phrases used in the book were printed in Roman letters.

Except for differences in spelling, punctuation, capitalization and paragraphing and a few differences which will be pointed out, the texts of 102, 103 and 103a are the same. The text was divided into more paragraphs in the Augsburg edition than in the Leipzig and Magdeburg editions, the paragraphing, with one exception (C_{I}) being the same in the latter two. Also, except for the recto of B_{III} and the verso of C_{II} these two are identical in division into pages. "Für" in the Augsburg edition is "vor" in the Leipzig and Magdeburg editions. In the Augsburg edition a verb is given completely, whereas in the others the auxilliary verb is omitted. In a few instances the word order in the editions is different and sometimes a number written out in one or two editions is given by numerals in the other one or two. Once the adjective "schöner" is used in the Leipzig and Magdeburg editions and omitted in the other. Once "Amen" appears in the Augsburg edition after the expression of a wish from God, but is omitted from the Leipzig and Magdeburg editions. Twice phrases are included in the Augsburg and Magdeburg editions but not in the other. In the first case the Augsburg edition (B_{I}) reads as follows (the italicized words do not appear in the Leipzig edition (B_{I}) but do in the Magdeburg edition): "... da ich wider auff den Abend vmb 5. vhr / die höhe oder erhebung des Cometen supra Horizontem occidentalem bey 43. Grad / *20. Minut. gefunden / vnd ist also in den 3. tagen fast bey 2. Grad* höher gegen Mitternacht zu kom̃en / der mittel Stern aber im Adeler ist eben vmb dieselbige zeit auff 30. Grad vnd 20. Minuten vber der Erden gestanden: ..." In the second instance, the Augsburg edition (B_{II}), speaking of Camerarius' book, reads as follows (the italicized words do not appear in the Leipzig edition (B_{II}) but do appear in the Magdeburg edition): "... vnnd vor kurtzer zeit solche verdeutscht ist worden / Damit aber ein jeder / auch gemeiner Mann möge wissen / vnd selbst lesen /

was für vnglück / jammer / elend / vnnd beschwerligkeit nach Cometen allzeit *erfolget seind / habe ich guter wolmeinunge fürnemer Cometen* / so von Christi vnsers einigen Erlösers vnd Seligmachers Geburt her / gesehen worden / hiebey setzen / . . ." In one instance (A_{iv}) a long phrase was omitted from the Magdeburg edition, the complete phrase reading as follows in the other two editions (copied from the Leipzig edition, the italicized words not appearing in the Magdeburg edition): ". . . vnd wie die obseruationes in den Instrumenten zeigten / war er im letzten grad des Steinbocks damals / *gegen mitternacht zu aber vmb 4. grad höher in den 3. tagen gestiegen / vñ haben* gegen mitternacht zu auff der seiten gleich gegen vber gestanden / . . ." Unfortunately, these differences do not show which edition was printed first.

On A_{iii} of the Magdeburg edition the phrase "des zeichen" was omitted although it appears in the other editions. Similarly the word "vhr" was omitted on the verso of A_{iii} of the Magdeburg edition although it was used in the other two. A_{iv} of the Magdeburg edition has "siebendehald" and "ein vnd viertzigste halbe" in place of "6½" and "40½" of the other editions, and the verso of A_{iv} similarly has "neundehalb" in place of "8½" and "andern" in place of "2." Also on the same page the Magdeburg edition has "75" in place of "57" of the other editions, which latter must be correct since the numbers relate to minutes. The verso of A_{iv} of the Leipzig edition has "1½" in place of "anderthalb" and "anderhalb" of the Magdeburg and Augsburg editions respectively. On the verso of B_{i} the Magdeburg edition has "14." in place of "24" in the other editions. B_{ii} of the Magdeburg edition has "auff den 1."' in place of "auff 1." of the other two editions. B_{iii} of the Leipzig edition reads "dreymal" where the other editions read "zweymal", and gives the date "703" in place of "603" as do the other editions, which are probably right. The verso of B_{iii} of the Magdeburg edition omits "wider", which is given in both the other editions, and on the same page gives the date "745" in place of "761" which is given, probably correctly, in the other editions. Similarly on the following page the year "557" was given in the Magdeburg edition, in place of "945" of the other two. The Magdeburg edition, on the verso of B_{iv}, used "worden" in place of "ist worden" of the other editions and on the following page "ben" in place of their "haben". On C_{iii} of the Magdeburg edition "allen" used in both the other editions is omitted and on the following page the Greek verse by Camerarius is omitted although it appears in the other editions.

Šubar Landškrounský, Valentin

103b Kázanij o Hrozné Kométě / kteráž se začala kazovati ten den po Swatém Martině s wečera / Léta Páně tč. LXXVII. Včiněné w Neděli XXIIII po Swaté Trogicy téhož Léta tč od Kněze Walentina Ssubara Landtsskronského / toho času Kazatele Slowa Božijho w Domažlicých. Wytisstěno w Starém Městě Pražském v Giřijho Melantrycha z Awentýnu: a M. Danyele Adama Pražského. M. D. LXXVIII°.

Not consulted. Bibliothèque Nationale et de Université, Prague 54 B 132, Tresor E 173, adl 2. The title and its German translation were furnished by Professor Quido Vetter.

German translation of title: Predigt von dem schrecklichen Komet, welcher sich zu zeigen begann den Tag nach St. Martin abends, des Jahres Gottes 77 d. Z., gemacht den 24. Sonntag nach der Hl. Dreifaltigkeit desselben Jahres d. Z. von dem Priester Walentin Šubar von Landskron, der Zeit Prediger des Wortes Gottes in Domažlice. Gedruckt in der Altstadt von Prag bei Georg Melantrych von Aventin und Mg. Adam von Prag, 1578.

Sueuus, Sigismundus (or Suevus, Siegmund)

103c Cometen, was sie fůr grosse Wunder vnd schreckliche Ding zu bedeuten, vnd anzukůndigen pflegen. Mit viel gedenckwirdigen Historien und Exempeln erkleret. Aus guten Chroniken vnd andern Büchern mit Fleiss zusammen getragen vnd mennigklichē zur trewen Warnung vnd ernsten Busspredigt fůrgestellet. Durch Sigismundum Sueuum Freistadiensem, Prediger zum Lauben. Oremus, emendemus, & vincemus. (at end: Gedruckt zu Görlitz, bey Ambrosio Fritsch. 1578)

Not located. Jöcher, IV, 931; Scheibel, 108

Description: The title is given as it was given by Scheibel, who said that the volume is an octavo of thirteen signatures. It was printed in Görlitz in 1578 by Ambrosius Fritsch. Jöcher said that the above item was included in the collected works of Suevus entitled *Spiegel des menschlichen Lebens* which was published in Breslau and in 1588 in Leipzig.

T., B. I.

See item 79b

T., L.

103d Vom Cometen, So jtzund in Latitudine Meridionali ascendente etc. zu einem zeichen Göttliches zorns etc. in Druck vorfertiget. L. T.

Not located. Weller (1857-8), 215

Description: Weller said that no place of publication was given but that the work was printed in 1577. He thought that the author might possibly be Thurneysser. Weller called the work a quarto. There remains the possibility that Weller or his source of information, F. Heerdegen's catalogue 228, was in error when citing this work and that the work in question is identical with item 3 of this bibliography.

Thurneysser, Leonard

[104] Discorso Natvrale Sopra xiij. Nouilunij dell' anno M. D. LXXXI. Calcvlato Per L'Eccell. Dottore, il Sig. Leonardo Thurneijssero Astrologo, & Medico del Brandeburg Elettor dell' Imperio. Tradotto di lingua Tedesca nella nostra Italiana. Con vn Trattato delle Comete. Cō Priuilegio Della Ces. Maestà. (at end: In Ferrara Per Vittorio Baldini, Con licenza de' Superiori.)

B.N. V 1210 Inv. Reserve.

Description: The tract is a quarto of four leaves. It is 195 mm. high. It was printed in Italian in Ferrara by Vittorio Baldini, probably in 1580, when a prognostication for 1581 would have been in demand. The title-page has a decorative frame and also a woodcut with the legend "Ytambien Meqvema" over the picture of a log fire. On the verso of the title-page is a figure showing the signs of the zodiac. The last leaf has a poem in Italian with the caption "Mario Napolitano, Ai Lettori", and a woodcut of a comet, clouds and stars, like the cut on the title-pages of items 41 and 71 and on the verso of B_4 of item 85. Item [104] does not deal with the comet of 1577, although part of the work may be a translation of part of item 105, no copy of which has been located.

105 Ein kurtzer und einfältiger Bericht Leonhart Thurneyssers zum Thurns, Churfl. Brandenb. bestallten Leib - Medici, ůber den 136 und in diesem lauffenden 77 Jar am 19 Octobris erstlich erschienenen Cometen, aller Welt zum Dienst und getreuer Warnung publicirt.

Not located. Carl, 53; Lalande, 104; Scheibel, 97; Schottenloher, IV, 378; Weller (1857-8), 323; Wolf, III, 32-3, note 59; Wolf (1877), 408, note

Description: The book was printed in Berlin by Michael Hentzsken in German in 1577. The title is given above as it was given by Scheibel, who called the work a quarto of eight signatures and said that it appeared in 1577. Weller said that there were four folio leaves and two celestial maps. Bassaeus, II, 308, listed the tract with the title *Leonhart Thurneyssers Bericht vber den 136. vnd in dem verloffenen 1577. Jahr am 19 Tag Octobris erschienen Cometen sampt derselben Tafeln vnnd Demonstration* and said that it was printed in Berlin in 1578. Bassaeus may have been in error or he may have been listing a second edition. Zedler, XLIII, 2009, listed a *Tractat, de Cometa, 1577* but the reference is too vague to be of use here.

T., T. (probably Thomas Twyne)

105a A View Of certain wonderful effects, of late dayes come to passe: and now newly conferred with the presignyfications of the Comete, or blasing Star, which appered in the Southwest vpō the. X. day of Nouem. the yere last past. 1577. Written by T. T. this. 28. of-Nouember. 1578. (at end: Finis. T.T. Imprinted at London by

Richarde Jhones, and are to be sould ouer against Saint Sepulchres Church without Newgate .1. Decem. 1578.)

B.M. 1395.c.3; photostatic copy, C.U.L. B523.6 T94

Description: The volume is a quarto with signatures A_i to C_{iv}. It is 17 cm. high, but the upper margin has certainly been cut, and the lower one probably has been cut. The pages are unnumbered. The verso of the title-page is blank. On the title-page there is a woodcut representing a comet. The recto and verso of A_{ii} contain the dedication, which is to Giles Lambert and is signed " T. T."

Vigenère, Blaise de

106 [and 107] Traicté Des Cometes, Ov Estoilles Chevelves, Apparoissantes Extraordinairement au ciel : Auec leurs causes & effects. Par Bl. de Vige^re^. A Paris, Chez Nicolas Chesneau, rue sainct Iacques, au Chesne verd. M. D. LXXVIII. Avec Privilege Dv Roy.

C.D.H. and B.N. V_8 1636, V 1660, V 21089

Description: The volume is a quarto with signatures A_i to Y_{ii}, there being no signatures J, U, or W. It is 158 mm. high (C.D.H. copy). The pages are numbered 2 to 171, the title-page and the verso of Y_{ii} being unnumbered. The book was printed in French prose in 1578. It ends on page 171 with a quotation, in French, from Seneca. The verso of page 171 contains an excerpt from the permission granted by the King, mentioning the title of the book, and dated from Paris, December 10, 1577. The title-page has a woodcut, probably the printer's device. The verso of the title-page, numbered 2, has a woodcut showing an eagle grasping a star in its beak, with streamers coming from the star, giving the semblance of a comet, the eagle at the same time holding fire and lightning with its claws. Alongside the eagle is an inverted arrow with four stars along its shaft and beams coming out of the upper part. Beneath the woodcut is a Latin poem signed by Io. Auratus, poet of the King. The poem represents the eagle telling that he is the minister of God, with fire at the head and feet, the former to warn humans, the latter to punish them if they do not mend their ways. Jo. Auratus is the Latin form of the name Jean Dorat, under which the poet was better known.[6]

The text begins on page 3. The B.N. copy V_8 1636 has two plates between pages 2 and 3. They are entitled as follows: I " Pleiades " and II " a) Cingvli et Ensisorionis Asterismus b) Nebvlosa Orionis c) Nebvlosa Praeslp:" They do not occur in the other three copies noted. The C. D. H. copy has an extra title-page and an extra leaf, pages 7 and 8, like an extra sheet folded around signature A. The extra pages 7 and 8 are like the true ones with the exception of the use of abbreviations, the division into lines, and the position on the page of the printed marginal notes. However, the title-pages are different, the extra one being earlier. It, too, contains the woodcut of the printer's device, but differently proportioned. The extra title reads " Traicté Des Cometes, Ov Excroissance De Lvmiere Apparoissant és estoilles. Par Bl. de Vig[re]. A Paris, Chez Nicolas Chesneau, rue sainct Iacques, au Chesne verd. M. D. LXXVII. Avec Privilege Dv Roy." The verso of the extra title-page carries a woodcut similar to the one on the verso of the 1578 title-page excepting that there is no arrow next to the eagle. Beneath the woodcut are lines in French, with a cadence, but not rimed. They are the translation of the Latin verse on the verso of the 1578 title-page, but they are unsigned.

The 1577 title-page might be given a separate number, 107, as was done in the original bibliography (Hellman), but it is doubtful that a book will ever be found to go with it. It seems that two title-pages were printed for the same book which, when issued, bore the 1578 title.

The C.D.H. copy is bound in vellum with Claude de Seissel's *Histoire Singvliere Dv Roy Loys xij* (Paris, 1558) and his *La Grand' Monarchie De France* ... *Auec la loy Salicque* . . . (Paris, 1557).

William IV, Landgrave of Hesse Cassel

See item [19]

6 Dorat was probably born in 1508. He died November 1, 1588. He was a member of the Pléiade. See Dorat.

[108] Coeli & siderum in eo errantium Observationes Hassiacae, Illustrissimi Principis Wilhelmi Hassiae Lantgravii auspicijs quondam institutae. Et Spicilegium biennale Ex Observationibvs Bohemicis V. N. Tychonis Brahe. Nunc primum publicante Willebrordo Snellio. R. F. Quibus accesserunt, Ioannis Regiomontani & Bernardi Walteri Observationes Noribergicae. Lvgdvni Batavorvm. Apud Iustvm Colstervm, Anno cIↃ IↃ CXVIII.

C.D.H.; C.U.L. 522.19 Sn 2 (missing leaves 65 to end) Description: The volume[7] is divided into two parts. Part I is a quarto of six leaves plus signatures a_1 to p_2 (without a signature j). The pages are numbered from 1 to 116 from the recto of a_1 to the end. Part II is a quarto with signatures A_1 to R_4 (without a signature J). The leaves are numbered from 1 to 68. The volume is 203 mm. high (C.D.H. copy, the C.U.L. copy having been cut). On the title-page there is printed a wreath enclosing the inscription " Homo ad immortalium cognitionem nimis est mortalis." The work contains several diagrams and numerous tables. The B.M. catalogue included the words " partim ab ipsomet . . . partim ab ipsius mathematicis " in the title, but those words seem to have been taken from page 1, the seventh leaf, where, however, the " m " of " mathematicis " is capitalized. The book was printed in Leyden in 1618 by Iustus Colsterus and was edited by Willebrord Snell. The verso of the title-page has a poem in Greek with the caption "Alexis." The preface dedicates the book to Mauritius, Landgrave of Hesse, the son of William IV. It is followed by a poem to William by Peter Cunaeus. The observations begin on page 1, starting with solar observations by William IV from 1561 to 1582, signed by both William and Rothmann, who, however, was not present at the early observations. On page 15 there begins a section of planet observations by Bürgi, made with the Cassel sextant, from 1590 to 1597. On page 69 begin selected Bohemian observations by Tycho, beginning in 1599, Gregorian calendar, and continuing into 1601, the year of his death,

7 Delambre (1819), 335 ff., analyzed this book quite carefully, but the analysis given below was made independently.

the observations of the last year being from Prague. These are followed by an eight-page explanatory notice by Snell and observations of the zodiacal obliquity, from Eratosthenes to Copernicus, and an explanation of them. There are also observations for a short interval when the solar quadrant pointed north of the ecliptic, and further solar observations by Bürgi and Tycho. The second part of the book, too, deals with solar observations, those of Regiomontanus and Bernard Walther. These are followed by an article by Johann Schöner on the construction and use of a rectangle or astronomical radius, and a section of observations by Regiomontanus, Peurbach, Walther and others of eclipses, comets, planets and fixed observations [stars], mentioning the comet of 1472, a section on observations by Walther from 1474 to 1504 and another technical section on observing by Schöner, a short note to a folio of Walther's observations, and a section with sixteen problems connected with observing a comet, by Regiomontanus, which section had been published separately in 1531 (see chapter II, above). The book has nothing to do with the comet of 1577 and was erroneously included in the original bibliography. The Landgrave seems never to have published a tract or treatise on the comet of 1577. A clew to a published work by the Landgrave on the comet of 1577 is contained in a foot-note in Motley, 343. Motley's text mentions the apprehension felt by the Landgrave because of the appearance of the comet, and the foot-note reads "'Summa, der comett und die grosse *prodigia* so diesz jahr gesehenn wordenn wollen ihre wirckung haben. Gott gebe dasz sie zu eynem guten ende lauffen.' Archives et Corresp., vi. 140—Compare Strada, ix. 463." The reference from Strada, (Strada, I, 334), provided that the corresponding citation was found in the available edition, merely mentions the comet as a sign of war and gives the dates of the comet's visibility. The available edition of volume VI of the *Archives ou Correspondance* was printed in Leyden in 1839 (Archives, VI). Page 140 gives part of a letter written before the appearance of the comet. However, on page 256 in a letter from the Landgrave to Count John of

Nassau written on November 29, 1577, the following sentence occurs, " Der grosse ungeheure Comet (1) so izo stehet und bey menschen gedencken keiner so grosz erschienen, bedeut fürwahr etwas grosz, darumb ist sich zu bessern undt zu Gott zu bekehren die höchste zeitt." The reference, (1), is to a passage in Archives, V, 34-5, which again accents the Landgrave's astrological and religious point of view, speaks of the nova of 1572, and, in a note, refers to Rommel. In Archives, VI, 268-9, in another letter from the Landgrave to Count John, dated December 18th of the same year, the comet is again spoken of as a terrible omen. The foot-note in Motley may be giving the title of a work or it may be quoting from an unidentified source.

The C.D.H. copy of item [108] is bound in vellum together with Snell's *Descriptio Cometae, qui anno 1618 mense Novembri primùm effulsit* ... (1619). The title page of item [108] has the not entirely legible signature, in red ink, of a previous owner, dated from Leipzig in April 1641.

Wincklerus, Nicolaus

109 Cometa Pogonias, Qvi Anno Labente 1577. Mense Novembri Et Decembri Apparvit, Demonstratus vnà cum parallaxi, distantia à Centro terrae, & significatione eius. Avtore D. Nicolao Vuincklero, Halae Sueuorum Physico. Cum gratia & Priuilegio Caesareae Maiestatis, ad annos Sex. Noribergae. cIↄ. Iↄ. Lxxviii. (at end: Excudebatur Noribergae, in Officina Typographica Catharinae Gerlachin, & Haeredum Iohannis Montani. M. D. LXXVIII.)

C.D.H.; B.M. 532. e. 60

Description: The volume is a quarto with signatures A_1 to C_4. It is 197 mm. high (C.D.H. copy). The pages are unnumbered. The verso of C_4 is blank. On the title-page is a woodcut showing a circular astronomical diagram labeled, at the center, " Cometa apparuit Anno dñj. M: D: LXXVII: Die XII: Nouemb: Hora V. Min: XXV. P.M. Halae. Súae: úorum ". The same figure appears on the recto of B_3. On the verso of the title-page is a Latin verse by Joannes Stechmannus. A_2 and its verso contain the dedication which begins " Reverendo, Genere Et Virtvte Nobili Viro, Erasmo Nevvstettero, Dicto Stvrmer, A Schönfeld, Bambergensis ac Vvürtzburgensis Ecclesiarum cathedralium Canonico D. D. Ioannis Baptistę in Haugis & Gangolffi ibidem praeposito, & Decano Collegij Comburgensis dignissimo, Domino, omni reuerentia obseruando, S. D.". This was dated from Schwäbisch-Hall, January 1, 1578 and signed by Winckler. The text, which is divided into three chapters, starts on the recto of A_3 and continues to the verso of C_3. Each chapter, as well as the dedication, begins with a woodcut initial. On the versos of A_4 and B_1 are two astronomical diagrams showing the comet's position. The recto of C_4 gives the place and date of printing and has a triangular woodcut design. The book was printed in Latin in Nuremberg in 1578 by Catherine Gerlach and the heirs of Joannes Montanus. Concerning Winckler's middle name, Eberhard, see Thorndike, VI, 133.

The C.D.H. copy is bound in boards, with the title in ink on a piece of paper pasted on the front cover. The inside of the front cover has the book-plate of J.L.E. Dreyer. The title-page has several numbers and letters written on it in red ink and several in pencil. In the third chapter several phrases have been underlined in ink and ink lines have been drawn vertically in the margins.

Woldstedt, Fridericus

110 De Gradu Praecisionis Positionum Cometae Anni Millesimi Quingentesimi Septuagesimi Septimi A Celeberrimo Tychone Brahe Per Distantias A Stellis Fixis Mensuratas Determinatarum Et De Fide Elementorum Orbitae Quae Ex Illis Positionibus Deduci Possunt, Specimen Academicum Quod Venia Amplissimi Ordinis Philosophorum Ad Imp. Alexand. In Fennia Univ. Praeside Nathan. Gerhardo Af Schultén Phil. Doctore, Mathem. Professore Publ. et Ord., Ordd. Imp. Reg. de S:to Stanislao in secunda et Imp. de S:to Wladimiro in quarta Classe Equite, &c. Pro Gradu Philosophiae Doctoris Obtinendo Publico Examini Modeste Subjicit Fridericus Woldstedt Philosophiae Licentiatus In Audi-

torio Philos. Die VIII Junii MDCCCXLIV. H. A. M. S. Helsingforsiae Ex Officina Typographica Frenckelliana.

C.U.L. 523.6 Z 17 Q (3)

Description: This tract consists of a title-page, its blank verso, 15 numbered pages, and the blank verso of the fifteenth. The book is 252 mm. high. There are numerous tables. The book was printed in Helsingfors at the Frenckel press in 1844.

Zeissius (or Zeisius or Zeise or Zeysius), Matthaeus

111 Beschreibung der Cometen, besonders dess von 1577.

Not located. Bassaeus, II, 310; Carl, 53; Gesner (1583), 593; Lalande, 107; Scheibel, 109; Weller (1857-8), 361.

Description: The title is given above as it was given by Carl. The book was printed in Frankfort-on-the-Oder in 1578, although Bassaeus gave the date 1577. Rasch, on the recto of A_{iii} of the 1584 edition of his *Gegenpractic* (see chapter VII, n. 5, above) listed, as number 19 of a list of prognostications, a *Beschreibung vnd erklärung der schröcklichen / vngewöhndlichen / harechtigen / feurigen Sternen / so man Cometen nennet / von ihren vrsachen / bedeuttung / vnd wirckung / durch M. Mattheum Zeysium,* printed in Frankfort in 1578, which is undoubtedly item 111. Rasch listed the work again in the 1588 edition of his *Gegenpractic*, there specifying Frankfort-on-the-Oder as the place of printing.

BIBLIOGRAPHY OF REFERENCES

This bibliography includes all references cited in the footnotes and in the appendix to this dissertation, and gives the abbreviations by which they are cited. Also included are the names of libraries from which copies of works listed in the appendix have been obtained. Separate entries have been made of the abbreviated forms used in citing short articles in this list.

A.D.B.

Allgemeine deutsche biographie; herausgegeben durch die Historische commission bei der K. [Beyer.] akademie der wissenschaften. Leipzig, Duncker & Humblot, 1875-1912. 56 v.

Aa

Aa, Abraham Jacob van der. Biographisch woordenboek der Nederlanden,... Haarlem, Brederode, 1852-78. 12 pts.

Abraham

Fracastoro, Girolamo. Fracastor Syphilis; or, The French disease; a poem in Latin hexameters... with a translation, notes, and appendix by Heneage Wynne-Finch... and an introduction by James Johnston Abraham . . . London, W. Heinemann medical books ltd., 1935. (Cited here only for the introduction, pp. 1-39.)

Adam (1615)

Adam, Melchior. Vitae Germanorvm svperiori, et qvod excvrrit, secvlo philosophicis et hvmanoribvs literis clarorum; sive, Literati in Germania seculi. Volumen I. . . . Frankfort, Hoffmann for Rosa, 1615.

Adam (1653)

Adam, Melchior. Vitae germanorum theologorum, qvi superiori secvlo ecclesiam Christi voce scriptisqve propagarunt et propugnarunt, congestae & ad annum usque cIↄ Iↄ c XVIII. deductae . . . Frankfort, for the widow of Rosa, 1653.

Adam (1705)

Adam, Melchior. Dignorum laude virorum, quos musa vetat mori, immortalitas, seu Vitae theologorum, jure-consultorum, & politicorum, medicorum, atque philosophorum, maximam partem germanorum, nonnullam quoque exterorum, . . . pluribus olim minoris formae tomis congestae cincinnataeque; . . . 3rd edition, Frankfort-on-the-Main, J. M. à Sande, 1705.

Albategni

Albategni. Al-Battanī sive Albatenii opus astronomicum. Ad fidem codicis escurialensis arabice editum Latine versum, adnotationibus instructum a Carolo Alphonso Nallino. Milan, 1903, 1907, 1899. 3 v. (Pubblicazioni del reale osservatorio di Brera in Milano. XL, pts. 1-3)

Albertus Magnus

Albertus Magnus. B. Alberti Magni . . . Opera omnia, ex editione Lugdunensi . . . castigata . . . etiam revisa et locupletata cura ac labore Augusti Borgnet. v. 4. Paris, Ludovicus Vivès, 1890.

Alfraganus (1669)

Alfraganus. . . . [Arabic title] Muhammedis Fil. Ketiri Ferganensis, qui vulgo Alfraganvs dicitur, elementa astronomica, Arabicè & Latine. . . . Amsterdam, Jansonius à Waasberge and Weyerstraet, 1669.

Alfraganus (1910)

Alfraganus. Il 'Libro dell' aggregazione delle stelle' . . . pubblicato con introduzione e note da Romeo Campani. Città di Castello, 1910. (Collezione di opuscoli Danteschi, v. 87-90)

Allen

Allen, Don Cameron. The star-crossed renaissance; the quarrel about astrology and its influence in England. Durham, North Carolina, Duke university press, 1941.

Allgemeine deutsche biographie . . . See A.D.B.

Allgemeines gelehrten-lexicon . . . See Jöcher

Andreas, Valerius. See Bib. Belg. Valerii Andreae

Angelus

Angelus, Jacobus, of Ulm. Tractatus de cometis. [Memmingen in Bavaria, ca. 1490]. (3 signatures, A and C being 3 sheets, each, folded in 2, B being 2 sheets folded in 2. Cornell university copy used)

Angiolgabriello

Angiolgabriello di Santa Maria [Calvi Paolo]. Biblioteca e storia di quei scrittori cosi delle città come dei territorio di Vicenza che pervennero fino ad ora a notizia . . . Vicenza, 1772-82. 6 v.

Annalen

Annalen der physik . . . 1st series, ed. by Ludwig Wilhelm Gilbert . . . v. 5 : 112, Halle, 1800. A "Nachricht" dealing with Scultetus' *Prognosticon Meteorographicum perpetuum.*

Annals of medical history, series 3, v. 2, New York, P. B. Hoeber, 1940.

Antoniadi

Antoniadi, Eugène Michel. L'astronomie égyptienne depuis les temps les plus reculés jusqu'à la fin de l'epoque alexandrine; . . . Paris, Gauthier-Villars, 1934.

Arago

Arago, François. Oeuvres complètes . . . publiées . . . sous la direction de M. J.-A. Barral . . . v. 3, Paris, Gide and J. Baudry, Leipzig, T. O. Weidel, 1855.

Arber

Arber, Edward, editor. A transcript of the registers of the Company of stationers of London; 1554-1640. . . . London, Privately printed, 1875-1877; Birmingham, 1894. 5 v.

Archives

[Groen van Prinsterer, Guillaume]. Archives ou correspondance inédite de la maison d'Orange-Nassau. . . . Series 1, Leiden, Luchtmans, 1835-1847, 8 v. and table and supplement. Series 2, Utrecht, Kemink, 1857-1861.

Aristotle (1914)

Aristotle. De Mundo [translated] by E. S. Forster, 1914. (In Works of Aristotle, translated into English under the editorship of J. A. Smith and W. D. Ross, v. 3. Oxford, Clarendon Press).

Aristotle (1923)

Aristotle. Meteorologica [translated] by E. W. Webster, 1923. (In Works of Aristotle, translated into English under the editorship of J. A. Smith and W. D. Ross. v. 3. Oxford, Clarendon Press).

L'Art Ancien

L'Art Ancien S.A., bookshop of Zurich, Switzerland. Catalogues.

Asclepiodotus

Asclepiodotus. The tactics tr. by C. H. and W. A. Oldfather. (In Aenaeus Tacitus, Asclepiodotus, Onasander; with an English translation by members of the Illinois Greek club. London, Heinemann, 1923: 227-340. [Loeb classical library]).

Aulus Gellius. See Gellius.

B.M.

British museum. Dept. of printed books.

B.M. catalogue

British museum. Dept. of printed books . . . Catalogue of printed books in the library of the British museum . . . London, Clowes, 1881-1900. 83 v. and Supplements, London, Clowes, 1900-1905. 10 v.

B.N.

Paris. Bibliothèque nationale.

B.N. catalogue

Paris. Bibliothèque nationale. Département des imprimés. Catalogue général des livres imprimés . . . Auteurs . . . v. 1-150.[1] Paris, Impr. nat., 1897-1938.

Bacon (1897)

Bacon, Roger. The 'Opus Majus' of Roger Bacon, edited with introduction and analytical table by John Henry Bridges. v. 1, Oxford, Clarendon press, 1897. (The "Introduction" was published separately. See Bridges.)

[1] v. 150 was the last volume used in the preparation of this dissertation.

Bacon (1920)

Bacon, Roger. Secretum secretorum cum glossis et notulis, tractatus brevis et utilis ad declarandum quedam obscure dicta Fratris Rogeri. Oxford, Clarendon press, 1920. (Opera hactenus inedita Rogeri Baconi . . . Edidit Robert Steele, fasc. 5)

Bailly

Bailly, Jean Sylvain. Histoire de l'astronomie moderne depuis la fondation de l'école d'Alexandrie, jusqu'à l'époque de M.D.CC.XXX . . . new ed. . . . Paris, De Bure, 1779-1785. 3 v.

Barber

Barber, G[odfrey] L[ouis]. The historian Ephorus. Cambridge, University press, 1935.

Barrettus

[Barrettus, Lucius, ed., (pseudonym of Albertus Curtius)]. Historia Coelestis. Ex libris commentariis manuscriptis observationum vicennalium viri generosi Tichonis Brahe Dani. . . . 2 pts. Augsburg, Simon Utzschneider, 1666.

Bassaeus

Bassaeus, Nicolaus. Collectio in vnvm corpvs, omnivm librorvm Hebraeorvm, Graecorvm, Latinorvm necnon Germanice, Italice, Gallicè, & Hispanicè scriptorum, qui in nundinis Francofurtensibus ab anno 1564 . . . ad . . . 1592 . . . venales extiterunt; . . . Frankfort-on-the-Main, Bassaeus, 1592, 3 v.

Bate

Bate, Henri, de Malines. Speculum divinorum et quorundam naturalium (étude critique et texte inédit) par G. Wallerand . . . Louvain, Institut supérieur de philosophie de l'université, 1931. (Les philosophes belges; textes et études . . . XI)

Baudrier

Baudrier, Henri Louis. Bibliographie lyonnaise. Recherches sur les imprimeurs, libraires, relieurs et fondeurs de lettres de Lyon au XVIe siècle. Par le président Baudrier. Publiées et continuées par J. Baudrier. Lyons, 1895-1921. 12 series.

Baumgärtel

Baumgärtel. Die älteste karte der Oberlausitz. (Oberlausitzisches magazin . . . Görlitz. v. 67: 247-250. 1891.)

Baur, editor

See Grosseteste. This reference refers only to the editor's writings

Baur (1917)

Baur, Ludwig. Die philosophie des Robert Grosseteste, bischofs von Lincoln († 1253) . . . Münster i. W., Aschendorff, 1917. (Beiträge zur Geschichte der Philosophie des Mittelalters . . . Band XVIII. Heft 4-6)

Bede (1843-4)

Bede, the Venerable. The complete works of Venerable Bede, in the original Latin, collated with the manuscripts, and various printed editions, accompanied by a new English translation of the historical works, and a life of the author. By the Rev. J. A. Giles . . . London, Whittaker, 1843-4. 12 v.

Bede (1896)

Bede, the Venerable. Venerabilis Baedae Historiam ecclesiasticam gentis Anglorum, Historiam abbatum, Epistolam ad Ecgberctum, . . . (Charles Plummer, editor) v. 1, Oxford, Clarendon press, 1896.

Bern (Mittheilungen)

Bern. Naturforschende gesellschaft. Mittheilungen. Bern, K. J. Wyss, 1843-1895.

Berry

Berry, Arthur. A short history of astronomy . . . London, Murray, 1898.

Bib. Belg. Gand

Bibliotheca Belgica. Bibliographie générale des Pays-Bas, par le bibliothécaire en chef et les conservateurs de la bibliothèque de l'université de Gand. Series 1-2, Gand, Camille Vyt; La Haye, Mart. Nijhoff, 1880-1890, 1891-1923.

Bib. Belg. Valerii Andreae

Andreas, Valerius. Bibliotheca belgica: De Belgis vita scriptisq. claris. Praemissa topographica Belgii totius sev Germaniae inferioris descriptione. new edition. Louvain, I. Zegers, 1643.

Bibliothèque Nationale et de Université, Prague

Biographie Nationale . . . de Belgique

Académie royale des sciences, des lettres et des beaux-arts de Belgique. Biographie nationale. Brussels, Bruylant-Christophe, 1866-1938. 27 v.

Biographisches lexicon der hervorragenden ärzte

Biographiches lexicon der hervorragenden ärzte aller zeiten und völker, unter mitwirkung der herren prof. E. Albert . . . prof. A. Anagnostakis . . . [u. a.] und unter spezial-redaktion von dr. E. Gurlt . . . und dr. A. Wernich . . . hrsg. von dr. August Hirsch . . . 2nd ed. . . . Berlin, Vienna, Urban & Schwarzenberg, 1929-1934. 5v.

Blumhof

Blumhof, Johann Georg Ludolph. Vom alten mathematiker Conrad Dasypodius. Ein literarischer versuch der Königlichen societät der wissenschaften zu Göttingen im September 1794. . . . Mit einer vorrede des herrn hofraths Kästner. Göttingen, J. C. D. Schneider, 1796.

Bodin

Bodin, Jean. Le theatre de la natvre vniverselle . . . auquel on peut contempler les causes efficientes & finales de toutes choses, desquelles l'ordre est continué par questions & responces en cinq liures . . . translated from the Latin by François de Fovgerolles . . . Lyons, Iean Pillehotte, 1597. (The Latin edition was published in the same year.)

Bodleian library catalogue

Oxford. University. Bodleian library. Catalogus librorum impressorum bibliothecae Bodleianae in Academia oxoniensi . . . Oxford, University press, 1843-1851. 4v.

Boffito

Boffito, Giuseppe. Un poeta della meteorologia. Gioviano Pontano. Memoria letta all' Accademia pontaniana . . . Naples, Stab. tip nella R. Università, 1899.

Bök

Bök, August Friedrich. Geschichte der herzoglich Wůrtenbergischen Eberhard Carls Universitåt zu Tůbingen im Grundrisse. Tübingen, Cotta, 1774.

Boll

Boll. F. Hephaistion 8) Hephaistion von Theben (Pauly-Wissowa, 8: 309-310)

Boll (1894)

Boll, Franz. Studien über Claudius Ptolemäus. Ein beitrag zur geschichte der griechischen philosophie und astrologie (Jahrbücher für classische philologie . . . supplement, 21: 49-244, 1894)

Boll (1918)

Boll, Franz. Antike beobachtungen farbiger sterne . . . .(K. Akademie der wissenschaften, Munich. Philosophisch-philologische und historische klasse. Abhandlungen . . . 30, no. 1, 1918)

Bonatti (1491)

Bonatus, Guido. [Liber astronomicus, ed. by Johannes Angelus], Augsburg, Ratdolt, 1491. (The title and the name of the editor are taken from the colophon.)

Borchert

Borchert, Ernst. Die lehre von der bewegung bei Nicolaus Oresme . . . Münster i. W., Aschendorff, 1934.

Bostock and Riley

See Pliny. This reference refers only to the editors' notes.

Bouché-Leclercq

Bouché-Leclercq, A[uguste]. L'astrologie grecque. Paris, Leroux, 1899

Brahe, Tycho. Historia coelestis. See Barrettus.

Brahe

Brahe, Tycho. Tychonis Brahe dani opera omnia, edidit I. L. E. Dreyer . . . Hauniae, in libraria Gyldendaliana, 1913-1929. 15v.

Brewster

Brewster, Sir David. The martyrs of science. Life of Galileo, Tycho Brahe, and Kepler. New edition. London, Chatto and Windus, 1874.

Bridges

Bridges, John Henry. The life & work of Roger Bacon, an introduction to the Opus majus . . . London, Williams & Norgate, 1914.

Broadbent

Broadbent, N. M., booksellers of London. List 3 (1940)

Brockhaus

Brockhaus' konversations-lexikon. Der grosse Brockhaus; handbuch des wissens . . . 15th edition. Leipzig, F. A. Brockhaus, 1928-1935. 20v.

Bruhns

Bruhns. Apianus: Peter A. (A.D.B., 1: 505-6)

Bruun

Bruun, Christian Walther. Bibliotheca danica. Systematisk fortegnelse over den danske literatur fra 1482 til 1830, efter samlingerne i det Store kongelige bibliothek i Kjøbenhavn. Kjøbenhavn, Gyldendal, 1877-1931. v. 1-4, suppl. and index

C.D.H.

C. Doris Hellman, library of

C.U.L.

Columbia university library

Calinich

Calinich, Robert. Aus dem sechszehnten jahrhundert. Culturgeschichtliche skizzen . . . Hamburg, Mauke, 1876.

Cambridge University library, manuscripts

Cambridge. University. Library. A catalogue of the manuscripts preserved in the library of the University of Cambridge. . . . Cambridge, University press; . . . 1856-67. 5 v.

v. Campenhausen

v. Campenhausen. Synesios. 1) S. von Kyrene (Pauly-Wissowa, new series, 4: 1362-1365)

Cantor (1892)

Cantor, Moritz. Vorlesungen über geschichte der mathematik; . . . v. 2, Leipzig, Teubner, 1892.

Cantor (1905)

Cantor, Moritz. Hieronymus Cardanus. Ein wissenschaftliches lebensbild aus dem XVI. jahrhundert (Neue Heidelberger jahrbücher . . . 13: 131-143, 1905)

Cantor (Gemma)

Cantor. Gemma-Frisius: Cornelis G. (A.D.B., 8:555)

Capelle (1905a)

Capelle, Wilhelm. Der physiker Arrian und Poseidonios (Hermes; zeitschrift für classische philologie 40: 614-635, 1905)

Capelle (1905b)

Capelle, Wilhelm. Die schrift von der welt. Ein beitrag zur geschichte der griechischen popularphilosophie (Neue jahrbücher für das klassische altertum, geschichte und deutsche literatur und für pädagogik . . . 15: 529-568, 1905)

Capelle (1908)

Capelle, Wilhelm. Erdbeben im altertum (Neue jahrbücher für das klassische altertum, geschichte und deutsche literatur und für pädagogik . . . 21: 603-633, 1908)

Capelle (1913)

Capelle, Wilhelm. Zur geschichte der meteorologischen litteratur (Hermes; zeitschrift für classische philologie . . . 48: 321-358, 1913)

Cardan (1554)

Cardan, Jerome, ed. Hieronymi Cardani . . . In Cl. Ptolemaei Pelvsiensis. IIII de astrorum iudicijs, aut, ut uulgò uocant, quadripartitae constructionis, libros commentaria . . . Nunc primùm in lucem aedita. Praeterea, eiusdem Hier. Cardani Geniturarum XII . . . exempla. Atque alia multa . . . Ac denique eclipseos, quam grauissima pestis subsecuta est, exemplum. Basle, Petri, 1554.

Cardan (1578)

Cardan, Jerome. Les livres de Hierome Cardanvs . . . intitulez de la subtilité, & subtiles inuentions, ensemble les causes occultes, & raisons d'icelles. Traduits de latin en françoys, par Richard le Blanc. Nouuellement reueuz, corrigez, & augmentez sur le dernier exemplaire latin de l'auteur, & enrichy de plusieurs figures necessaires. Paris, Beys, 1578.

Cardan (1663)

Cardan, Jerome. . . . Opera omnia: . . . Curâ Caroli Sponii . . . Lyons, Huguetan and Ravaud, 1663. 10v.

Carl

Carl, Ph[ilipp]. Repertorium der cometen-astronomie. Munich, Rieger, 1864.

Carl (Ms.)

[Carl, Philipp]. Vorlesungen ueber astronomie. Geschichte der astronomie. Manuscript, 19th century (after 1860)

Cartier

Cartier, Alfred. . . . .Bibliographie des éditions des de Tournes, imprimeurs lyonnais, mise en ordre avec une introduction et des appendices par Marius

Audin et une notice biographique par E. Vial . . . Paris, Editions des Bibliothèques nationales de France [1937-8]. 2 v. paged continuously. (Numbers cited are item numbers.)

Castellanus

Castellanus, Petrus. Vitae Illvstrivm Medicorvm Qui toto orbe, ad haec vsque tempora floruerunt . . . Antwerp, Gulielmus à Tongris, 1617.

Cat. Belg.

[Houzeau, Jean Charles.] Catalogue des ouvrages d'astronomie et de météorologie qui se trouvent dans les principales bibliothèques de la Belgique, préparé . . . . à l'Observatoire royal de Bruxelles; suivi d'un Appendice qui comprend tous les autres ouvrages de la bibliothèque de cet établissement. Brussels, F. Hayez, 1878. (Numbers cited are item numbers.)

Catholic encyclopedia

Catholic encyclopedia; an international work of reference on the constitution, doctrine, discipline and history of the Catholic church. New York, Appleton, [c 1907-1922], 17v.

Celoria

Celoria, Giovanni. Sulle osservazioni di comete fatte da Paolo dal Pozzo Toscanelli e sui lavori astronomici suoi in generale . . . Milan, Hoepli, 1921. (Milan, R. Osservatorio . . . di Brera. Publication no. 55)

Chalmers

Chalmers, Alexander. General biographical dictionary. London, 1812-1817. 32v.

Chiaramonti, Scipione. See Claramontius.

Cicero

Cicero, Marcus Tulius. Cicero: De natura deorum. Academica. With an English translation by H. Rackham . . . London, Heinemann, 1933. (Loeb classical library)

Cicogna

Cicogna, Emmanuele Antonio. Saggio di bibliografia veneziana . . .Venice, G. B. Merlo, 1847.

Claramontius (1619)

Chiaramonti, Scipione. Discorso della cometa pogonare dell' anno MDCXVIII. Di Scipione Chiaramonti Gentilhuomo Cesenate. Aggiuntaui la riposta della cometa prossima antecedente dell' istesso. Venice, Pietro Farri, 1619. (The C.U.L. copy was used.)

Claramontius (1621)

Claramontius, Scipio. Antitycho Scipionis Claramontii Caesenatis in qvo contra Tychonem Brahe, & nonnullos alios rationibus eorum ex opticis, & geometricis principijs solutis demonstratur cometas esse svblvnares non coelestes. Venice, Deuchinus, 1621. (There is a copy in the N.Y.P.L.)

Claramontius (1636)

Claramontius, Scipione. Scipionis Claramontii Caesenatis de sede svblvnari cometarvm opuscula tria, in supplementum Anti-Tychonis cedentia: . . . Amsterdam, Janssonius, 1636. (The H.C.L. copy was used.)

Clarke

See Seneca. This reference refers only to the editor's notes or introduction.

Clavius

Clavius, Christopher. Novi Calendarii Romani Apologia, Aduersus Michaelem Maestlinum Goeppingensem, in Tubingensi Academia Mathematicum, Tribvs Libris Explicata. Rome, Sanctius, 1588.

Collard

Collard, Auguste. Magister Jacobus de Ulma et son "Tractatus De Comeis" (2me Congrès National des Sciences... Organisé per la Fédération Belge des Sociétés Scientifiques — Bruxelles, 19-23 juin 1935 — Comptes Rendus, 1: 82-88. Brussels, Secrétaire Général: J.-P. Bosquet, [Marcel Hayez, printer], 1935.)

Coote

Coote, C. H., ed. Johann Schöner, professor of mathematics at Nuremberg. A reproduction of his globe of 1523 long lost, his dedicatory letter to Reymer von Streytperck and the 'De Molvccis' of Maximilianus Transylvanus, with new translations and notes on the globe by Henry Stevens . . . edited with an introduction and bibliography by C. H. Coote . . . London, Henry Stevens & Son, 1888.

Crawford library

Edinburgh. Royal observatory. Crawford library.

Crawford library catalogue

Edinburgh. Royal observatory. Catalogue of the Crawford library of the Royal observatory, Edinburgh. Edinburgh, Pub. by authority of Her Majesty's government, 1890.

Curtze (1870)

Curtze, Maximilian. Die mathematischen schriften des Nicole Oresme . . . ein mathematisch-bibliographischer versuch . . . Berlin, S. Calvary, 1870.

Curtze (1878)

Curtze, Maximilian, ed. Inedita Coppernicana. Aus den handscriften zu Berlin, Frauenburg, Upsala und Wien herausgegeben von Maximilian Curtze (Coppernicus-verein für wissenschaft und kunst zu Thorn. Mitteilungen... 1, Leipzig, 1878)

D.N.B.

Dictionary of national biography. . . . From the earliest times to 1900 . . . London, Published since 1917 by the Oxford university press [1921-1922] 22v. (v. 22 is 1st supplement)

Darmstaedter

Darmstaedter, Ludwig. . . . Handbuch zur geschichte der naturwissenschaften und der technik. In chronologischer darstellung. 2nd edition, . . . Berlin, J. Springer, 1908.

Davis and Orioli

Davis and Orioli, bookseller of London. Catalogues.

Degeorge

Degeorge, Léon. . . . La maison Plantin à Anvers; monographie complète de cette imprimerie célèbre aux XVIe et XVIIe siècles; . . . 2nd edition . . . Brussels, Gay and Doucé, 1878.

Delambre (1817)

Delambre, Jean Baptiste Joseph. Histoire de l'astronomie ancienne; . . . Paris, Ve Courcier, 1817. 2v.

Delambre (1819)

Delambre, [Jean Baptiste Joseph]. Histoire de l'astronomie du moyen âge; . . . Paris, Ve Courcier, 1819.

Delambre (1821)

Delambre, [Jean Baptiste Joseph]. Histoire de l'astronomie moderne; . . . Paris, Ve Courcier, 1821. 2v.

Dibelius

Dibelius, Franz. Zur geschichte und charakteristik Nikolaus Selneckers. (Dibelius, Franz and Lechler, Gotthard. Beiträge zur sächsischen kirchengeschichte . . . Leipzig, Barth, 4: 1-20, 1888.)

Dictionary of national biography . . . See D.N.B.

Diodorus (1793-1807)

Diodorus Siculus. Διόδωρος. Diodori Siculi Bibliothecae historicae libri qui supersunt, e recensione Petri Wesselingii, cum interpretatione latina Laur. Rhodomani atque annotationibus variorum integris . . . Nova ed., cum commentationibus III Chr. Gottl. Heynii et cum argumentis disputationibusque Ier. Nic. Eyringii . . . Zweibrücken, typ. Societatis, 1793-1807. 11v.

Diodorus (1933-9)

Diodorus Siculus. Diodorus of Sicily; with an English translation by C. H. Oldfather . . . v. 1-3, London, Heinemann, 1933-9. (Loeb classical library)

Döllinger

Döllinger, J[ohann Joseph Ignaz von]. Die reformation, ihre innere entwicklung und ihre wirkungen im umfange des Lutherischen bekenntnisses. 2nd ed. Arnheim, Witz, 1853-4. 3 v.

Doppelmayr

Doppelmayr, Johann Gabriel. Historische nachricht von den nürnbergischen mathematicis und künstlern . . . Nuremberg, Monath, 1730.

Dorat

Dorat, Jean. Oevvres poétiqves de Iean Dorat, poète et interprète dv roy, avec une notice biographique et des notes, par Ch. Marty-Laveaux. Paris, A. Lemerre, 1875.

Drecker

Drecker, Professor Dr. [Joseph]. Zeitmessung und sterndeutung in geschichtlicher darstellung. Berlin, Borntraeger, 1925.

Dreyer, editor

See Brahe. This reference refers only to the editor's notes.

Dreyer (1890)

Dreyer, John Louis Emil. Tycho Brahe, a picture of scientific life and work in the sixteenth century . . . Edinburgh, Adam and Charles Black, 1890.

Dreyer (1906)

Dreyer, John Louis Emil. History of the planetary systems from Thales to Kepler . . . Cambridge, University press, 1906.

Druon

Druon, H., ed. Oeuvres de Synésius . . . tr. . . . en français et précédées d'une étude biographique et littéraire, par H. Druon. Paris, Hachette, 1878.

Duhem

Duhem, Pierre. Le système du monde; histoire des doctrines cosmologiques de Platon à Copernic. Paris, A. Hermann and son, 1913-1917. 5v.

Du Verdier

Du Verdier, Antoine. La bibliotheqve d'Antoine Dv Verdier, seignevr de-Vavprivas, contenant le catalogue de tous ceux qui ont escrit, ou traduict en françois, & autres dialectes de ce royaume . . . Et à la fin vn supplement de l'epitome de la bibliotheque de Gesner. Lyons, Barthelemy Honorat, 1585.

Edelstein

Edelstein, Ludwig. The philosophical system of Posidonius. (The American journal of philology . . . 57: 286-325, 1936)

Emerson

Emerson, Edwin. Comet Lore; Halley's comet in history and astronomy. New York, Schilling Press, [1910].

Emmanuel college catalogue

Cambridge. University. Emmanuel college. Library. A hand-list of English books in the library of Emmanuel college, Cambridge, printed before MDCXLI. [Cambridge] Printed for the Bibliographical society at the Cambridge university press, 1915.

Enciclopedia Italiana

Enciclopedia italiana di scienza, lettere ed arti. v. 29, Rome, Istit. della encic. ital., fondata da Giovanni Treccani, 1936.

Enciclopedia vniversal

Enciclopedia vniversal ilvstrada europeo-americana . . . Barcelona, Espasa-Calpe, [1905-c1930], 70 v. in 71.

Engelbrecht

Engelbrecht, August [Gottfried]. Hephaestion von Theben und sein astrologisches compendium. Ein beitrag zur geschichte der griechischen astrologie . . . Vienna, 1887. (Jahres-Bericht über das Gymnasium der k.k. Theresianischen Akademie in Wien, . . . 1887.)

Ephorus

Ephorus. Ephori Cumaei Fragmenta. Collegit atque illustravit Meier Marx . . . Praefatus est Frid. Creuzer . . . Karlsruhe, Marx, 1815.

Ersch and Gruber

Allgemeine encyclopädie der wissenschaften und künste von genannten schriftstellern bearb. sec. 1, pt. 23. Leipzig, Brockhaus, 1832.

Estreicher

Estreicher, Stanisław. Bibliographia polska. Crakow, Czcionkami Drukarni Uniwersytetu Jagiellońskiego, 1872-1938.

Fabricius

Fabricius, Johann Albert. Jo. Alberti. Fabricii . . . Bibliotheca graeca, sive, notitia scriptorum veterum graecorum, quorumcunque monumenta integra, aut fragmenta edita exstant: tum pleorumque è mss. ac deperditis. . . . Hamburg, Liebezeit [etc.], 1708-1728 [irregular]. 6 books in 14v.

Favaro, editor

See Galilei.

Favaro (1876)

Favaro, Antonio. Copernicus und die entwickelung seines systems in Italien. Recension von Berti's festrede an der Universität Rom,... Dresden, B. G. Teubner, 1876.

Favaro (1881)

Favaro, Antonio. Die hochschule Padua zur zeit des Coppernicus. . . . tr. by Maximilian Curtze. (Coppernicus-verein für wissenschaft und kunst zu Thorn. Mitteilungen . . . 3, pt. I: 1-60, 1881)

Fernel

Fernelius, Ioannes . . . Cosmotheoria, libros duos complexa . . . Paris, Colinaeus, 1527.

Fiedler

Fiedler, J. Peurbach und Regiomontanus. Eine biographische skizze. (Jahresbericht des königlichen katholischen gymnasiums zu Leobschütz, womit zu der öffentlichen prüfung aller klassen am 15. august c. und zu der feierlichen entlassung der abiturienten am 16. august c. ergebenst im namen des lehrer-collegiums einladet professor und erster oberlehrer dr. J. Fiedler . . . pt. 1: 1-27, Leobschütz, Schiffmann, 1870)

FitzGerald

See Synesius. This reference refers only to the editor's notes.

Foppens

Foppens, Jean François. Bibliotheca belgica, sive Virorum in Belgio vitâ, scriptisque illustrium catalogus, librorumque nomenclatura; continens scriptores à clariss. viris Valerio Andrea, Auberto Miraeo, Francisco Sweertio, aliisque, recensitos, usque ad annum M.D.C.LXXX . . . Cura & studio Joannis Francisci Foppens . . . Brussels, Petrus Foppens, 1739. 2v. paged continuously

Fowler

Fowler, Harold N[orth], ed. Panaetii et Hecatonis librorvm fragmenta collegit praefationibvs illvstravit . . . Bonn, Charles George, [1885].

Fracastoro

Fracastoro, Jerome. Hieronymi Fracastorii Veronensis opera omnia, in vnum proxime post illius mortem collecta: . . . Accessit index locvpletissimvs. 2nd edition . . . Venice, Ivntas, 1574.

Frank

Frank, Gustav. Geschichte der protestantischen theologie . . . v. 1, Leipzig, Breitkopf and Härtel, 1862.

Freher

Freher, Paulus. D. Pauli Freheri . . . Theatrum virorum eruditione clarorum . . . Nuremberg, Hofmannus & the Knorzius press, 1688. 4v. paged continuously

Friedrich

Friedrich, Johann. Astrologie und reformation; oder, Die astrologen als prediger der reformation und urheber des bauernkrieges; ein beitrag zur reformationsgeschichte . . . Munich, Rieger, 1864.

Frisch, editor

See Kepler. This reference refers only to the editor's notes.

G. S.

George Sarton

Galileo

Galilei, Galileo. Le opere di Galileo Galilei. Ed. nazionale sotto gli auspicii di Sua Maestà il rè d'Italia. (Edited by Antonio Favaro) Florence, G. Barbèra, 1890-1909. 20v.

Galileo (1623)

Galilei, Galileo. Il Saggiatore nel quale con bilancia esquisite e giusta si ponderano le cose contenute nella libra astronomica e filosophica di Lotario Sarsi Sigensano . . . Rome, Giacomo Mascardi, 1623.

Gassendi

Gassendi, Pierre. Tychonis Brahei, equitis Dani, astronomorum coryphaei, vita, authore Petro Gassendo . . . Accessit Nicolai Copernici, Georgii Peurbachii, & Joannis Regiomontani . . . vita. 2nd edition. Haag, Netherlands, Adrian Vlacq, 1655.

Gellius

Gellius, Aulus. The Attic nights of Aulus Gellius, with an English translation by John C. Rolfe . . . London, Heinemann, 1927-8. 3v. (Loeb classical library)

Gesner (1583)

Gesner, Konrad. Bibliotheca institvta et collecta, primum a Conrado Gesnero: deinde in epitomen redacta, & nouorum librorum accessione locupletata, tertiò recognita, & in duplum post priores editiones aucta, per Iosiam Simlerum: iam verò postremò aliquot mille, cùm priorum tùm nouorum authorum opusculis, ex instructissima Viennensi Austriae imperatoria bibliotheca amplificata, per Iohannem Iacobum Frisium . . . Zurich, Froschovervs, 1583.

Ghilini

Ghilini, Girolamo. Teatro d'hvomini letterati. Aperto dall' abbate Girolamo Ghilini . . . Venice, Guerigli, 1647. 2v. in 1.

Gilbert

Gilbert, William. De mundo nostro sublunari philosophia nova. Opus posthumum ab authoris fratre collectum . . . Amsterdam, Elzevir, 1651.

Gloriosi

Gloriosi, Giovanni Camillo. De Cometis Dissertatio Astronomico-Physica Pvblice Habita In Gymnasio Patavino Anno Domini MDCXIX. A Ioanne Camillo Glorioso Gifonensi. Publico tunc temporis eiusdem Gymnasij Mathematico. In qua per triplices, easque celebriores hypotheses vltro citroque disputatur. Venetiis, MDCXXIV. Ex Typographia Varisciana. Svperiorvm Permissv Et Privilegiis. (The title is given here according to the rules adopted for the appendix.)

Goldschmidt

Goldschmidt, E. P. and Co., bookseller of London. Catalogues.

Gräve

Gräve. M. Bartholomäus Scultetus, bürgermeister zu Görlitz. (Oberlausitzische gesellschaft der wissenschaften. Neues lausitzisches magazin. Görlitz, 3: 455-505, 1824)

Green, editor

Whitney, Geoffrey. Whitney's "Choice of Emblemes". A Fac-simile reprint. Edited by Henry Green, M. A. with an introductory dissertation, essays literary and bibliographical, and explanatory notes. London, Reeve; Chester, Minshull & Hughes; Nantwich, Griffiths; 1866.

Greiff, Sebastian. See Marlishusanus

Greswell

Greswell, William Parr. Memoirs of Angelus Politianus, Joannes Picus of Mirandula, Actius Sincerus Sannazarius, Petrus Bembus, Hieronymus Fracastorius, Marcus Antonius Flaminius, and the Amalthei: translations from their poetical works: and notes and observations concerning other literary characters of the fifteenth and sixteenth centuries. 2nd edition . . . London, Cadell and Davies, 1805.

Grosseteste

Grosseteste, Robert. Die philosophischen werke des Robert Grosseteste, bischofs von Lincoln. Zum erstenmal vollständig in kritischer ausg. besorgt von dr. Ludwig Baur . . . Münster i. W., Aschendorff, 1912. (Beiträge zur Geschichte der Philosophie des Mittelalters . . . Band IX)

Gruterus

Janus Gruterus. Delitiae C. Poetarvm Belgicorvm . . . , v. 2, pt. 1, Frankfort, Hoffmann, 1614.

Guides-Joanne: Lyonnais

Collection des guides-Joanne. Bourgogne, Morvau, Jura, Lyonnais. Paris, Hachette, 1912.

Guillemin

Guillemin, Amédée. The world of comets, by Amédée Guillemin . . . Tr. and ed. by James Glaisher . . . London, S. Low, Marston, Searle, & Rivington, 1877.

Gundel

Gundel. Kometen (Pauly-Wissowa, 11: 1143-1193)

Gundel, Wilhelm (editor). See Hermes (1936).

Günther (1880)

Günther, Siegmund. Der Wapowski-brief des Coppernicus und Werner's tractat über die präcession. (Coppernicus-verein für wissenschaft und kunst zu Thorn. Mitteilungen . . . 2, pt. 1, Thorn, 1880)

Günther (1882)

Günther, Siegmund. Peter und Philipp Apian, zwei deutsche mathematiker u. kartographen. Ein beitrag zur gelehrten-geschichte des XVI. jahrhunderts . . . (Abhandlungen der königl. böhmischen gesellschaft der wissenschaften. Series 6. v. 11. Mathematisch-naturwissenschaftliche Classe, no. 4, Prague, 1882.)

Günther (1887)

Günther, Siegmund. Ein stück meteorologie und astrologie aus alt-München . . . (Jahrbuch für Münchener geschichte . . . year 1 (1887): 75-92.)

Günther (1890)

Günther, Siegmund. Münchener Erdbeben - und Prodigien - litteratur in älterer Zeit. (Jahrbuch für Münchener geschichte . . . year 4 (1890): 233-256.)

Günther (1901)

Günther, Siegmund. Die kompromiss-weltsysteme des XVI, XVII und XVIII jahrhunderts. (Annales internationales d'histoire; Congrès de Paris 1900; 5e section; histoire des sciences. Paris, Colin, 1901. 121-145.)

Günther (Maestlin)

Günther. Maestlin: Michael M. (A.D.B., 20: 575-580)

Günther (Regiomontanus)

Günther. Müller: Johannes M. (A.D.B., 22: 564-581)

Günther (Voegelin)

Günther. Voegelin: Johannes V. (A.D.B., 40: 142-3)

H.C.L.

Harvard college library

Haag

Haag, Eugène and Émile. La France Protestante . . . Paris, Cherbuliez, 1846-1859. 10v.

Hagecius (1574)

Hájek, Tadeáš. Dialexis De Novae Et Privs Incognitae Stellae Invsitatae Magnitvdinis & splendissimi luminis apparitione, & de eiusdem stellae vero loco constituendo. Adiuncta est ibidem ratio inuestigandae parallaxeos cuiuscunque Phaenomeni, eiúsque à centro terrae distantia, Meteorologicam doctrinam mirificè illustrans: nunc primum conscripta & edita, Per Thaddaeum Hagecium ab Hayck, Aulae Caesareae Maiestatis Medicum. Accesserunt aliorum quoque doctissimorum virorum de eadem stella scripta: & quaedam alia, quae versa pagella cognosces. Francofvrti Ad Moenvm. M.D.LXXIIII.

Hain

Hain, Ludwig [Friedrich Theodor]. Repertorium bibliographicum . . . ad annum MD. . . . Stuttgart, Cotta, 1826-1838. 2v. in 4 (Numbers cited are item numbers.)

Haly

Haly, Albohazen. Albohazen Haly filii Abenragel, scriptoris arabici, De ivdiciis astrorvm libri octo, doctorvm aliqvot virorvm opera in latinvm sermonem conuersi, . . . Basle, Henricpetrina, [1571].

Hantzsch

Hantzsch, Viktor. Die ältesten gedruckten karten der sächsisch-thüringischen länder (1550-1593), hrsg. und erläutert von Viktor Hantzsch . . . Leipzig,

Teubner, 1905. (Saxony. Sächsische kommission für geschichte. Aus den schriften . . . [12])

Haselbach

Haselbach, Karl. Ueber Johann Rasch's Weinbuch und die Weinkultur in Niederösterreich, vornemlich im XVI. Jahrhundert. (Blätter der vereines für landeskunde von Niederösterreich. new series, 15: 161-186, 1881.)

Haskins (1922)

Haskins, Charles H[omer]. Michael Scot and Frederick II. (Isis, 4 (2): 250-275, 1922.)

Haskins (1927)

Haskins, Charles Homer. Studies in the history of mediaeval science . . . 2nd edition, Cambridge, Harvard university press, 1927.

Haton

Haton, Claude. Mémoires de Claude Haton contenant le récit des événements accomplis de 1553 à 1582, principalement dans la Champagne et la Brie, publiés par M. Félix Bourquelot, . . . Paris, Imprimerie impériale, 1857. 2v. (Coll. de doc. inéd. sur l'hist. de France . . . 1e série. Histoire politique)

Haym

Haym, Nicola Francesco. Biblioteca Italiana, o sia Notizia de' libri rari italiani . . . Milan, Galeazzi, 1771-3. 2v.

Hazlitt (1867)

Hazlitt, William Carew. Handbook to the popular, poetical and dramatic literature of Great Britain, from the invention of printing to the restoration. London, J. R. Smith, 1867.

Hazlitt (1876-1903)

Hazlitt, William Carew. Bibliographical collections and notes on early English literature. 1474-1700. London, Quaritch, 1876-1903. 4 ser., suppl. 1-2 to ser. 3.

Hazlitt (1893)

Hazlitt, William Carew. General index to Hazlitt's Handbook and his Bibliographical collections, by G. J. Gray. London, Quaritch, 1893.

Hegi

Hegi, Friedrich. Neues zur lebensgeschichte Dr. Konrad Türsts. (urkundlich 1466-1503 (?). (Anzeiger für schweizerische geschichte. Hrsg. von der Allgemeinen geschichtforschenden gesellschaft der Schweiz. neue folge. 11 (1910-13): 280-289, Bern, 1912.)

Hellman

Hellman, C. Doris . . . A bibliography of tracts and treatises on the comet of 1577 . . . Bruges (Belgium) The Saint Catherine press, ltd. [1934] (Reprinted from Isis, no. 63, v. 22, 1, December, 1934)

Hellmann (1883)

Hellmann, Gustav. Repertorium der deutschen Meteorologie. Leistungen der Deutschen in Schriften, Erfindungen und Beobachtungen auf dem Gebiete der Meteorologie und des Erdmagnetismus von den ältesten Zeiten bis zum Schlusse des Jahres 1881. Leipzig, W. Engelmann, 1883.

Hellmann (1891)

Hellmann, Gustav. . . . Meteorologische volksbücher. Ein beitrag zur geschichte der meteorologie und zur kulturgeschichte . . . Berlin, H. Paetel, 1891. (Sammlung populärer Schriften herausgegeben von der Gesellschaft Urania zu Berlin. No. 8.)

Hellmann (1899)

Hellmann, Gustav. . . . Wetterprognosen und wetterberichte des XV. und XVI. jahrhunderts . . . Berlin, A. Asher & Co., 1899. (Neudrucke von schriften und karten über meteorologie und erdmagnetismus . . . no. 12)

Hellmann (1917)

Hellmann, Gustav. Entwicklungsgeschichte des meteorologischen Lehrbuches. (Beiträge zur Geschichte der Meteorologie, II, no. 6, Berlin, Behrend, 1917)

Hellmann (1924)

Hellmann, Gustav. . . . Versuch einer geschichte der wettervorhersage im XVI. jahrhundert . . . Berlin, Akademie der wissenschaften, 1924. (Abhandlungen der Preussischen akademie der wissenschaften . . . 1924. Physikalisch-mathematische klasse, . . . 1)

Heppe

Heppe, Heinrich. Geschichte des deutschen Protestantismus in den jahren 1555-1583. . . . Frankfort-on-the-Main, Völcker, 1865, 4v.

Hermes (1936)

Hermes *Trismegistus*. Neue astrologische texte des Hermes Trismegistos; . . . von Wilhelm Gundel . . . Munich, 1936. (Abhandlungen der Bayerischen akademie der wissenschaften. Philosophisch-historische abt. n.f. hft. 12)

Hevelius (1665)

Hevelius, Johannes. . . . Prodromus Cometicus, Quo Historia, Cometae Anno 1664 exorti . . . exhibetur . . . Dantzic, 1665.

Hevelius (1666)

Hevelius, Johannes. . . . Descriptio Cometae Anno Aerae Christ. M. DC. LXV. exorti, . . . Dantzic, 1666.

Hevelius (1668)

Hevelius, Johannes. . . . Cometographia, Totam Naturam Cometarum; . . . Dantzic, 1668.

Hiersemann

Hiersemann, Karl W., bookseller of Leipzig. Catalogues.

Hind

Hind, John Russell. The comet of 1556; being popular replies to every-day questions, referring to its anticipated re-appearance, with some observations on the apprehension of danger from comets. By J. Russell Hind . . . London, Parker, 1857.

Hoefer

Hoefer, Ferdinand, ed. Nouvelle biographie générale depuis les temps les plus reculés jusqu'à nos jours, avec les renseignements bibliographiques et l'indication des sources à consulter; publiée par MM. Firmin Didot frères, sous la direction de M. le Dr. Hoefer . . . Paris, Firmin Didot frères, 1853-70 [v. 1, 1855], 46v.

Hoek

Hoek, Martin. De kometen van de jaren, 1556, 1264 en 975, en hare vermeende identiteit . . . 's Gravenhage, Van Cleef, 1857.

Holmes

Holmes, Thomas James. Increase Mather; a bibliography of his works . . . with an introduction by George Parker Winship and supplementary material by Kenneth Ballard Murdock and George Francis Dow . . . Cleveland, O., 1931 [Harvard University Press] 2v.

Houzeau

Houzeau, J. C. and Lancaster, A. Bibliographie générale de l'astronomie . . . ou catalogue methodique des ouvrages, des mémoires et des observations astronomiques publiés depuis l'origine de l'imprimerie, jusqu'en 1880. 2v., v. 1 in 2 pts., Brussels, F. Hayez, 1887 and 1889; v. 2, Brussels, Xavier Havermans, 1882. (Unless otherwise indicated, numbers cited are item numbers.)

Houzeau (Vade-mecum)

Houzeau, J. C. Vade-mecum de l'astronomie. Brussels, F. Hayez, 1882. (Numbers cited are item numbers.)

Houzeau, J. C. See Cat. Belg.

Isis

Isis; international review devoted to the history of science and civilization. v. 1 —, Bruges, Saint Catherine press ltd. . . . March, 1913–

Jacobus Angelus. See Angelus.

Jancke (1861a)

Jancke, J. C. O., Memorabilia scholastica Gorlicensia, ex ore M. Barthol. Sculteti in Diario ejus manuscripto, ab anno: 1567-1594. posteritati data, . . . (Oberlausitzische gesellsschaft der wissenschaften. Neues lausitzisches magazin . . . 38: 265-280, 1861.)

Jancke (1861b)

Jancke, J. C. O. M. Bartholomäi Sculteti lusatische reisen und seine mappa Lusatiae. (Oberlausitzische gesellschaft der wissenschaften. Neues lausitzisches magazin . . . 38: 280-285, 1861.)

Jancke (1868)

Jancke, [J. C. O.] Memorabilien aus Sculteti Diarium. (Oberlausitzische gesellschaft der wissenschaften. Neues lausitzisches magazin . . . 44: 267-273, 1868.)

Jancke (1870)

Jancke, [J. C. O.] Eine Scultetus'sche Inschriften - Collection. (Oberlausitzische gesellschaft der wissenschaften. Neues lausitzisches magazin . . . 47: 119-120, 1870.)

Janssen

Janssen, Johannes. Geschichte des deutschen volkes seit dem ausgang des mittelalters . . . Freiburg in Breisgau, Herder, 1883-1894. 8v.

Jastrow

Jastrow, Morris, Jr. Die religion Babyloniens und Assyriens . . . Giessen, Ricker, 1905-12. 2v. in 3.

Jöcher

Jöcher, Christian Gottlieb. Allgemeines gelehrten-lexicon, darinne die gelehrten aller stände sowohl männ-als weiblichen geschlechts, welche vom anfange der welt bis auf ietzige zeit gelebt, . . . Leipzig, Gleditsch, 1750-1. 4v.—Fortsetzung und ergänzungen . . . Leipzig, Gleditsch, 1784-87; Delmenhorst, Jöntzen, 1810; Bremen, Heyse, 1813-19; Leipzig, Selbstverlag der deutschen gesellschaft, 1897. 7v. (The "Fortsetzung und ergänzungen" were begun by Johann Christoph Adelung, who died in 1806. In the copy which was used for this dissertation, v. 3-6 of these supplements were missing.)

John Crerar Library (of Chicago)

John of Damascus

John of Damascus. Exposition of the Orthodox Faith, translated by the Rev. S. D. F. Salmond (A select library of Nicene and post-Nicene fathers of the Christian church. 2nd series . . . 9, pt. 2, New York, Charles Scribner's sons, 1899.)

Johnson (1934)

Johnson, Francis R. and Larkey, Sanford V. Thomas Digges, the Copernican system, and the idea of the infinity of the universe in 1576 (Huntington Library Bulletin, no. 5: 69-117, Cambridge, Mass., Harvard university press, April, 1934.)

Johnson (1936)

Johnson, Francis R. The influence of Thomas Digges on the progress of modern astronomy in sixteenth-century England. (Osiris, 1: 390-410, January, 1936.)

Johnson (1937)

Johnson, Francis R. Astronomical thought in renaissance England; a study of the English scientific writings from 1500 to 1645 . . . Baltimore, The Johns Hopkins press, 1937.

Jourdain

Jourdain, Charles. Nicolas Oresme et les astrologues de la cour de Charles V. (Revue des questions historiques; . . . 18: 136-159, 1875.)

Kästner

Kästner, Abraham Gotthelf. Geschichte der mathematik seit der wiederherstellung der wissenschaften bis an das ende des achtzehnten jahrhunderts, . . . Göttingen, Rosenbusch, 1796-1800. 4v.

Kauffmann

Kauffmann. Artemidoros 35) Von Parion (Pauly-Wissowa, 2: 1332-4)

Kaussen

Kaussen, Joseph. Physik und ethik des Panätius . . . Bonn, S. Foppen, 1902.

Kepler

Kepler, Johann. Joannis Kepleri astronomi opera omnia. Edidit Dr. Ch. Frisch . . . Frankfort-on-the-Main and Erlanger, Heyder & Zimmer, 1858-71. 8v.

Kepler (1596)

Kepler, Johannes. Prodromus Dissertationvm Cosmographicarvm, Continens Mysterivm Cosmographicvm, De Admirabili Proportione Orbivm Coelestivm, . . . Demonstratvm Per Qvinqve regularia corpora Geometrica . . . Addita est erudita Narratio M. Georgii Ioachimi Rhetici, de Libris Reuolutionum . . . Tvbingae Excudebat Georgius Gruppenbachius, Anno M.D.XCVI.

Kestner

Kestner, Christian Wilhelm. Medicinisches gelehrten - lexicon; darinnen die leben der berühmtesten aerzte . . . beschrieben worden . . . Jena, successors to Johann Meyers, 1740.

Klatt

Klatt, Detloff. David Chytraeus als geschichtslehrer und geschichtschreiber . . . Rostock, successors of Adler, 1908.

Kluckhohn

Kluckhohn, August. Der sturz der kryptocalvinisten in Sachsen 1574. (Historische zeitschrift hrsg. von . . . Sybel . . . Munich, Literarisch-artistische anstalt, 18: 77-127, 1867.)

Koch (1907-1910)

Koch, Ernst. Moskowiter in der Oberlausitz und M. Bartholomäus Scultetus in Görlitz. Kulturbilder aus der zweiten Hälfte des XVI. Jahrhunderts. . . . (Oberlausitzische gesellschaft der wissenschaften . . . Neues lausitzisches magazin, 83: 1-90; 84: 41-109; 85: 255-290; 86: 1-80, Görlitz, 1907-1910.)

Koch (1916)

Koch, Ernst. Scultetica. (Oberlausitzische gesellschaft der wissenschaften . . Neues lausitzisches magazin, 92: 20-58, Görlitz, 1916.)

Krabbe

Krabbe, Otto. Die Universität Rostock im funfzehnten und sechzehnten jahrhundert. Rostock, successors of Adler, 1854.

Kroker

Kroker, Ernst. Bartholomäus Scultetus und seine Karte von Sachsen (1568). (Neues archiv für Sächsische geschichte und altertumskunde . . . 42: 270-277, Dresden, 1921.)

Kugler

Kugler, Franz Xavier. Sternkunde und sterndienst in Babel.. . . . Münster i. W., Aschendorff, 1907-35. 4 v. in 5.

Lalande

Lalande, [Joseph] Jérôme [Le Français] de. Bibliographie astronomique; avec l'histoire de l'astronomie depuis 1781 jusqu'a à 1802: par Jérôme de La Lande . . . Paris, De l'Imprimerie de la République, an XI. = 1803.

Lancaster, A. See Houzeau.

Larkey, Sanford V. See Johnson (1934).

Lauffer

Lauffer, Otto. Der komet im volksglauben. (Verein für volkskunde. Zeitschrift, 27: 13-35, Berlin, 1917.)

Legré

Legré, Ludovic. Un philosophe provençal au temps des Antonins, Favorin d'Arles, sa vie—ses oeuvres—ses contemporains. Marseilles, Aubertin and Rolle, 1900.

Le Long (1709)

Le Long, Jacques. Bibliotheca sacra sev syllabvs omnivm ferme sacrae scriptvrae editionvm ac versionvm secundum seriem . . . v. 2, Leipzig, Gleditsch & Weidmann, 1709.

Le Long (1719)

Le Long, Jacques. Bibliotheque historique de la France; contenant le catalogue de tous les ouvrages, tant imprimez que manuscrits, qui traitent de l'histoire de ce roïaume, ou qui y ont rapport: avec des notes critiques & historiques. . . . Paris, Gabriel Martin, 1719.

Leovitius

Leovitius, Cyprianus. De Nova Stella. Ivdicivm Cypriani Leovitii à Leonicia, Mathematici, de noua Stella siue Cometa, viso mense Nouembri ac Decembri, Anni Domini 1572. Item mense Ianuario & Februario, Anni Domini 1573. Cum gratia & priuilegio Caesareae Maiestatis. Impressum Lavingae ad Danubium. M. D. LXXIII. (On A_3 v: Elaboratum die 20. Februarij, Anno 1573.) This is a quarto of one signature. The title is given in the manner used for the items in the appendix. The C.D.H. copy was used.

Leupold

Leupold "ducatus Austrie filii". Compilatio de astrorum scientia. Augsburg, Ratdolt, January 9, 1489. (The copy belonging to the Plimpton library of Columbia university was used.)

Lindesiana

[Crawford, James Ludovic Lindsay, 26th earl of]. Bibliotheca Lindesiana . . . Catalogue of the printed books preserved at Haigh Hall, Wigan, co. pal. Lancast. . . . [Aberdeen], Aberdeen university press, 1910. 4v.

Little

Little, Andrew George, ed. Roger Bacon essays, contributed by various writers on the commemoration of the seventh centenary of his birth, . . . Oxford, Clarendon press, 1914.

Littrow

Littrow, Karl v. The observations by Fabricius and Heller of the comet of 1556. Being the substance of a paper by Karl v. Littrow in Sitzungsberichte Wiener Akad. der Wissenschaften, vol. xx. p. 301. Translated by . . . A. S. Herschel . . . (Royal Astronomical Society, London. Monthly notices . . . 77: 633-643, London, 1917.)

Loth

Loth, Otto. Al-kindi als Astrolog (Morgenländische Forschungen. Festschrift Herrn Professor Dr. H. L. Fleischer . . . gewidmet von Seinen Schülern . . . Leipzig, Brockhaus, 1875. pt. VIII: 263-309.)

Lozzi

Lozzi, Carlo. Biblioteca istorica della antica e nuova Italia; saggio di bibliografia, analitico, comparato e critico, compilato sulla propria collezione con un discorso proemiale da Carlo Lozzi . . . Imola, Galeati, 1886-7. 2v.

Lubienski

Lubienski, Stanislaw. Stanislai de Lubienietz . . . Theatrum cometicum, duabus partibus constans, quarum altera . . . cometas anni 1664 & 1665 variis vivorum per Europam clariss. cum quibus auctor de hoc argumento contulit, observationibus . . . descriptos . . . exhibit . . . altera continet Historiam CCCCXV. cometarum . . . ad ann. 1665 . . . Et Theatri cometici exitus, sive De significatione cometarum . . . Amsterdam, Franciscus Cuperus, 1666-68, v. 2.

Ludendorff

Ludendorff, H. Die kometen-flugschriften des XVI. und XVII. jahrhunderts. (Zeitschrift für bücherfreunde . . . Jahrg. 12: 501-506, Berlin, 1908-9.)

Lydiat

Lydiat, Thomas. Praelectio Astronomica De Natvra Coeli & conditionibus elementorum: . . . London, Ioannes Bill, 1605.

Lydus

Ioannes Lydus. Ioannes Lydus ex recognitione Immanuelis Bekkeri, Bonne, 1837. (Corpus scriptorum historiae byzantinae . . . 31.)

Mädler

Mädler, J[ohann] H[einrich] von. Geschichte der himmelskunde von der ältesten bis auf die neueste zeit. Braunschweig, Westermann, 1873. 2v.

Maestlin (1588)

Maestlin, Michael. Epitome Astronomiae, Qva Brevi Explicatione Omnia, Tam Ad Sphaericam quàm Theoricam eius partem pertinentia, ex ipsius scientiae fontibus deducta, perspicuè per quaestiones traduntur Conscripta per M. Michaelem Maestlinvm Goeppingensem, Matheseos in Academia Tubingensi Professorem. Iam nunc ab ipso Autore diligenter recognita. Cum Priuilegio Caesareae Maiestatis. Tvbingae, Excudebat Georgius Gruppenbachius. Anno 1588. (This title is given in the manner used for the appendix.)

Maestlin (1610)

Maestlin, Michael. Epitome Astronomiae, . . . Jam Nunc Ab Autore denuo diligenter recognita: Additis insuper, Iuniorum gratiâ, breuibus ex doctrina Triangulorum Sphaericorum praeceptis, computandi praecipuas primi Motus Tabulas: alijsq́; nonnullis, ad Astronomiae cognitionem scitu partim necessarijs, partim iucundis . . . Tübingen, Philip Gruppenbach, 1610.

Malchin

Malchin, Franz. De avctoribvs qvibvsdam qvi Posidonii libros meteorologicos adhibvervnt, dissertatio inavgvralis . . . Rostock, Charles Boldtius, 1893.

Manilius

Manilius, Marcus. M. Manili AstronomicΩn Libri Qvinqve. Iosephvs Scaliger Ivl. Caes. F. Recensvit, ac pristino ordini suo restituit. Eiusdem Ios. Scaligeri Commentarius in eosdem libros, & Castigationum explicationes. Lvtetiae, Apud Mamertum Patissonium Typographum Regium, in officina Roberti Stephani. M. D. LXXIX. Cvm Privilegio Regis.

Marcellinus

Ammianus Marcellinus. The Roman history of Ammianus Marcellinus, during the reigns of the emperors Constantius, Julian, Jovianus, Valentinian, and Valens. Translated by C. D. Yonge . . . with a general index. London, Bell, 1911. (Bohn's classical library.)

Marí

Marí, Antoni Quintana i. Tablas cronologicas para Cataluña (Archeion, Archivo de Historia de la Ciencia, XXV - 1943 - N. 2/3: 204-214. Santa Fe, Argentina, 1943.)

Marlishusanus

Greiff, Sebastian, (of Marlishausen?). Vom Cometen Anno 1596. Im anfang des Monats Julii erschienen. Aus dess Hocherleuchtē Astronomi / Philosophi vnd Medici D. Philippi Theo. Paracelsi scriptis / Colligiret vnd beschrieben. Durch / Sebastianum Greiffen Marlishusanum / Bůrgern inn Erffurdt . . . Erfurt, Martin Wittel, [1596].

Marx

Marx, Meier, ed. See Ephorus. This reference applies to all phrases not by Ephorus.

Meunier

Meunier, Francis. Essai sur la vie et les ouvrages de Nicole Oresme . . . Paris, Typ. de C. Lahure, 1857.

Michaud

. . . Biographie universelle ancienne et moderne. . . . sous la direction de MM. Michaud. 2nd edition. Paris, Vivès [1880]. 45v.

Monatliche correspondenz

Monatliche correspondenz zur beförderung der erd- und himmels-kunde hrsg. von F. von Zach. 12: 267-303 (article on "Landraf Wilhelm IV"), Gotha, Becker, 1805.

Montucla

Montucla, Jean Étienne. Histoire des mathématiques dans laquelle on rend compte de leur origine jusqu'à nos jours; . . . new ed., Paris, H. Agasse, [1799] - 1802. 4v.

Morley

Morley, Henry. Jerome Cardan. The life of Girolamo Cardano, of Milan, physician . . . London, Chapman and Hall, 1854. 2v.

Motley

Motley, John Lothrop. The rise of the Dutch republic, a history, by John Lothrop Motley . . . v. 3, New York and London, Harper, 1901.

Müller, A.

Müller, Adolf. Galileo Galilei und das kopernikanische weltsystem, . . . Freiberg in Breisgau, St. Louis, Mo. etc., Herder, 1909.

Müller, K. K.

Müller, K. K. Asklepiodotos 10) (Pauly-Wissowa, 2: 1637-1641.)

Mützell

Mützell, Julius. Geistliche lieder der evangelischen kirche aus dem sechszehnten jahrhundert. . . . . Berlin, Enslin, 1855. 3v.

McColley

McColley, Grant. An early friend of the Copernican theory: Gemma Frisius. (Isis, 26(2) : 322-325, March, 1937.)

N. Y. P. L.

New York Public Library

Nationalbibliothek Wien

Nicéron

[Nicéron, Jean Pierre.] Mémoires pour servir à l'histoire des hommes illustres dans la re'publique des lettres, avec un catalogue raisonne' de leurs ouvrages . . . Paris, Briasson, 1729-45. 43v. in 44.

Nielsen

Nielsen, Lauritz [Martin]. Dansk bibliograpfi, 1551-1600, med saerligt hensyn til dansk bogtrykkerkunsts historie . . . Copenhagen, Gyldendal, 1931-3. (Numbers cited are item numbers.)

Nijhoff

Nijhoff, Martin, bookseller of the Hague, Holland. Catalogues.

Noback

Noback, Friedrich. Münz-, maas- und gewichtsbuch. . . . 2nd ed., newly augmented. Leipzig, Brockhaus, 1879.

Nolthius (1572)

Nolthius, Andreas. Observatio vñd Beschreibuñg des Newen Cometen / so vmb das ende des 1572. vnd noch in diesem 73. Jar erschienen / Geschehen vnd gestellet / Durch Andream Nolthium Mathematicum. Gedruckt zu Erffurdt / zum bundten Lawen bey Sanct Paul. (no date; a photostatic copy was furnished by the Royal Astronomical Society.)

Nourry

Nourry, Emile, bookseller of Paris. Catalogues.

Ockenden, editor

See Twyne, Thomas. This reference refers only to the editor's notes.

Olbers

Olbers, Heinrich Wilhelm Matthias. Ueber den von Apian im jahr 1533 beobachteten kometen. (Berliner astronomisches jahrbuch für 1800 . . . J. E. Bode, ed.,: 126-131, Berlin, 1797.)

Origen

Origen. The writings of Origen, tr. by Frederick Crombie . . . Edinburgh, Clark, 1869-1872. 2v. (Anti-Nicene Christian library, 10, 23.)

Ortroy (1901)

Ortroy, Fernand Gratien van. Bibliographie de l'oeuvre de Pierre Apian. (Le bibliographe moderne; courrier international des archives et des bibliothèques . . . 5th year: 89-156, 284-333, Paris, 1901.)

Ortroy (1920)

Ortroy, Fernand Gratien van. Bio-bibliographie de Gemma Frisius, fondateur de l'École belge de géographie, de son fils Corneille et de ses neveux les Arsenius . . . [Brussels, Lamertin, 1920.] (Acad. royale des sciences, des lettres et des beaux arts, Brussels. Mém. Collection in 8°. Series 2, v. 11, pt. 2.)

Osiris

Osiris; studies on the history and philosophy of science, and on the history of learning and culture . . . 1–, Bruges (Belgium), The Saint Catherine press ltd., 1936–

Pachymerēs

Pachymerēs, Geōrgius. Historia. v. 1, [Rome, 1666] (Corpus Byzantinae historiae. 18, pt. 1.)

Palau y Dulcet

Palau y Dulcet, Antonio. Manual del librero hispano-americano; inventario bibliográfico de la producción cientifica y literaria de España y de la América latina desde la invención de la imprenta hasta nuestros dias, con el valor comercial de todos los artículos descritos. Barcelona, Libreria anticuaria, 1923-27. 7v.

Paracelsus

Paracelsus. Avr. Philip. Theoph. Paracelsi Bombast ab Hohenheim . . . Opera omnia: medico-chemico-chirvrgica, tribvs volvminibvs comprehensa. Ed. novissima et emendatissima, ad germanica & latina exemplaria accuratissimè collata: . . . Geneva, I. Antonius and Samuel De Tournes, 1658. 3v. in 1.

Paulsen

Paulsen, Peter. David Chyträus als historiker. Ein beitrag zur kenntnis der deutschen historiographie im reformationsjahrhundert. . . . Rostock, Hinstorff, 1897.

Pauly-Wissowa

Pauly, August Friedrich von. Paulys real-encyclopädie der classischen altertumswissenschaft; neue bearbeitung . . . unter mitwirkung zahlreicher fachgenossen hrsg. von G. Wissowa . . . Stuttgart, Metzler, 1894-1940. v. 1–

Peignot

Peignot, L. G. Dictionnaire historique et bibliographique, abrégé des personnages illustres, célèbres ou fameux de tous les siècles et de tous les pays du monde, avec les dieux et les héros de la mythologie, . . . Paris, Haut-Coeur and Gayet j[e], 1815-1822. 3v.

Pérez Pastor

Pérez Pastor, Christóbal. Bibliografía madrileña, ó Descripción de las obras impresas en Madrid . . . por . . . Don Cristóbal Pérez Pastor . . . Obra premiada por la Biblioteca nacional . . . é impresa á expensas del estado. v. 1, Madrid, Tip. de los huérfanos, 1891.

Petit

Petit, Louis David. Bibliotheek van Nederlandsche pamfletten. Verzamelingen van de bibliotheek van Joannes Thysius en de bibliotheek der Rijks-universiteit te Leiden. . . . v. 1-2, s'-Gravenhage, M. Nijhoff, 1882-4.

Phares

Phares, Simon de. Recueil des plus célèbres astrologues et quelques hommes doctes, faict par Symon de Phares, du temps de Charles VIIIe, publié d'après le manuscrit unique de la Bibliothèque nationale, par le dr. Ernest Wickersheimer. Paris, Champion, 1929.

Picatoste y Rodriguez

Picatoste y Rodríguez, Felipe. Apuntes para una biblioteca científica española del siglo XVI; estudios biográficos y bibliográficos de ciencias exactas, física y naturales y sus inmediatas aplicaciones en dicho siglo . . . Madrid, M. Tello, 1891.

Pingré

Pingré, [Alexandre Guy.] Cométographie; ou Traité historique et théorique des comètes . . . Paris, Imprimerie royale, 1783-4. 2v.

Planck

Planck, G. J. Geschichte der protestantischen theologie von Luthers tode bis zu der einführung der konkordienformel. . . . v. 2, 3. (Geschichte der entstehung, der verånderungen und der bildung unseres protestantischen lehrbegriffs von anfang der reformation bis zu der einführung der konkordienformel. 5, 6, Leipzig, Crusius, 1798-1800.)

Plantin Museum, Antwerp

Pliny

Pliny, the Elder. The natural history of Pliny. Tr., with copious notes and illustrations, by the late John Bostock . . . and H. T. Riley . . . London, H. G. Bohn, 1855-7. 6v.

Pliny (1784)

Pliny, the Elder. Caii Plinii Secundi Historiae naturalis libri XXXVII, ex recognitione Joannis Harduini et Gabrielis Broterii, cum notis selectioribus . . . v. 1, Venice, Bettinelli, 1784.

Plutarch (1883)

Plutarch's Morals. Tr. from the Greek by several hands. Cor. and rev. by William W. Goodwin . . . With an introduction by Ralph Waldo Emerson . . . Boston, Little, Brown, and company, 1883. 5 v.

Plutarch (1893)

Plutarch. Plutarchi Chaeronensis Moralia recognovit Gregorius N. Bernardakis . . . v. 5, Leipzig, Teubner, 1893.

Poggendorff

Poggendorff, J. C. Biographisch-Literarisches Handwörterbuch zur Geschichte der exacten wissenschaften . . . v. 1-2, Leipzig, Johann Ambrosius Barth, 1863.

Pogo (1934)

Pogo, A[lexander]. Earliest diagrams showing the axis of a comet tail coinciding with the radius vector. (Query no. 35 in Isis, 20 (2): 443-4, January, 1934.)

Pogo (1935)

Pogo, A[lexander]. Gemma Frisius, his method of determining differences of longitude by transporting timepieces (1530), and his treatise on triangulation (1533). (Isis, 22 (2): 469-485, February, 1935.)

Pontano (1519)

Pontano, Giovanni Gioviano. Ioannis Ioviani Pontani opera omnia solvta oratione composita. pt. 3. [Colophon: Venice, Aldus, 1519].

Pontano (1902)

Pontano, Giovanni Gioviano. Ioannis Ioviani Pontani carmina. Testo fondato sulle stampe originali, e riveduto sugli autografi, introduzione bibliografica, ed appendice di poesie inedite; a cura di Benedetto Soldati . . . Florence, Barbèra, 1902, 2v.

Prandtl

Prandtl, Wilhelm. Die Bibliothek des Tycho Brahe. Vienna, Reichner, 1933.

Preger

Preger, Wilhelm. Matthias Flacius Illyricus und seine zeit . . . Erlangen, Bläsing, 1859-61. 2v.

Pressel

Pressel, Theodor. David Chyträus . . . Elberfeld, Friderichs, 1862. (Leben und ausgewählte schriften der väter und begründer der lutherischen kirche, . . . 8.)

Prins

Prins, Anthonij Winkler. Winkler Prins' algemeene encyclopaedie. New edition, edited by J. de Vries. Amsterdam, Elsevier, 1932-8. 16 v.

Proctor and Crommelin

Proctor, Mary and Crommelin, A. C. D. Comets; their nature, origin, and place in the science of astronomy . . . London, Technical press, 1937.

Proctor, Robert

Proctor, Robert, ed. [British Museum. Dept. of printed books.] Catalogue of books printed in the XVth century now in the British Museum . . . London, Printed by order of the Trustees, 1908-1935. v. 1-7.

Prowe

Prowe, Leopold. Nicolaus Coppernicus . . . Berlin, Weidmann, 1883-4. 2v., v. 1 in 2 pts.

Pruckner

Pruckner, Hubert . . . Studien zu den astrologischen schriften des Heinrich von Langenstein. Leipzig / Berlin, B. G. Teubner, 1933.

Ptolemy (1541)

Ptolemy, Claudius. Clavdii Ptolemaei pelvsiensis alexandrini omnia, qvae extant, opera, . . . Basle, Henry Petrus, 1541.

Ptolemy (1822)

Ptolemy, Claudius. Ptolemy's Tetrabiblos, or Quadripartite: being four books of the influence of the stars. Newly translated from the Greek paraphrase of Proclus. With a preface, explanatory notes, and an appendix, containing extracts from the Almagest of Ptolemy, and the whole of his Centiloquy; together with a short notice of Mr. Ranger's Zodiacal planisphere, and an explanatory plate. By J. M. Ashmand . . . London, Davis and Dickson, 1822.

Quételet

Quételet, Ad[olphe]. Histoire des sciences mathématiques et physiques chez les Belges . . . Brussels, Hayez, 1864.

R.A.S.

Royal Astronomical Society, London

Ranke

Ranke, Leopold von. Zur deutschen geschichte. Vom religionsfrieden bis zum dreissigjährigen krieg. . . . 2nd edition. Leipzig, Duncker and Humblot, 1874. (Ranke, Leopold von. Sämmtliche Werke. 2nd ed. . . . 7)

Rasch, Johann. Gegenpractic, wider etliche aussgangen Weissag, Prognostic und Schrifften, sonderlich des Misocaci, uber das 84. und 88. Jare, . . . Munich, A. Berg. 1584.

Rasch, Johann. Gegenpractic. Urthail und allgemainer khurtzer bericht wider etliche aussgangene weissag, prognostic, practic und troeschrifften, auss den zuefällen des 84. unnd 88. wunderjaren, sunderlich des Misocacs, von undergang hoches Geschlechts und der Röm-Clerisey, von änderung der Reich und Religion, Von Antichrist, von lester zeit und end der weld. Munich, Adam Berg, 1588

Rauscher

Rauscher, Julius. Der Halleysche komet im jahre 1531 und die reformatoren. (Zeitschrift für kirchengeschichte. . . . 32: 259-276, June 24, 1911.)

Realencyklopädie

Herzog, Johann Jakob. Realencyklopädie für protestantische theologie und kirche, begründet von J. J. Herzog; in 3. verb. und verm. aufl. . . . hrsg. von Albert Hauck. Leipzig, Hinrichs, 1896-1913. 24v.

Rehm (1907)

Rehm, Albert. Anlage und buchfolge von Senecas Naturales Quaestiones. (Philologus; zeitschrift für das classische alterthum . . . 66: 374-395. Leipzig, 1907.)

Rehm (1922)

Rehm, Albert. Das siebente buch der Naturales Quaestiones des Seneca und die kometentheorie des Poseidonios. (K. Akademie der wissenschaften, Munich. Philosophisch-philologische und historische klass. Sitzungsberichte, jahrg. 1921, 1. abhandlung, Munich, 1922)

Reinhardt (1921)

Reinhardt, Karl. Poseidonios . . . Munich, Beck, 1921.

Reinhardt (1926)

Reinhardt, Karl. Kosmos und sympathie; neue untersuchungen über Poseidonios . . . Munich, Beck, 1926.

Reisacher

Reisacher, Bartholomew. De Mirabili Nouae ac splendidiss: Stellae, Mense Nouembri anni 1572; primùm conspectae, ac etiam nunc apparentis, Phoenomeno, Ivdicivm, Et Prognosticon Scriptvm: Ad Sereniss: Ac Potentiss: Principem ac Dominum, Dominum Rhodolphvm Dei gratia, Regem Hungariae, Archiducem Austriae, &c. per Barptolemaevm Reisacherum, Medicinae Doctorem, Xenodochij Caes: Physicum, ac Mathematicum Viennensem. Adivncta Est Brevis Commentatio De Eadem Stella, Thaddaei Ab Hayck, A: Rom: Caes: Maie: Medici. Viennae Avstriae ex Officina Typographica Caspari Stainhoferi. Anno M. D. LXXIII. (The H. C. L. copy was used.)

Renouard

Renouard, Ph[ilippe]. Bibliographie des éditions de Simon de Colines, 1520-1546, . . . Paris, Paul, Huard and Guillemin, 1894.

Riccardi

Riccardi, Pietro. Biblioteca matematica italiana dalla origine della stampa ai primi anni del secolo XIX; . . . Modena, Società tipografica, 1873-1893. 3v.

Riccioli

Riccioli, Giovanni Battista. Almagestvm novvm astronomiam veterem novamqve complectens observationibvs aliorvm, et propriis nouisque theorematibus, problematibus, ac tabulis promotam, in tres tomos distribvtam . . . Avctore P. Ioanne Baptista Ricciolo Societatis Iesu Ferrariensi . . . Bononiae, ex typographia haeredis Victorij Benatij, 1651. 1v. in 2. (The references to I and II are to parts 1 and 2.)

Richel

Richel, Arthur. Astrologische Volksschriften der Aachener Stadtbibliothek. (Aachen. Stadtbibliothek. Festschrift aus anlass der eröffnung des bibliothekgebäudes der stadt Aachen . . . Herausgegeben von Dr. Emil Fromm . . . Aachen, Cremerschen Buchhandlung, 1897, pt. 1, 49-93 [Zeitschrift des Aachener Geschichtsvereins, 19.])

Rieger

Rieger, František Ladislav. Slovník naučný. Praha, Kober a Markgraf, 1860-1873. 10v.

Ringshausen

Ringshausen, Karl Wilhelm. Poseidonios—Asklepiodot—Seneca und ihre anschauungen über erdbeben und vulkane . . . Borna-Leipzig, Noske, 1929.

Rixner

Rixner, Thaddeus Anselm and Siber, Thaddeus. Leben und lehrmeinungen berühmter physiker am ende des XVI. und am anfange des XVII. jahrhunderts, als beyträge zur geschichte der physiologie in engerer und weiterer bedeutung; hrsg. von Thaddä Anselm Rixner . . . und Thaddä Siber . . . v. 1-2. Sulzbach, Seidel, 1829, 1820. (The first volume is called a second edition.)

Robinson

Robinson, [James] Howard. The great comet of 1680; a study in the history of rationalism . . . Northfield, Minn. [Press of the Northfield news] 1916.

Rommel

Rommel, Christoph v. Neuere geschichte von Hessen . . . v. 1, 2. Cassel, Friedrich Perthes of Hamburg, 1835-7. (pt. 4, sections 1 and 2 of Geschichte von Hessen . . .)

Rosen

Rosen, Edward. Three Copernican treatises: The Commentariolus of Copernicus, The Letter against Werner, The Narratio prima of Rheticus, trans. with introd. and notes . . . New York, Columbia university press, 1939. (Records of civilization: sources and studies, edited under the auspices of the Dept. of history, Columbia university . . . no. XXX.)

Rosenthal

Rosenthal, Ludwig, bookseller of the Hague, Holland, formerly of Munich. Catalogues.

Rossi

Rossi, Giuseppe. Girolamo Fracastoro in relazione all' Aristotelismo e alle scienze nel rinascimento . . . Pisa, Spoerri, 1893.

Roth

Roth, F. W[ilhelm] E. Zur bibliographie des Henricus Hembuche de Hassia, dictus de Langenstein . . . [Leipzig, O. Harrassowitz, 1888.] (Beihefte zum Centralblatt für bibliothekswesen, 1, pt. 2: 97-118)

Rothschild

Rothschild, James, baron de. Catalogue des livres composant la bibliothèque de feu M. le baron James de Rothschild . . . Paris, D. Morgand, 1884-1920. 5v.

Ruge

Ruge, S. Geschichte der sächsischen kartographie im 16. jahrhundert. (Zeitschrift für wissenschaftliche geographie, . . . II jahrgang: 89-94, 143-145, 223-235, Lahr, 1881.)

Rumor

Rumor, Sebastiano. Bibliografia storica della citta' e provincia di Vicenza. Vicenza, Giuseppe, 1916-39. 2v.

Russell, Dugan and Stewart

Russell, Henry Norris. Astronomy; a revision of Young's Manual of astronomy. v. 1, The solar system. Boston, New York, etc., Ginn and company, [1926].

Salembier

Salembier, Louis. Le cardinal Pierre d'Ailly, chancelier de l'Université de Paris, évêque du Puy et de Cambrai, 1350-1420 . . . Tourcoing, Georges, 1932.

Sánchez Pérez

Sánchez Pérez, José A. Las Matemáticas en la Biblioteca del Escorial . . . Madrid, Maestre, 1929.

Santucci

Santucci, Antonio. Trattato Nvovo Delle Comete, Che Le Siano Prodotte in Cielo, e non nella regione dell'aria, come alcuni dicono, D'Antonio Sanctvcci Da Ripomaranci Lettore publico delle Matematiche nello Studio di Pisa, e Cosmografo del Gran Duca di Toscana. Con l'aggiunta che le Sfere del Fuoco, e dell' Aria non si muouino di moto circolare delle 24. hore, dedicato à S.A.S. In Fiorenza, Appresso Gio. Antonio Caneo 1611. Con licenza de' SS. Superiori. (The title is given here according to the rules adopted for the appendix.)

Sarton

Sarton, George. Introduction to the history of science. Baltimore, pub. for Carnegie inst. by Williams & Wilkins, 1927-31. v. 1-2.

Sarton (1936)

Sarton, George. Early editions of the Deutsche Sphaera of Conrad of Megenberg. (Query no. 63 in Isis, 25 (2) : 455, September, 1936.)

Schaff-Herzog

Schaff-Herzog encyclopedia. The new Schaff-Herzog encyclopedia of religious knowledge, . . . , based on the third edition of the Realencyklopädie . . . New York and London, Funk and Wagnalls, 1908-1914. 12v. and index.

Scheibel

[Scheibel, Johann Ephraim]. Einleitung zur mathematischen bücherkentnis . . . Breslau, J. E. Meyer, 1775-1787. 17 pts. in 3 v. Volume I is dated 1781 and is called "a new edition". (Unless otherwise indicated, references to v. 3, pts. 15 and 16.)

Schenck

Schenck, Johann Georg. Biblia iatrica; siue, Bibliotheca medica macta, continvata, consvmmata, qva velvt favissa, avctorum in sacra medicina scriptis cluentium, reique medicae monumentorum, ac diuitiarum thesaurus cluditur. . . . Frankfort, Published by A. Hummius, 1609.

Scheuchzer

Scheuchzer, Johann Jacob. Bibliotheca scriptorum historiae naturalis . . . Accessit . . . Jacobi Le Long, . . . de scriptoribus historiae naturalis Galliae. . . . Zurich, Heideggeri & Soc., 1751.

Schmidt

Schmidt, Wilhelm. Heron von Alexandria, Konrad Dasypodius und die Strassburger astronomische muensteruhr. . . . (Abhandlungen zur geschichte der mathematik. 8: 175-194, Leipzig, Teubner, 1898.)

Schoepflin

Schoepflin, Johann Daniel. Alsatia illustrata. Colmar, 1751-61. v. 2.

Schöner

Schöner, Johann. Coniectur odder abnemliche ausslegung Joannis Schoners vber dē Cometen so im Augstmonat / des M.D.XXXj. jars erschinen ist / zu ehren einem erbern Rath / vnd gmainer Burgerschafft der stat Nuremberg aussgangen. Nuremberg, Friderich Peypus. (no date)

Schotel

Schotel, G[illes] D[ionysius] J[acobus]. Vaderlandsche volksboeken en volkssprookjes van de vroegste tijden tot het einde der 18e eeuw; . . . Haarlem, A. C. Kruseman, 1873-4. 2 v. in 1.

Schottenloher

Schottenloher, Karl. Bibliographie zur deutschen geschichte im zeitalter der glaubensspaltung, 1517-1585; im auftrag der Kommission zur erforschung der geschichte der reformation und gegenreformation hrsg. von Karl Schottenloher . . . Leipzig, K. W. Hiersemann, 1933-9. 5v.

Scultetica (1837)

Altes urkundenverzeichniss. (Oberlausitzische gesellschaft der wissenschaften. Neues lausitzisches magazin . . . 15: 320, Görlitz, 1837.)

Scultetica (1842)

Inhalt der von M. Bartholom. Scultetus gesammelten urkundenbücher (Coll. II.) (Oberlausitzische gesellschaft der wissenschaften. Neues lausitzisches magazin . . . 20: 135-8, Görlitz, 1842.)

Scultetica (1915)

B. Sculteti e libris rerum gestarum Gorlicensium. Abschnitt III, Ex libro expeditionum bellicarum 1404-1479, bearbeitet von Dr. W. v. Boetticher. (Oberlausitzische gesellschaft der wissenschaften. Neues lausitzisches magazin . . . 91: 161-197, Görlitz, 1915.)

Seneca

Seneca, Lucius Annaeus. Physical science in the time of Nero, being a translation of the Quaestiones naturales of Seneca, by John Clarke, with notes on the treatise by Sir Archibald Geike. London, Macmillan, 1910.

Seyn

Seyn, Eugène de. Dictionnaire des ecrivains belges, bio-bibliographie . . . Bruges, "Excelsior", 1930-1. 2v.

Singer

Singer, Charles and Dorothea. The Scientific position of Girolamo Fracastoro . . . (Annals of medical history, 1: 1-34, New York, P. B. Hoeber, Spring, 1917.)

Smith (1917)

Smith, David Eugene. Medicine and mathematics in the sixteenth century. (Annals of medical history, 1: 125-140, New York, P. B. Hoeber, Summer, 1917.)

Sotheran

Sotheran, Henry and Co., bookseller of London. Catalogues.

Spunda

Spunda, Franz. Paracelsus. Vienna and Leipzig, Karl König, [1925]. (Menschen Völker Zeiten Eine Kulturgeschichte In Einzeldarstellungen Herausgegeben Von Max Kemmerich. 6.)

Stevens, Henry. See Coote.

Stimson

Stimson, Dorothy. The gradual acceptance of the Copernican theory of the universe, . . . Hanover, N. H., 1917.

Stobaeus

Stobaeus, Joannes. **Ιωάννου Στοβαίου 'Εκλογῶν βιβλία β.** Ioannis Stobaei Eclogarum physicarum et ethicarum libri duo. Accedit Hieroclis Commentarius in Aurea carmina Pythagoreorum. Ad mss. codd. recensuit Thomas Gaisford . . . Oxford, University press, 1850. 2v.

Stoddart

Stoddart, Anna M. The life of Paracelsus, Theophrastus von Hohenheim, 1493-1541. London, John Murray, 1911.

Stolpe

Stolpe, P[eter] M[atthias]. Dagspressen i Danmark, dens vilkaar og personer indtil midten af det attende aarhundrede. . . . v. 1, Copenhagen, Samfundet til den danske literaturs fremme, 1878.

Strada

Strada, Famian. De bello belgico . . . Rome, Schevs, Corbelletti, 1640-1647. 2v.

Strauss

Strauss, David Friderich. Leben und schriften des dichters und philologen Nicodemus Frischlin. Ein beitrag zur deutschen culturgeschichte in der zweiten hälfte des sechszehnten jahrhunderts. . . . Frankfort-on-the-Main, Literarische anstalt, 1856.

Strieder

Strieder, Friedrich Wilhelm. Grundlage zu einer hessischen gelehrten- und schrifsteller-geschichte. Von der reformation bis 1806 . . . v. 17, Marburg, 1819.

Struve

Struve, Otto [Wilhelm], ed. Pulkovo Astronomicheskaia observatoriia. Librorum in bibliotheca Speculae pulcovensis anno 1858 exeunte contentorum catalogus systematicus. Edendum curavit et praefatus est Otto Struve . . . St. Petersberg, Acad. imper. of sciences, 1860-1880. 2v.

Sudhoff

Sudhoff, Karl. Bibliographia Paracelsica — Besprechung der unter Theophrast von Hohenheim's Namen 1527-1893 erschienenen Druckschriften. Berlin, Georg Reimer, 1894. (Versuch einer Kritik der Echtheit der Paracelsischen Schriften I. Theil.)

Synesius

Synesius Cyrenaeus. The essays and hymns of Synesius of Cyrene, including the address to the Emperor Arcadius and the political speeches; tr. into English with introduction and notes by Augustine FitzGerald . . . London, Oxford university press, 1930. 2v.

Tallarigo

Tallarigo, Carlo Maria. Giovanni Pontano e i suoi tempi; monografia del prof. Carlo Maria Tallarigo, con la ristampa del dialogo Il Caronte e del testo delle migliori poesie latine, colla versione del prof. Pietro Ardito. Naples, Morano, 1874. 2v. paged continuously.

Thomas Aquinas (1886)

Thomas Aquinas, St. Sancti Thomae Aquinatis Doctoris Angelici opera omnia iussu impensaque Leonis XIII P. M. edita. Tomus tertius, Commentaria in libros Aristotelis de caelo et mundo, de generatione et corruptione et meteorologicum ad codicis manuscriptos exacta cura et studio Fratrum Ordinis Praedicatorum. Rome, 1886.

Thomas Aquinas (1921)

Thomas Aquinas, St. The "Summa Theologica" . . . literally translated by Fathers of the English Dominican Province. v. 3[6]. London, Burns Oates & Washbourne, Ltd., 1921.

Thompson

Thompson, R. Campbell. The reports of the magicians and astrologers of Nineveh and Babylon in the British Museum. London, Luzac, 1900. 2 v.

Thomson

Thomson, S. Harrison. The text of Grosseteste's de cometis. (Isis, 19 (1): 19-25, April, 1933.)

Thorndike

Thorndike, Lynn. A history of magic and experimental science . . . New York, v. 1-2 Macmillan; v. 3-6 Columbia university press, 1929-41. 6v.,

v. 1-2 (During the first thirteen centuries of our era), v. 3-4 (Fourteenth and fifteenth centuries), v. 5-6 (The sixteenth century). Volumes 1-2 are now published by the Columbia university press.

Thorndike (1916)

Thorndike, Lynn. The true Roger Bacon. (Reprinted from the American historical review, 21, no. 3: 237-257, 468-480, New York, Macmillan, January and April, 1916.) (Pages 238-245 have been incorporated in Thorndike, II, 619-628.)

Thorndike (1929)

Thorndike, Lynn. Science and thought in the fifteenth century; studies in the history of medicine and surgery, natural and mathematical science, philosophy and politics . . . New York, Columbia university press, 1929.

Thou-Teissier

Thou, Jacques Auguste de. Les éloges des hommes savans, tirez de l'histoire de M. de Thou, avec des additions contenant l'abbrégé de leur vie, le jugement & le catalogue de leurs ouvrages, par Antoine Teissier . . . v. 4, 4th ed., Leyden, Haak, 1715.

Tiraboschi

Tiraboschi, Girolamo. Storia della letteratura italiana . . . Rome, Salvioni, 1782-5. 9v. in 12.

Twyne

Twyne, Thomas. Thomas Twyne's discourse on the earthquake of 1580, edited by R. E. Ockenden, M.A. Oxford, Pen-in-hand publishing co., 1936.

Ungerer

Ungerer, Alfred. Les horloges astronomiques et monumentales les plus remarquables de l'antiquité jusqu'à nos jours. Strasburg, Chez l'auteur, 1931.

Uranophilus

Uranophilus, Christianus. Cometisther [sic] Gedenckzettel / Darinnen nicht allein die Cometen kůrtzlich vor Augen gestellet werden / welche der Hőchste Gott / Von Christi Geburt / biss auff gegenwårtiges 1677ste Jahr / der sůndigen Welt am Himmel gezeiget; Sondern auch Die jenigen Begebenheiten so auff einen und den andern erfolget. Herauss gegeben von Christiano Uranophilo. Leipzig, Christian Kirchner, [1677].

Uzielli

Uzielli, Gustavo. La vita e i tempi di Paolo dal Pozzo Toscanelli, ricerche e studi. Con un capitolo (VI) sui lavori astronomici del Toscanelli, di Giovanni Celoria. Rome, Ministero della pubblica istruzione, 1894.

Van Ortroy, F. See Ortroy, Fernand Gratien van.

van Someren

Utrecht. Rijks-universiteit. Bibliotheek. . . . Pamfletten niet voorkomende in afzonderlijk gedrukte catalogi der verzamelingen in andere openbare Nederlandsche bibliotheken, beschreven door J. F. van Someren . . . v. 1, Utrecht, A. Oosthoek, 1915.

Vetter

Vetter, Quido. Tadeáš Hájek z Hájku. (Ke čtyřsetletému výročí jeho narození.) (Zvláštní otisk z časopisu "Říše hvězd", ročník VI., číslo 6.) A special reprint from the periodical "The Empire of the Stars", year VI [1926], no. 6. (The translation used for the present dissertation was made by Dr. J. Novak.)

Vetter (1926)

Vetter, Quido. O Tadeáši Hájkovi z Hájku. (Ke čtyřstaletému výročí jeho narození.) A special reprint from "Zeměměřičského Věstníku", 2, 1926.

Vetter (1928)

Vetter, Q[uido]. Rapporto sulle lettere indirizzate al Dottor Taddeo Hagecius da Hayck, astronomo, medico e matematico Ceko conservate a Breslavia. (Reprint from Atti del congressso internazionale dei matematici, 499-501, Bologna, September 3-10, 1928 - VI.)

Vetter (1937)

Vetter, Q[uido]. L'Histoire des sciences en Tchécoslovaquie. (Reprint from Lychnos 1937: 243-7, Upsala, 1937.)

Vollbehr

U. S. Library of Congress . . . Exhibit of books printed during the XVth century and known as incunabula, selected from the Vollbehr collection purchased by act of Congress 1930; list of books. Washington, U. S. Govt. print. off., 1930.

Vossius

Vossius, Gerardus Joannes. Gerardi Ioannis Vossii De vniversae mathesios natvra & constitvtione liber; cui subjungitur chronologia mathematicorvm . . . Amsterdam, Blaev, 1650.

Wackernagel

Wackernagel, Philipp. Das deutsche kirchenlied von der ältesten zeit bis zu anfang des XVII. jahrhunderts. v. 4, Leipzig, Teubner, 1874.

Wallerand

See Bate. This reference refers only to the editor's notes.

Waters

Waters, W[illiam] G[eorge]. Jerome Cardan; a biographical study. London, Lawrence & Bullen, 1898.

Watt

Watt, Robert. Bibliotheca Britannica; or, A general index to British and foreign literature. Edinburgh, Constable, 1824. 4v.

Wedel

Wedel, Theodore Otto. ... The mediaeval attitude toward astrology, particularly in England ... New Haven, Yale university press ..., 1920.

Weidler

Weidler, Johann Friedrich ... Historia astronomiae, sive De ortv et progressv astronomiae liber singvlaris. Wittenberg, Schwartz, 1741.

Weller (1857-8)

Weller, E[mil]. Cometen-Literatur [1572-7]. (Anzeiger für kunde der deutschen vorzeit. new series, v. 4 (1857), columns 321-4, 359-362; v. 5 (1858), column 215.)

Weller (1860)

Weller, Emil. Die deutschen zeitungen des sechzehnten jahrhunderts. [continuations] (Serapeum. Zeitschrift für bibliothekwissenschaft, handschriftenkunde und ältere litteratur ... 21: 14-16, 30-32, 60-64, 77-80, 109-112, 127-128, 157-160. [1572-1582.] Leipzig, Weigel, 1860.)

Weller (1862-4)

Weller, Emil. Annalen der poetischen national-literatur der Deutschen im XVI. und XVII. jahrhundert. Nach den quellen bearbeitet ... Freiburg im Breisgau, Herder, 1862-4. 2v.

Weller (1872)

Weller, Emil. Die ersten deutschen zeitungen; herausgegeben mit einer bibliographie (1505-1599) von Emil Weller. Tübingen, H. Laupp, 1872. (Bibliothek des litterarischen vereins in Stuttgart. 111.)

White

White, Andrew Dickson. A history of the warfare of science with theology in christendom ... v. 1, New York, D. Appleton, 1899.

Wickersheimer

Wickersheimer, Ernest. Les médecins de la nation anglaise (ou allemande) de l'Université de Paris aux XIVe et XVe siècles ... (Bulletin de la Société française d'histoire de la médecine ... 12 (1913): 285-344, 537-8. Paris, 1913.)

Wierzbowski

Wierzbowski, Teodor. Bibliographia polonica XV ac XVI ss. ... Varsoviae in officina typ. C. Kowalewski, 1889-94. 3v.

Wilamowitz (1902)

Wilamowitz-Moellendorff, Ulrich von. Griechisches lesebuch ... 2nd impression, v. 1, pt. 2, Berlin, Weidmann, 1902.

Wilamowitz (1906)

Wilamowitz-Moellendorff, U. v. Der physiker Arrian (Hermes; zeitschrift für classische philologie ... 41: 157-8, 1906.)

Will

Will, Georg Andreas. Nůrnbergisches gelehrten-lexicon oder Beschreibung aller nurnbergischen gelehrten beyderley geschlechtes nach ihrem leben, verdiensten und schrifften zur erweiterung der gelehrten geschichts-kunde und verbesserung vieler . . . fehler . . . Nuremberg and Altdorf, L. Schůpfel, 1755-8. 4v.

Witte

Witte, Henning. Diarium biographicum, in quo scriptores seculi post natum Christum XVII. praecipui, . . . juxta annum diemque cujusvis emortualem, consisè descripti magnô adducuntur numerô. . . . Dantzic, Published by Martin Hallervordius, 1688.

Wohlwill

Wohlwill, Emil. Galilei und sein kampf für die copernicanische lehre. . . . Hamburg and Leipzig, L. Voss, 1909-26. 2v.

Wolf, J. C.

Wolf, Johann Christoph. . . . Bibliotheca hebraea; sive, Notitia tvm avctorvm hebr. cvjvscvnqve aetatis, tvm scriptorvm, qvae vel hebraice primvm exarata vel ab aliis conversa svnt, ad nostram aetatem dedvcta. . . . v. 1. Hamburg and Leipzig, Liebezeit, 1715.

Wolf

Wolf, Rudolf. Biographien zur kulturgeschichte der Schweiz. Zurich, Orell, Füssli & Co., 1858-62. 4v.

Wolf (1845)

Wolf, Rudolf. Conrad Dasypodius. (Bern. Mittheilungen der naturforschenden gesellschaft in Bern, 1845, no. 56: 137-8. There is a brief supplement, 1854: 69.)

Wolf (1849)

Wolf, Rudolph. Ueber die älteste Cometen-Litteratur der Schweiz. (Bern. Mittheilungen der naturforschenden gesellschaft, 1849, nos. 156-7: 102-5.)

Wolf (1852)

Wolf, Rudolph. Christian Wursteisen von Basel. (Bern. Mittheilungen der naturforschenden gesellschaft, 1852, nos. 237-8: 105-111.)

Wolf (1877)

Wolf, Rudolf. Geschichte der astronomie . . . Munich, Oldenbourg, 1877. (K. Acad. der wissenschaften, Munich. Historische commission, ed. Geschichte der wissenschaften in Deutschland. Neuere zeit. 16.)

Wood.

Wood, Anthony à. Athenae oxoniensis. An exact history of all the writers and bishops who have had their education in the University of Oxford . . . new edition, 5v., London, F. C. and J. Rivington, 1813-1820.

Woolhouse

Woolhouse, W. S. B. Measures, weights & moneys of all nations. London, Crosby, Lockwood, 1881.

Zedler

Zedler, Johann Heinrich. Grosses vollständiges universal lexikon aller wissenschaften und künste . . . Leipzig and Halle, Zedler, 1732-1750. 64v.

Zeiller

[Zeiller, Martin] M. Z. Topographia Bohemiae, Moraviae et Silesiae; das ist, Beschreibung vnd eigentliche abbildung der vornehmsten vnd bekandtisten stätte, vnd plätze, in dem königreich Boheim vnd einverliebten landern, Mähren, vnd Schlesien. Frankfort, Merian, 1650.

Zeumer

Zeumer, Johann Caspar. Vitae professorum ivrivm, qvi in illvstri academa jenensi . . . vixervnt, . . . Jena, Printed for Tobias Oehrlingus. no date.

Ziegler

Ziegler, Alexander. Regiomontanus, (Joh. Müller aus Königsberg in Franken) ein geistiger vorläufer des Columbus . . . Dresden, Höckner, 1874.

Zinner (1934)

Zinner, Ernst. Die Fränkische sternkunde im 11. bis 16. jahrhundert. (Bericht der naturforschenden gesellschaft in Bamberg, 27: 1934.)

Zinner (1938)

Zinner, Ernst. Leben und wirken des Johannes Müller von Königsberg genannt Regiomontanus. (Schriftenreihe zur bayerischen Landesgeschichte. Hrsg. v. d. Kommission f. bayerische Landesgeschichte bei der bayerische akademie der wissenschaften. 31, Munich, C. H. Beck, 1938.)

INDEX

Index of Printers Cited in Appendix

Index of Persons

Names in *italics* are references cited in bibliography.

VITA

Clarisse Doris Hellman (Mrs. Morton Pepper) was born in New York City on August 28, 1910. She received her B.A. degree from Vassar College in 1930 and her M.A. degree from Radcliffe College in 1931. She is a member of Phi Beta Kappa and was graduated from Vassar with honors. She was a Vassar College Fellow, in mathematics and astronomy, at Radcliffe in 1930-1, a Columbia University Scholar in history in 1931-2 and a Columbia University Fellow in history in 1932-3.

At Vassar College she did independent study in mathematics under the late Professor Henry S. White and independent study in astronomy under the late Professor Caroline E. Furness. At Radcliffe, her work consisted mainly of research in the history of science under Doctor George Sarton. At Columbia University she attended Professor Lynn Thorndike's seminar entitled *Studies in the intellectual history of the closing medieval and early modern centuries* and the late Professor Frederick Barry's seminar entitled *Special problems in the history of science.*

In addition to reviews and an answer to a query in *Isis,* Miss Hellman's publications include:

George Graham: Maker of Horological and Astronomical Instruments (*Popular Astronomy,* 1931; *Vassar Journal of Undergraduate Studies,* 1931).

An Unpublished Diary of Edward Jenner (1810-1812) (*Annals of Medical History,* 1931).

Jefferson's Efforts towards the decimalization of U. S. weights and measures (*Isis,* 1931).

John Bird (1709-1776) *Mathematical Instrument - Maker in the Strand* (*Isis,* 1932).

A Bibliography of tracts and treatises on the comet of 1577 (*Isis,* 1934).

Legendre and the French Reform of weights and measures (*Osiris,* 1936).

Bei Fragen zur Produktsicherheit wenden Sie sich bitte an:
If you have any questions regarding product safety,
please contact:

Walter de Gruyter GmbH
Genthiner Straße 13
10785 Berlin
productsafety@degruyterbrill.com